Springer Series in Operations Research and Financial Engineering

The Springer Series in Operations Research and Financial Engineering publishes monographs and textbooks on important topics in theory and practice of Operations Research, Management Science, and Financial Engineering. The Series is distinguished by high standards in content and exposition, and special attention to timely or emerging practice in industry, business, and government. Subject areas include:

Linear, integer and non-linear programming including applications; dynamic programming and stochastic control; interior point methods; multi-objective optimization; Supply chain management, including inventory control, logistics, planning and scheduling; Game theory Risk management and risk analysis, including actuarial science and insurance mathematics; Queuing models, point processes, extreme value theory, and heavy-tailed phenomena; Networked systems, including telecommunication, transportation, and many others; Quantitative finance: portfolio modeling, options, and derivative securities; Revenue management and quantitative marketing Innovative statistical applications such as detection and inference in very large and/or high dimensional data streams; Computational economics.

Johannes O. Royset

Optimization Concepts

Springer

Johannes O. Royset
Daniel J. Epstein Department
of Industrial and Systems Engineering
University of Southern California
Los Angeles, CA, USA

ISSN 1431-8598 ISSN 2197-1773 (electronic)
Springer Series in Operations Research and Financial Engineering
ISBN 978-3-032-36818-8 ISBN 978-3-032-36819-5 (eBook)
https://doi.org/10.1007/978-3-032-36819-5

Mathematics Subject Classification: 90C46, 90C15, 65K05, 90-01, 90C90

This Springer imprint is published by the registered company Springer Nature Switzerland AG
The registered company address is: Gewerbestrasse 11, 6330 Cham, Switzerland

to aspiring optimizers

Preface

These notes serve as an introduction to the broad field of optimization, one of the main technologies of the modern information age. Companies want to forecast the demand for their products and plan their supply chain accordingly. Airlines have to assign crews and aircraft to specific routes. Engineers develop transportation networks while accounting for a growing population. Governments aim to optimally incentivize citizens and different sectors of the economy. All these challenges can be addressed by optimization technology. In the age of massive data and capable computers, optimization-based, automated approaches to prediction, planning, and decision making are becoming increasingly integral to nearly all human activity. Intelligent systems such as food delivery robots, military drones, and driverless cars make complicated decisions autonomously. Chat bots and AI agents help us to sort through data and complete increasingly complex tasks. Data about the location, purchase history, and interests of potential customers guide decisions about online advertisement. Optimization is the interdisciplinary field that supplies a wide range of applications with models and algorithms for making the best possible predictions and decisions.

Applications require skills in building mathematical models of real-world situations. But models are most useful when they're tractable and can be "solved," i.e., there are algorithms that produce from the models numerical estimates, predictions, and recommended courses of action. The underlying theory guides the development of models and algorithms, establishes their correctness, and informs us about computational efficiency and limitations. In these notes, we teach how to formulate optimization models and construct the necessary algorithms, but also introduce the underlying mathematical theory so the reader can justify a solution strategy and be prepared to tackle emerging challenges. We step through the stages

$$\text{real-world problem} \implies \text{optimization model} \implies \text{algorithm} \implies \text{analysis of results}$$

so that the reader sees the whole picture as it emerges in practical applications.

The notes support a one-semester introductory course in optimization for undergraduate students of engineering or for master students in an analytics program. We expect that the reader has a foundation in linear algebra and an introductory knowledge of calculus. No knowledge of probability and statistics is prerequisite. We summarize basic concepts when needed, including multivariate calculus. Formal mathematical reasoning is largely avoided, and the reader is referred to [30] for proofs and more advanced topics. Throughout, we take inspiration from material in [30], adapt and augment it with illustrations and explanations.

The notes include five chapters, intended to be read sequentially, and many examples and exercises with solutions. The first chapter focuses on minimizing smooth functions, provides gradient-based algorithms, introduces convexity, and gives examples from data fitting and inventory management. Chapter 2 addresses constraints, how to use them in modeling and how to solve the resulting formulations. It discusses how the fundamental optimality condition "gradient equal to zero" extends to nonsmooth and constrained problems, resulting in the KKT condition and various algorithms. Chapter 3 focuses on optimization under uncertainty and addresses recourse problems, risk minimization, and stochastic gradient descent for machine learning. Chapter 4 teaches how to model logical relations using integer variables and introduces the branch-and-bound algorithms, possibly enhanced with cutting planes. It also mentions network problems, network algorithms, and extensions to stagewise decision making and reinforcement learning. The last chapter guides the reader through sensitivity analysis and duality using the general approach of Rockafellians. Examples show how duality theory helps with reformulation and solution of difficult optimization problems. In this age of AI, keeping exercise solutions away from students is futile and we include fully worked out solutions in the appendix.

A unique feature of these notes relative to other introductory books is the attempt to connect elementary calculus to the celebrated KKT condition and more general characterization of optimality. Thus, subgradients emerge as the central concept connecting smooth functions to constrained problems. We also shy away from simply presenting dual problems as the result of algebraic operations and instead trace their origin back to sensitivity analysis and thereby stress the great multitude of dual problems. We hope these perspectives will give the reader a deeper understanding of the fundamental concepts in optimization and the many possibilities afforded by variational analysis.

Mathematical background material is highlighted in gray-shaded boxes. An active use of the index should settle most questions about definitions and notation. Definitions and theorems are stated in italic font. Greek letters typically denote scalars, lower-case roman letters represent vectors, and upper-case roman letters represent matrices. Vectors to be optimized are usually denoted by x. Multipliers are written as y and z. Functions are denoted by f, g, h and sometimes φ (the Greek letter "phi") and ψ (psi). Boldface letters indicate random variables, random vectors, and random matrices. In particular, $\boldsymbol{\xi}$ (ksi) is a random vector representing uncertain parameters and $\mathbb{E}[\boldsymbol{\xi}]$ is its expected value.

We use common set notation. We write $\mathbb{R}$ for the real line and $\mathbb{R}^n$ for the n-dimensional space. A point x in the set C is denoted by $x \in C$. A set C_1 is a subset of another set C_2 when $C_1 \subset C_2$, which doesn't rule out the possibility that the sets are identical. The intersection of C_1 and C_2 is denoted by $C_1 \cap C_2$. The union of the sets is $C_1 \cup C_2$.

Subscripts routinely designate components of a vector or coordinates of a point. For example, x_j is typically the jth component of a vector x, i.e., $x = [x_1 \ x_2 \ \dots \ x_n]$, or the jth coordinate of a point, i.e., $x = (x_1, \dots, x_n)$. The distinction between an n-dimensional vector and a point in n-dimensional space ($\mathbb{R}^n$) can be blurry with the same object sometimes being represented either way. We tend to interpret a point as a column vector and adopt the convention

$$(x_1, \ldots, x_n) = \begin{bmatrix} x_1 \\ \vdots \\ x_n \end{bmatrix} = [x_1 \ x_2 \ \ldots \ x_n]^\top,$$

where $^\top$ denotes the transpose of a vector. Brackets and parentheses indicate vectors and points but also intervals:

$(0, 1)$ is the *open* interval between 0 and 1, excluding the end points

$[0, 1]$ is the *closed* interval between 0 and 1, including the end points

$\{0, 1\}$ is just the two points 0 and 1.

This means that $(0, 1) \cap \{0, 1\} = \emptyset$ (the empty set). Since we use parentheses for points and column vectors, the context has to clarify what's meant with (0,1). It could be the open interval between 0 and 1, or the two-dimensional column vector of length one pointing straight up, or the point in $\mathbb{R}^2$ with first coordinate 0 and second coordinate 1.

The use of superscripts can be confusing at first; x^ν probably doesn't mean x to power ν (the Greek letter "nu") but rather iteration number ν of an algorithm.

The maturity of the field of optimization (see, e.g., [23]) makes it impossible to cover all the important subjects in an introductory course. Among many resources, we mention a few books that supplement these notes especially well. Our textbook "An Optimization Primer" [30] covers a wide array of topics at a more advanced level and would be the natural next step after mastering these notes. With a focus on applications, [9] provides a foundation in linear algebra and convex optimization. For a classical exposition of linear optimization, we refer to [7]. A concise review of convex analysis appears in [6]. Optimization models under uncertainty is the topic of [16]. Algorithmic details for linear and quadratic optimization problems as well as those involving smooth objective and constraint functions are laid out in [22]. First-order methods and their applications in machine learning are addressed by [2, 18]. Further treatment of discrete optimization at the graduate level appears in [37], while [34] describes heuristics and [1] discusses network flow. Comprehensive developments of multi-stage optimization under uncertainty appear in [3, 4, 24].

There are many open-source packages with optimization capabilities. A reputable list is found at `coin-or.org`. To solve an optimization problem, we would need a suitable *solver*, i.e., a computer implementation of an algorithm, but can also benefit from an *algebraic modeling language*, which allows us to specify objective functions and constraints easily as well as manage subproblems, batch runs, tolerances, and output formats. Pyomo (`pyomo.org`) is a Python-based, open-source modeling language that supports a diverse set of problems including those involving uncertainty [13].

Los Angeles, California *Johannes O. Royset*
June, 2026

Acknowledgements

I'm grateful for having written papers with Terry Rockafellar, Roger Wets, and Kevin Wood. Their rejection of the misguided maxim "it was hard to prove so it should be hard to read" helped me realize the importance of good technical writing. Make every word count. No unnecessary fluff. No cryptic notation and unnecessary complexity. While it's doubtful that the present notes meet these standards fully, I hope they can provide an accessible entry point to the fascinating and practically important world of mathematical optimization.

I'm extremely thankful for the many stimulating discussions with colleagues and students over the years. Faculty at the Naval Postgraduate School taught me invaluable lessons on what it takes to address real-world problems. The enthusiastic support of Jong-Shi Pang, John Gunnar Carlsson, and other USC colleagues encouraged me to undertake this project. The input from the ISE330 class at USC during Fall 2025 was instrumental in improving the initial draft of these notes. Andreas Royset critically reviewed the introduction to calculus and provided valuable examples. Many thanks go to Donna Chernyk at Springer and the series editors for again giving me the opportunity to put thoughts to paper.

Funding from the Office of Naval Research has been essential in developing this material over many years. In particular, I'm grateful to David Phillips, Kip Krebs, and Marc Steinberg for seeing the importance of optimization technology in a broad range of application areas. I'm also thankful for the flexibility and support provided by the Daniel J. Epstein Department of Industrial and Systems Engineering at USC.

Lastly, I'm indebted to my family for their patience and unwavering support. Their generosity in granting me endless "timeouts" is much appreciated.

Contents

Chapter 1
The Optimization Landscape

The task of finding a "best" decision when there's a trade-off between different quantities of interest such as cost and performance reduces to an optimization problem. Likewise, minimizing prediction error in a neural network by changing its trainable parameters is an optimization problem. Regardless of the application, optimization problems can be unified mathematically. We can formulate an optimization problem in terms of a function and the goal becomes to find the function's minimum. In this chapter, we'll identify the main ideas and challenges, and also introduce fundamental algorithms and supporting theory.

1.A Calculus: Our Secret Weapon

An optimization problem arises usually from modeling a real-world situation. It results in a function f that we then seek to minimize, i.e., find an argument x such that $f(x)$ is as low as possible. There might be none, one, or multiple such *minimizers* with Figure 1.1 illustrating some possibilities.

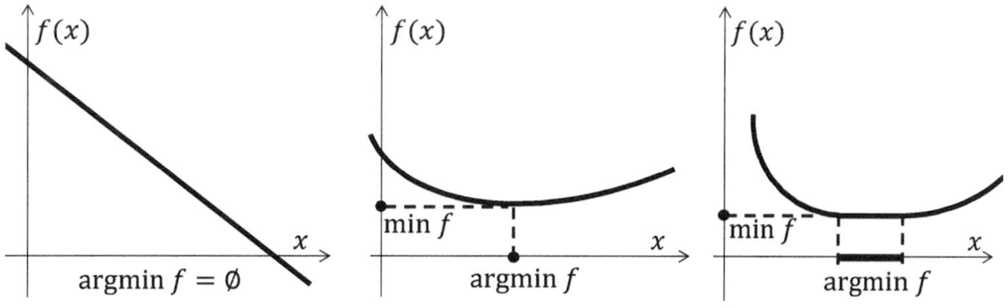

Fig. 1.1: Functions with no minimizers (left), one minimizer (middle), and multiple minimizers (right).

J. O. Royset, Optimization Concepts, *Springer Series in Operations Research and Financial Engineering*,
https://doi.org/10.1007/978-3-032-36819-5_1

Formally, the *set of minimizers* of f is denoted by

$$\operatorname{argmin} f = \left\{x^\star \mid f(x^\star) \le f(x)\ \forall x\right\}.$$

Here, the right-hand side reads: the set of those $x^\star$ such that $f(x^\star) \le f(x)$ for all x. When $x^\star \in \operatorname{argmin} f$, the number $f(x^\star)$ is called the *minimum value* of f and is denoted by $\min f$.

A function f to be minimized is an *objective function* and x is a *decision variable*. An example shows how a function and its decision variable arise in practice.

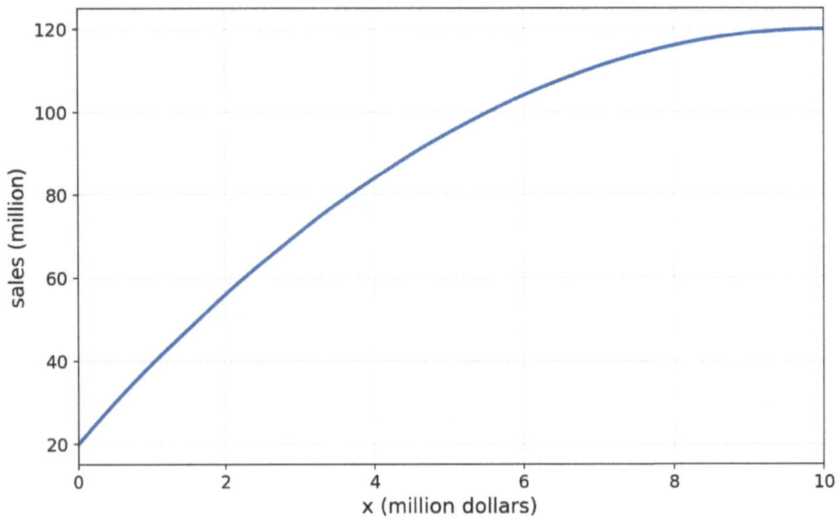

Fig. 1.2: Sales in millions as a function of advertisement expenditure in millions of dollars (x) using the formula $120 - (x - 10)^2$.

Example 1.1 (optimizing advertisement). We seek to determine the amount of advertisement dollars, denoted by x, that we should spend to

minimize advertisement expenditure minus profit from sales.

Historical data shows that at least for expenditures between zero and 10 million dollars, the number of sales varies with x according to the formula:

$$\text{millions of sales given } x \text{ million advertisement dollars} \;=\; 120 - (x - 10)^2.$$

This formula indicates a diminishing return as x increases; see Figure 1.2. The diminishing return causes a delicate trade-off between advertisement expenditure and profit from sales. If the profit per million sales is 0.1 million dollars, then the profit from sales becomes $0.1(120 - (x - 10)^2)$. Thus, the function to minimize is given by

$$f(x) = x - 0.1\big(120 - (x - 10)^2\big). \tag{1.1}$$

Figure 1.3 shows the graph of f. Reading off the graph, it appears that f has a unique minimizer near 5. The corresponding minimum value is $f(5) = -4.5$. Thus, the optimal decision would be to spend 5 million dollars on advertisement and this decision produces a profit from sales of 9.5 million. We make $9.5 - 5 = 4.5$ million dollars! Figure 1.3 shows that any advertisement amount between zero and 10 million dollars would result in a negative value of f, i.e., we would make money. However, the amount is significantly less as we deviate from the optimal decision. □

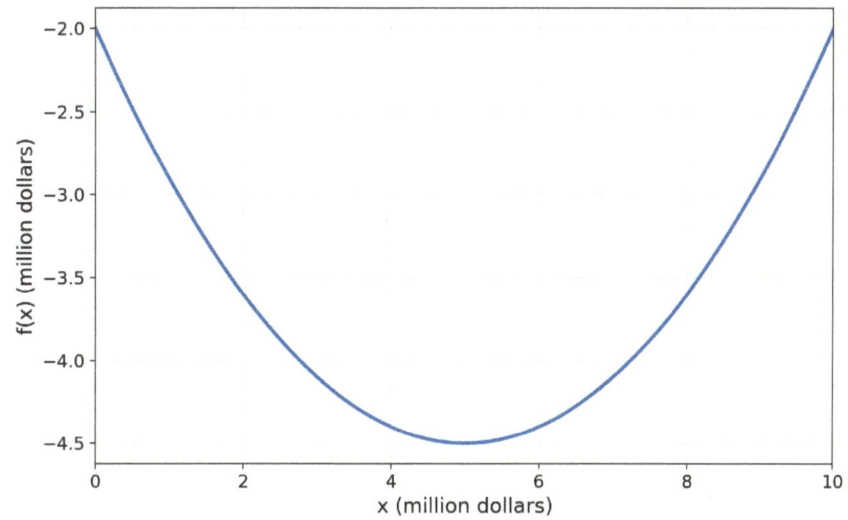

Fig. 1.3: Graph of f from (1.1) representing advertisement expenditure minus profit from sales as a function of x.

We solve the optimization problem in Example 1.1 by simply looking at the graph of f. Is there a more principal way? We could check a large number of different x-points and compute $f(x)$ for each one of them. The x-point with the lowest $f(x)$ could serve as our estimate of a minimizer. While this is doable in Example 1.1, it becomes problematic as we encounter real-world problems with many decision variables to optimize. For instances, an extension of Example 1.1 might involve different kinds of advertisement, say on social media, on TV, on radio, and per email. This would produce four variables to optimize and checking different possibilities becomes computationally costly. Recall, checking 1000 different values in each of four dimensions would requires $1000^4 = 10^{12}$ checks. In practice, we often face problems with thousands or millions of decision variables which rules out such naive checking. Beyond two variables, a graphical approach is also impossible.

A fundamentally different and scalable approach relies on centuries of mathematical developments. Pierre de Fermat (1607-1665), a French mathematician and lawyer, observed that for $x^\star$ to be a minimizer of f the tangent to the graph of f at the point $(x^\star, f(x^\star))$ must be "flat" [15]; see Figure 1.4. We would now express this statement using differential calculus and write

$$\text{if } x^\star \in \operatorname{argmin} f, \quad \text{then } f'(x^\star) = 0,$$

where $f'(x^\star)$ is the *derivative* of f at the point $x^\star$. Generally, the derivative at a point $\bar{x}$ is defined as the ratio of "rise" $f(\bar{x}+\tau) - f(\bar{x})$ over "run" τ, as the scalar τ vanishes. Specifically, one has

$$f'(\bar{x}) = \lim_{\tau \to 0} \frac{f(\bar{x}+\tau) - f(\bar{x})}{\tau}. \tag{1.2}$$

The derivative $f'(\bar{x})$ therefore conveys the slope of the tangent line of the graph of f at the point $(\bar{x}, f(\bar{x}))$.

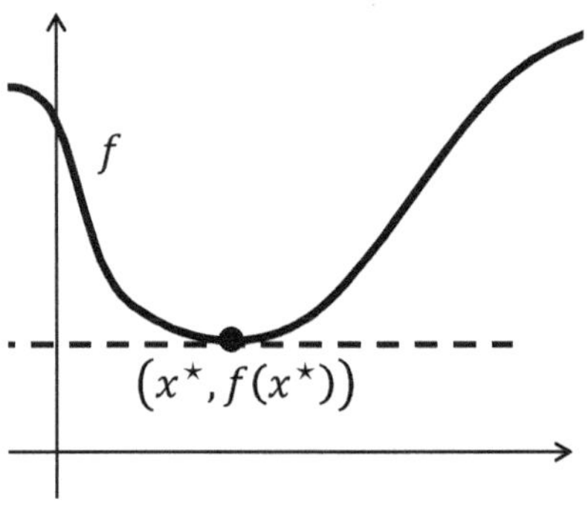

Fig. 1.4: Horizontal tangent to the graph of a function at a minimizer.

While a derivative isn't well-defined for all functions—kinks or jumps in the value of f near $\bar{x}$ could make the limit in (1.2) undefined—many common functions have easily available derivatives.

Differential Calculus. For any scalar α, some common functions and their derivatives include:

$$f(x) = x^\alpha \qquad f'(x) = \alpha x^{\alpha-1} \qquad \text{when } x > 0 \text{ or } \alpha \geq 1$$

$$f(x) = \exp(\alpha x) \qquad f'(x) = \alpha \exp(\alpha x).$$

The derivative is always zero for a constant function.

If the functions h and g are smooth, i.e., their derivatives are well-defined and they vary continuously, then the following rules are useful to compute derivatives of composite functions:

$$f(x) = h(g(x)) \qquad f'(x) = g'(x)h'(g(x))$$
$$f(x) = h(x)g(x) \qquad f'(x) = h'(x)g(x) + h(x)g'(x).$$

In particular, the derivative of a sum of functions is the sum of the derivatives of the functions.

For the function in Example 1.1, $f'(x) = 1+0.2(x-10)$, which vanishes at $x = 5$. This confirms the solution we identified graphically in the example.

Example 1.2 (a hiker's dilemma). Consider a hiker that starts in the northwest corner of the map in Figure 1.5 and seeks a minimum-time route to the southeast corner. Suppose that the map covers ten-by-ten miles and the upper half of the area has relatively firm ground on which the hiker estimates a walking speed of four miles per hour. The lower half of the area is marshy and walking speed reduces to two miles per hour. While walking diagonally across the map is an option, it might be faster to cross into the marshy area further east to reduce the distance walked at lower speed. Using x as the cross-over point (see Figure 1.5) and these assumptions, finding the best route amounts to minimizing the function

$$f(x) = \tfrac{1}{4}\sqrt{5^2 + x^2} + \tfrac{1}{2}\sqrt{5^2 + (10 - x)^2}.$$

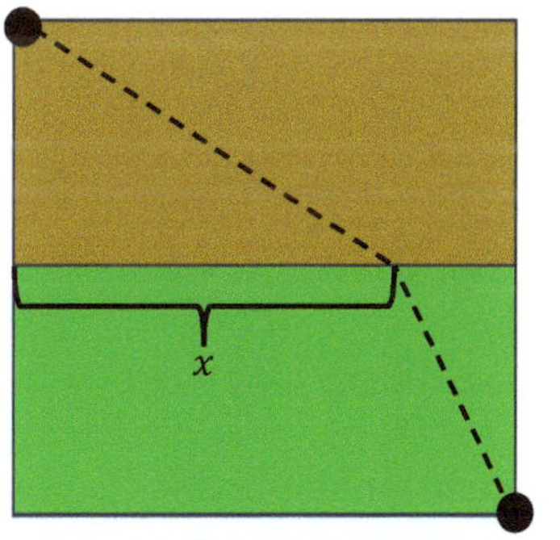

Fig. 1.5: A hiker seeks a time-minimizing route from the northwest corner to the southeast corner of a map while passing over firm (brown) and marshy (green) terrain.

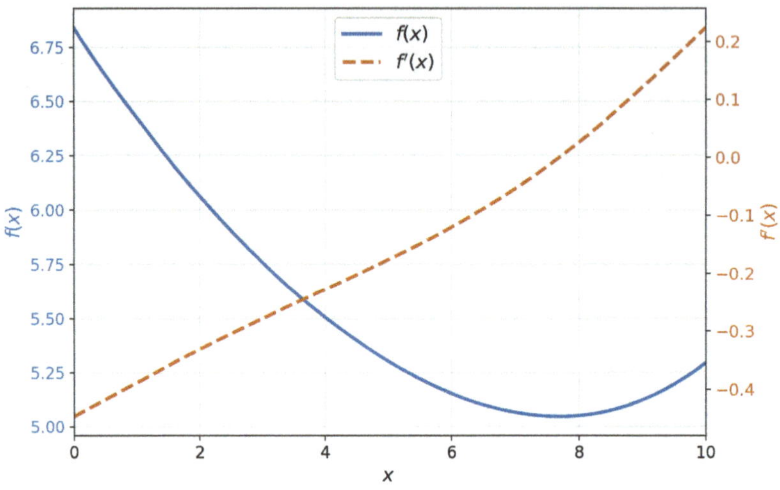

Fig. 1.6: Time as function of x (solid blue line) and its derivative (dashed orange line).

We compute the derivative of f as

$$f'(x) = \frac{x}{4\sqrt{5^2 + x^2}} + \frac{x - 10}{2\sqrt{5^2 + (10 - x)^2}}.$$

In contrast to Example 1.1, there's no easy analytical approach to solving the equation $f'(x) = 0$. Later, we'll learn about algorithms for solving such equations numerically. Figure 1.6 shows the graph of both f and f', with the latter vanishing at approximately $x = 7.7$. The best route takes $\min f \approx f(7.7) = 5.05$ hours. In contrast, the diagonal route (corresponding to $x = 5$) would take 5.30 hours, i.e., 15 minutes longer. □

Let's proceed with multivariate functions defined in terms of n variables, which then can represent optimization problems involving n decision variables. Typically, we write a vector of such variables as $x = (x_1, \ldots, x_n)$. Figure 1.7 shows the graph of a function f in two variables so that $x = (x_1, x_2)$, with

$$f(x) = 100(x_2 - x_1^2)^2 + (1 - x_1)^2. \tag{1.3}$$

An alternative view of such 2-dimensional functions is provided by their contour plot (see Figure 1.8), which is similar to a topographic map with darker colors indicating lower points in the "terrain." It's now clear that minimizing a function can be thought of as finding the bottom in a landscape defined by the graph of the function. As the number of dimensions grow beyond two, illustrations like Figures 1.7 and 1.8 aren't possible. Still, it's helpful to think of minimizing a function as being equivalent to finding your way to the bottom of a high-dimensional landscape.

Differential calculus again enters as the central tool, but now in its multivariate incarnation. For a function f in n variables, we define its *gradient* at the point $\bar{x}$ as

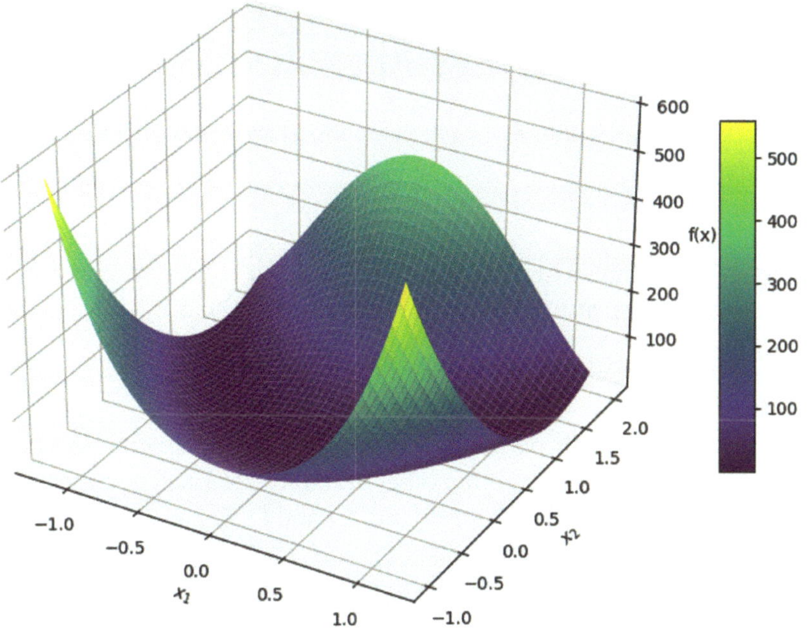

Fig. 1.7: Graph of the two-dimensional function from (1.3).

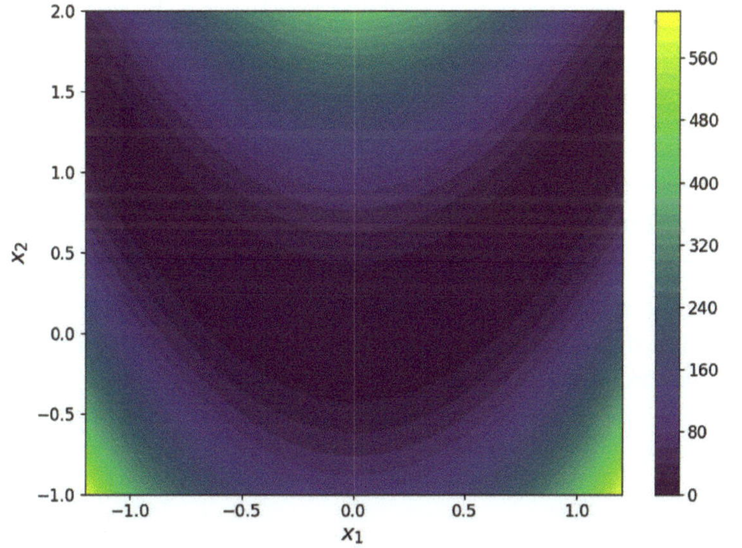

Fig. 1.8: Contour plot of the two-dimensional function from (1.3).

$$\nabla f(\bar{x}) = \begin{bmatrix} f'_{x_1}(\bar{x}) \\ f'_{x_2}(\bar{x}) \\ \vdots \\ f'_{x_n}(\bar{x}) \end{bmatrix},$$

where the jth *partial derivative* $f'_{x_j}(\bar{x})$ at $\bar{x}$ is defined as follows. Let g_j be the univariate function given as

$$g_j(x_j) = f(\bar{x}_1, \ldots, \bar{x}_{j-1}, x_j, \bar{x}_{j+1}, \ldots, \bar{x}_n).$$

This means that all variables of f except x_j are fixed at values specified by $\bar{x}$. Then, the jth partial derivative of f at $\bar{x}$ becomes the derivative of g_j at $\bar{x}_j$:

$$f'_{x_j}(\bar{x}) = g'_j(\bar{x}_j).$$

Informally, we obtain the jth partial derivative of f by computing the derivative with respect to x_j and pretending that all the other variables are fixed parameters.

As an example, the function f from Figure 1.7 and (1.3) has

$$f'_{x_1}(\bar{x}) = -400(\bar{x}_2 - \bar{x}_1^2)\bar{x}_1 - 2(1 - \bar{x}_1)$$
$$f'_{x_2}(\bar{x}) = 200(\bar{x}_2 - \bar{x}_1^2).$$

These partial derivatives then produce the gradient at $\bar{x}$ as

$$\nabla f(\bar{x}) = \begin{bmatrix} -400(\bar{x}_2 - \bar{x}_1^2)\bar{x}_1 - 2(1 - \bar{x}_1) \\ 200(\bar{x}_2 - \bar{x}_1^2) \end{bmatrix}. \tag{1.4}$$

A function f in n variables is *smooth* if all its partial derivatives are well-defined and they vary continuously. For such functions, Fermat's statement about optimality now gets translated to

The Fermat rule: if $x^\star \in \operatorname{argmin} f$, then $\nabla f(x^\star) = 0$

We note that 0 on the right-hand side is actually the n-dimensional *zero-vector*. Thus, the Fermat rule states that each partial derivative needs to vanish at a minimizer. Again returning to the function from (1.3) and its gradient in (1.4), the choice $\bar{x} = (\bar{x}_1, \bar{x}_2) = (1, 1)$ produces

$$\nabla f(\bar{x}) = \begin{bmatrix} -400(1 - 1^2)1 - 2(1 - 1) \\ 200(1 - 1^2) \end{bmatrix} = \begin{bmatrix} 0 \\ 0 \end{bmatrix}.$$

The implication in the Fermat rule goes from left to right. This means that there are functions with vanishing gradient at points other than minimizers. An example is $f(x) = -x^2$, with $f'(x) = -2x$. The point $x = 0$ isn't a minimizer, but actually a maximizer. Similarly, $f(x) = x^3$ produces $f'(x) = 3x^2$ and again $x = 0$ isn't a minimizer despite $f'(0) = 0$. In fact, $\operatorname{argmin} f = \emptyset$ (the empty set) in both cases. In view of these examples, we refer to the Fermat rule as a *necessary optimality condition*. However, it's certainly not sufficient for verifying that a point is a minimizer.

Although the Fermat rule only delivers a necessary condition for optimality, it leads to a practically important surrogate problem: Find x such that

$$\nabla f(x) = 0 \quad \text{or, equivalently,} \quad f'_{x_j}(x) = 0, \ \ j = 1, \dots, n.$$

The Fermat rule is therefore a means to transform, in some sense, an optimization problem into a problem of solving equations. It's a principal way of addressing optimization problems, one that avoids naively checking an exponentially large number of x-points.

1.B Data Fitting

Given the values $y_1, y_2, \dots, y_m$ of an unknown univariate function at the distinct points $x_1, x_2, \dots, x_m$, we're interested in finding a function of the form

$$p(x) = c_1 x + c_0$$

with values at $x_1, \dots, x_m$ as close as possible to the observed values. Figure 1.9 illustrates a particular case with $m = 11$ data points; see Exercise 1.5 for the actual numbers.

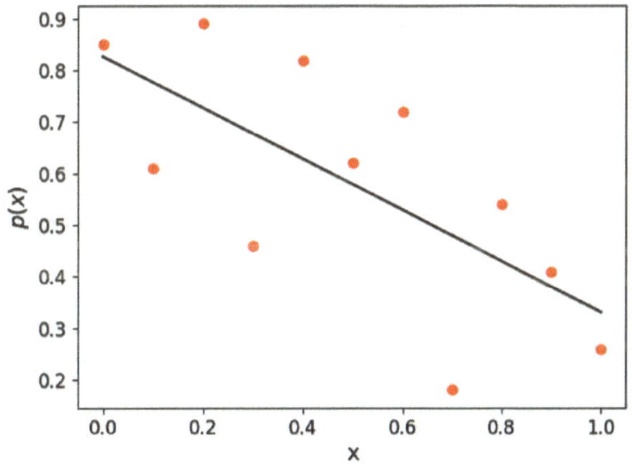

Fig. 1.9: Data points to be fitted (red dots) and the graph of a candidate function p.

When proximity is measured in the *least squares* sense, a formulation of this fitting problem is

$$\underset{c \in \mathbb{R}^2}{\text{minimize}}\, f(c) = \sum\nolimits_{i=1}^{m} (c_1 x_i + c_0 - y_i)^2,$$

where $c = (c_0, c_1)$. The notation $c \in \mathbb{R}^2$ under "minimize" is our way to specify that the vector c to be optimized is two-dimensional, i.e., c contains two decision variables. A formulation of an optimization problem is called an *optimization model* to stress its role in supporting decision making via modeling, which typically involves assumptions and simplifications. While p is a univariate function, the objective function f is bivariate as seen in Figure 1.10 when the data is taken from Figure 1.9. We say that f is defined on $\mathbb{R}^2$, the space of two-dimensional vectors.

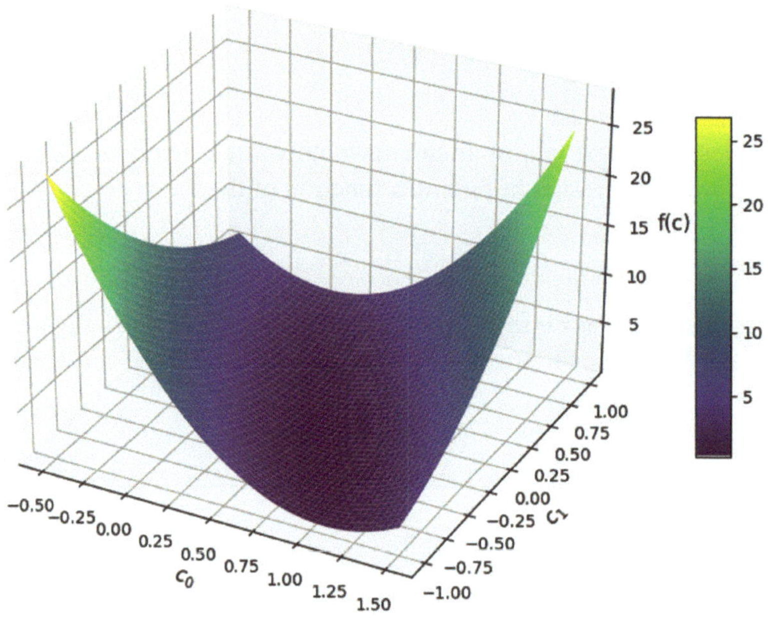

Fig. 1.10: Graph of the function f with variables c_0 and c_1.

Graphically, we identify a minimizer of f near $(0.8, -0.5)$; this might be seen most easily from the contour plot in Figure 1.11. A more principled approach relies on the Fermat rule and redirects the focus to solving

$$\nabla f(c) = \sum_{i=1}^{m} \begin{bmatrix} 2(c_1 x_i + c_0 - y_i) \\ 2(c_1 x_i + c_0 - y_i) x_i \end{bmatrix} = \begin{bmatrix} 0 \\ 0 \end{bmatrix}.$$

This vector equation amounts to two linear equations in the two unknown variables c_0 and c_1, with (x_i, y_i), $i = 1, \ldots, 11$, being the known data points from Figure 1.9. After some algebra, we find the unique solution $(0.82636, -0.49636)$, which actually corresponds to the function p graphed in Figure 1.9, i.e., $p(x) = 0.82636 - 0.49636x$. The solution produces a minimum value of f equal to 0.28495. No other coefficient values for c_0 and c_1 produce a smaller least-squares error. The minimum value amounts to the squared vertical distance between the data points and the line in Figure 1.9, summed up over all the data points, as we see from the definition of f.

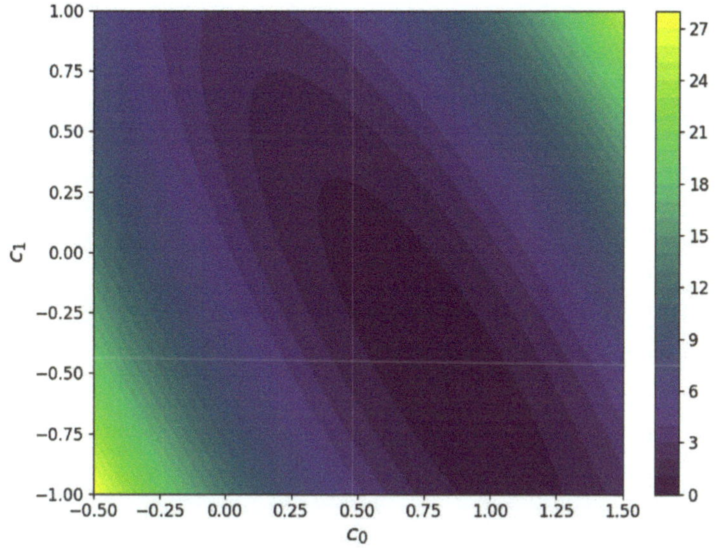

Fig. 1.11: Contour plot of the function f with variables c_0 and c_1.

We can generalize the problem by fitting a polynomial p_n of degree n, i.e.,

$$p_n(x) = c_n x^n + \cdots + c_2 x^2 + c_1 x + c_0,$$

which involves $1+n$ decision variables to be optimized. Setting $c = (c_0, c_1, \ldots, c_n)$ and recalling its interpretation as a column vector, the more general problem takes the form

$$\underset{c \in \mathbb{R}^{1+n}}{\text{minimize}}\, f(c) = \sum\nolimits_{i=1}^{m} \left(c_n x_i^n + \cdots + c_2 x_i^2 + c_1 x_i + c_0 - y_i\right)^2 = (Dc-y)^\top (Dc-y), \quad (1.6)$$

where D is the $m \times (1+n)$-matrix

$$D = \begin{bmatrix} 1 & x_1 & x_1^2 & \cdots & x_1^n \\ 1 & x_2 & x_2^2 & \cdots & x_2^n \\ \vdots & \vdots & \vdots & \ddots & \vdots \\ 1 & x_m & x_m^2 & \cdots & x_m^n \end{bmatrix} \quad \text{and} \quad y = \begin{bmatrix} y_1 \\ y_2 \\ \vdots \\ y_m \end{bmatrix}.$$

We stress that the optimization is now over $1+n$ variables by writing $c \in \mathbb{R}^{1+n}$ under "minimize."

Matrices and vectors help us to represent complex expressions succinctly but can sometimes be tricky to set up. It might not be immediately clear how the long sum defining $f(c)$ equals $(Dc-y)^\top(Dc-y)$. The best way to confirm the formula is to simply carry out the matrix and vector multiplications.

Matrices and Vectors. For $m \times n$-matrix A and $n \times q$-matrix B, with

$$A = \begin{bmatrix} A_{11} & A_{12} & A_{13} & \cdots & A_{1n} \\ A_{21} & A_{22} & A_{23} & \cdots & A_{2n} \\ \vdots & \vdots & \vdots & \ddots & \vdots \\ A_{m1} & A_{m2} & A_{m3} & \cdots & A_{mn} \end{bmatrix} \qquad B = \begin{bmatrix} B_{11} & B_{12} & B_{13} & \cdots & B_{1q} \\ B_{21} & B_{22} & B_{23} & \cdots & B_{2q} \\ \vdots & \vdots & \vdots & \ddots & \vdots \\ B_{n1} & B_{n2} & B_{n3} & \cdots & B_{nq} \end{bmatrix},$$

the multiplication AB is well-defined because the number of columns in A matches the number of rows in B. Specifically, AB is an $m \times q$-matrix with (i, j)-element

$$\sum_{k=1}^{n} A_{ik} B_{kj}.$$

The *transpose* of B is the $q \times n$-matrix

$$B^\top = \begin{bmatrix} B_{11} & B_{21} & B_{31} & \cdots & B_{n1} \\ B_{12} & B_{22} & B_{32} & \cdots & B_{n2} \\ \vdots & \vdots & \vdots & \ddots & \vdots \\ B_{1q} & B_{2q} & B_{3q} & \cdots & B_{nq} \end{bmatrix},$$

which means that $AB^\top$ is well-defined if $q = n$. Generally, $(AB)^\top = B^\top A^\top$. We may view an n-dimensional vector as an $n \times 1$-matrix, i.e., a *column vector*. In parallel, it can be viewed as a point in n-dimensional space ($\mathbb{R}^n$).

In (1.6), Dc is well defined because D is an $m \times (1 + n)$-matrix and c is an $(1+n)$-column vector, and the multiplication Dc produces an m-dimensional column vector. Subtraction yields the column vector $Dc - y$ so that $(Dc - y)^\top$ becomes an m-dimensional *row vector*. Viewing $(Dc - y)^\top (Dc - y)$ as the multiplication of a $1 \times m$-matrix by an $m \times 1$-matrix, we obtain that $f(c)$ is a 1×1-matrix, i.e., a *scalar*.

Applying the Fermat rule to (1.6), we observe that the objective function

$$f(c) = (Dc - y)^\top (Dc - y) = (Dc)^\top Dc - 2y^\top Dc + y^\top y = c^\top D^\top Dc - 2y^\top Dc + y^\top y.$$

We return to differentiation of functions of this form in Example 1.10 and simply note that if D and c were scalars then the derivative of $c^\top D^\top Dc = D^\top Dc^2$ would be $2D^\top Dc$. This "rule" yields

$$\nabla f(c^\star) = 2D^\top Dc^\star - 2D^\top y = 0 \tag{1.7}$$

for any minimizer $c^\star$ of f.

Checking dimensions in (1.7), the first term multiplies $(1 + n) \times m$ and $m \times (1 + n)$ matrices, and a $(1+n)$-dimensional column vector and this produces a $(1+n)$-dimensional column vector. The second term multiplies a $(1 + n) \times m$ matrix with an m-dimensional column vector to produce a $(1+n)$-dimensional column vector. Thus, $\nabla f(c^\star)$ is a $(1+n)$-dimensional column vector, which matches the dimension of c as expected.

The vector equation (1.7) is equivalent to

$$D^\top D c^\star = D^\top y. \tag{1.8}$$

This is a linear system of $n + 1$ equations with $n + 1$ unknowns in the vector $c^\star$.

Linear Algebra. The n-dimensional vectors $\{v^i, i = 1, \ldots, m\}$ are *linearly independent* when $\sum_{i=1}^m \alpha_i v^i = 0$ implies that $\alpha_i = 0$ for all i. For example, the 3-dimensional vectors $v^1 = (1, 0, 0)$ and $v^2 = (0, 1, 0)$ are linearly independent because the vector equation

$$\alpha_1 v^1 + \alpha_2 v^2 = \alpha_1 \begin{bmatrix} 1 \\ 0 \\ 0 \end{bmatrix} + \alpha_2 \begin{bmatrix} 0 \\ 1 \\ 0 \end{bmatrix} = \begin{bmatrix} \alpha_1 \\ \alpha_2 \\ 0 \end{bmatrix} = \begin{bmatrix} 0 \\ 0 \\ 0 \end{bmatrix}$$

necessarily implies that α_1 and α_2 are zero. In contrast, if $v^2 = (3, 0, 0)$, then

$$\alpha_1 v^1 + \alpha_2 v^2 = \alpha_1 \begin{bmatrix} 1 \\ 0 \\ 0 \end{bmatrix} + \alpha_2 \begin{bmatrix} 3 \\ 0 \\ 0 \end{bmatrix} = \begin{bmatrix} \alpha_1 + 3\alpha_2 \\ 0 \\ 0 \end{bmatrix} = \begin{bmatrix} 0 \\ 0 \\ 0 \end{bmatrix}$$

for nonzero α_1 and α_2.

The *rank* of an $m \times n$-matrix A is the number of linearly independent columns, or rows, in A. If $m = n$ and A has *full rank* (i.e., the rank is n), then A is *invertible* and its *inverse* A^{-1} satisfies $AA^{-1} = A^{-1}A = I$, where I is the *identity matrix* with ones along the diagonal and zeros elsewhere.

For any $m \times n$-matrix A, the $n \times n$-matrix $A^\top A$ has full rank if and only if the rank of A is n.

Assuming that $m \geq 1+n$, the columns of D are linearly independent because $x_1, \ldots, x_m$ are distinct points. Hence, $D^\top D$ is invertible and the system of equations in (1.8) is uniquely solved by

$$c^\star = (D^\top D)^{-1} D^\top y.$$

This is also the unique solution of (1.7).

Instead of computing the inverse, it's usually more efficient to solve a system of equations by Gauss elimination, iterative methods, or factor methods such as Cholesky decomposition. Figure 1.12 shows the optimal polynomial when $n = 5$. The minimum value of f equals 0.25265. This value is lower than 0.28495 obtained at the beginning of the section under the assumption of $n = 1$; see Figure 1.9. Since the case with $n = 1$ can be viewed as optimizing $c = (c_0, c_1, c_2, c_3, c_4, c_5)$ but with c_2, c_3, c_4, and c_5 forced to zero, it's natural that lifting this restriction provides a lower least-squares error. A comparison between Figures 1.9 and 1.12 indeed shows that the higher-order polynomial gets closer to the data points than the straight line.

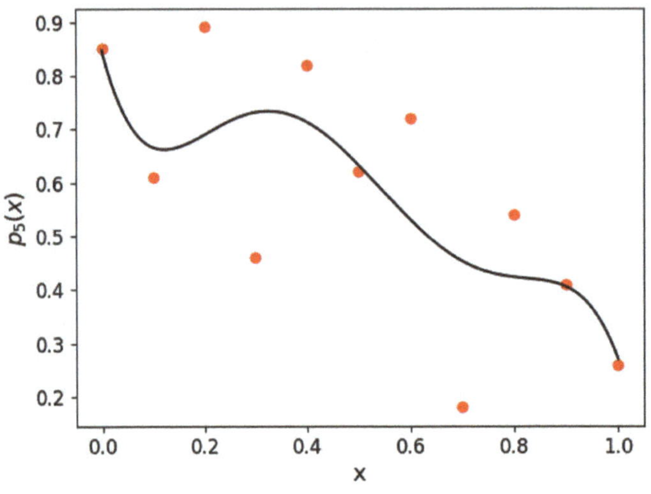

Fig. 1.12: An optimal fit of a polynomial of degree $n = 5$.

1.C Optimization under Uncertainty

In applications, problems are rarely as clear-cut as in the preceding sections. A frequent issue is that of data uncertainty. A versatile approach for dealing with a parameter to which we can't assign a specific value, perhaps because it's intrinsically varying or represents an unknown condition, is to model the parameter as a *random variable*. This allows us to leverage the tools of probability theory and statistics and brings us to *stochastic optimization problems*.

Random Variables. A random variable $\boldsymbol{\xi}$ (the Greek letter "ksi") takes on different values (outcomes) according its *distribution function* P, also called a cumulative distribution function (cdf). The probability that $\boldsymbol{\xi}$ takes a value no greater than ξ is then $P(\xi)$. (We denote random variables such as $\boldsymbol{\xi}$ using boldface, returning to regular font, such as ξ, for their values).

If P can be expressed by a *density function* p (also called a probability density function (pdf)), then

$$P(\xi) = \int_{-\infty}^{\xi} p(\xi_0)\, d\xi_0.$$

The *expectation* of $\boldsymbol{\xi}$ (or average, or expected value) is then

$$\mathbb{E}[\boldsymbol{\xi}] = \int_{-\infty}^{\infty} \xi p(\xi)\, d\xi.$$

Generally, a random variable $\boldsymbol{\xi}$ is *integrable* if $\mathbb{E}[\boldsymbol{\xi}]$ is finite; see [30, Section 3.B].

If $\boldsymbol{\xi}$ has a *finite distribution* with outcomes $\xi_1, \dots, \xi_m$ occurring with probabilities $p_1, \dots, p_m$, then

$$P(\xi) = \sum\nolimits_{i \mid \xi_i \le \xi} p_i \quad \text{and} \quad \mathbb{E}[\boldsymbol{\xi}] = \sum\nolimits_{i=1}^m \xi_i p_i.$$

Here the sum on the left is only over indices i with outcome $\xi_i \le \xi$.

Example 1.3 (newsvendor problem). In a contractual agreement, a newsvendor places an order for the daily delivery of a fixed number of newspapers to meet an uncertain daily demand $\boldsymbol{\xi}$. We think of $\boldsymbol{\xi}$ as a random variable with known distribution function P. The newsvendor is charged o cents per paper ordered and sells each for s cents, with $s > o$. Unsold papers can't be returned and are worthless at the end of the day. The goal of the newsvendor is to choose an order quantity that minimizes the expected (average) loss. This is a problem from the area of *inventory management*.

Detail. When ordering x newspapers and ξ is the demand, the loss (expense minus income) turns out to be

$$\begin{cases} ox - sx & \text{if } \xi \ge x \\ ox - s\xi & \text{otherwise,} \end{cases}$$

with negative values implying a profit. Thus, the loss can be written concisely as

$$\max\{ox - sx,\ ox - s\xi\}.$$

If the distribution function P can be expressed by a density function p, then an order of x newspapers would yield an expected loss of

$$\begin{aligned}
f(x) &= \mathbb{E}\big[\max\{ox - sx,\ ox - s\boldsymbol{\xi}\}\big] \\
&= \int_0^\infty \max\{ox - sx,\ ox - s\xi\} p(\xi)\, d\xi \\
&= \int_0^x (ox - s\xi) p(\xi)\, d\xi + \int_x^\infty (o - s) x p(\xi)\, d\xi \\
&= ox \int_0^x p(\xi)\, d\xi - s \int_0^x \xi p(\xi)\, d\xi + ox \int_x^\infty p(\xi)\, d\xi - sx \int_x^\infty p(\xi)\, d\xi \\
&= ox \int_0^\infty p(\xi)\, d\xi - s \int_0^x \xi p(\xi)\, d\xi - sx \int_x^\infty p(\xi)\, d\xi \\
&= ox \int_0^\infty p(\xi)\, d\xi - s \int_0^x \xi p(\xi)\, d\xi - sx \int_0^\infty p(\xi)\, d\xi + sx \int_0^x p(\xi)\, d\xi \\
&= (o - s)x + s \int_0^x (x - \xi) p(\xi)\, d\xi.
\end{aligned}$$

Here we consider only positive demand and set $p(\xi) = 0$ for negative ξ and then $\int_0^\infty p(\xi)\, d\xi = 1$.

The Fermat rule requires us to compute the derivative of f. Since f is given in terms of an integral, we can bring in the Leibniz rule for differentiation under an integral[1]. Applying the Leibniz and Fermat rules to the newsvendor problem, we establish that a minimizer $x^\star$ of f satisfies

$$f'(x^\star) = o - s + s(x^\star - x^\star)p(x^\star) + s\int_0^{x^\star} p(\xi)\, d\xi = o - s + sP(x^\star) = 0.$$

This is equivalent to having $P(x^\star) = 1 - o/s$; see Figure 1.13. In the figure, $x^\star$ is unique, but this isn't always the case because the graph of P could be "flat" when P has value $1 - o/s$.

In this problem the Fermat rule specifies not only a necessary but also a sufficient condition for a minimizer. Thus, the order quantity computed from the above expression actually minimizes the expected loss. A formal argument for this relies on the development in the following sections. □

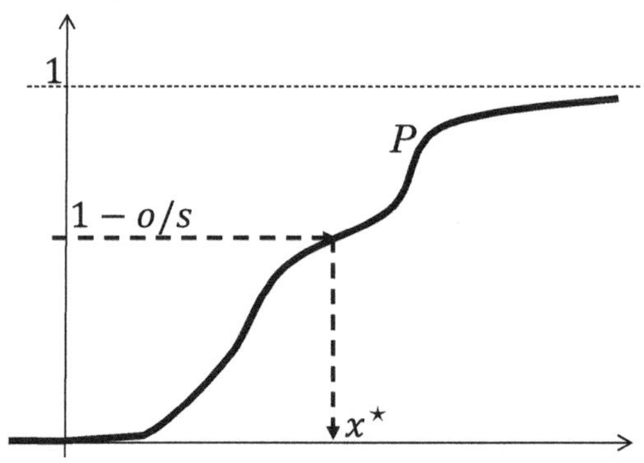

Fig. 1.13: An optimal order quantity $x^\star$ of newspapers under uncertain demand given by the distribution function P.

Usually, we construct a stochastic optimization problem from a *deterministic version*, where uncertainty about parameters is ignored. For the newsvendor problem in Example 1.3, we might want to minimize loss (expense minus income) subject to a requirement that an *assumed* demand $\hat{\xi}$ is met. This results in the (trivial) model

$$\underset{x\in\mathbb{R}}{\text{minimize}}\ (o - s)x \ \text{ subject to } \ x = \hat{\xi},$$

which specifies that the order quantity should be the assumed demand $\hat{\xi}$. However, it's unlikely that the demand will exactly match $\hat{\xi}$. Since the demand isn't known at the time

[1] Under some regularity conditions, if $f(x) = \int_0^x g(x,y)dy$, then $f'(x) = g(x,x) + \int_0^x g'_x(x,y)dy$.

of ordering newspapers, the actual loss has to be evaluated after the decision is made. Examining the decision and evaluation process more closely, we see it follows the pattern:

$$\text{decision: } x \rightsquigarrow \text{ observation: } \xi \rightsquigarrow \text{ evaluation: } \xi - x.$$

Moreover, the expected loss f in Example 1.3 is made up of two terms. The first one, $(o - s)x$, corresponds to the loss accrued when ordering and selling x newspapers. It's a negative value reflecting a positive margin for each sale. The second term evaluates the order quantity based on the disparity between the demand (ξ) and the available newspapers (x), which can only be measured *after* the demand is observed. The second term is the *expected recourse cost*, which can be expressed concisely as

$$\mathbb{E}\big[h(x - \boldsymbol{\xi})\big] = \int_0^\infty h(x - \xi)p(\xi)\, d\xi,$$

with recourse cost function

$$h(y) = \max\{0, sy\} = \begin{cases} 0 & \text{if } y < 0 \\ sy & \text{otherwise;} \end{cases}$$

see Figure 1.14. The expected recourse cost captures the fact that there's no additional cost if the demand turns out to be above the order quantity x. If the demand is lower, however, there's the additional cost $s(x - \xi)$ associated with the $x - \xi$ newspapers never sold but prematurely included in the first term. The recourse cost function h expresses how we evaluate different outcomes of $x - \boldsymbol{\xi}$, and the exact form of the function could vary depending on the application. For example, both $x > \xi$ and $x < \xi$ might be associated with a positive cost. Regardless of the form, the name "recourse" alludes to the possibility of adjustments beyond a simple evaluation after the demand has been observed. In the newsvendor problem, there's no real adjustment possible, but in Chapter 3 we'll consider models involving a "second-stage" decision.

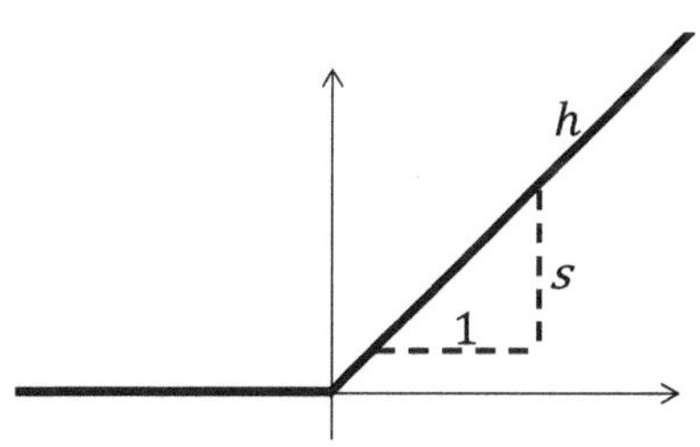

Fig. 1.14: The recourse cost function given by $h(y) = \max\{0, sy\}$.

In view of this discussion, we can formulate the newsvendor problem by starting from the deterministic version and adding the expected recourse cost to obtain

$$\underset{x \in \mathbb{R}}{\text{minimize}}\ (o - s)x + \mathbb{E}\big[h(x - \boldsymbol{\xi})\big]. \tag{1.9}$$

The resulting objective function is merely an alternative expression for f from Example 1.3 but illustrates two features of stochastic optimization problems: Decisions are modeled relative to the arrival of information and minimizers are obtained in a trade-off between initial (first-stage) and expected recourse (second-stage) costs.

1.D Convexity

In optimization, convexity plays the role of a fault line, both theoretically and practically. On the "convex side," we're on solid ground to engineer efficient numerical procedures whereas on the other side one may have to exert an exponentially growing amount of computational effort as the problem size increases or settle for algorithms that produce solutions satisfying an optimality condition derived from the Fermat rule. In this section, we'll lay out conditions for convexity and examine the consequences.

Definition 1.4 (convex set). *A set C is* convex *if it's either empty or for any $x^0, x^1 \in C$ and $\lambda \in [0,1]$, one has $(1-\lambda)x^0 + \lambda x^1 \in C$.* □

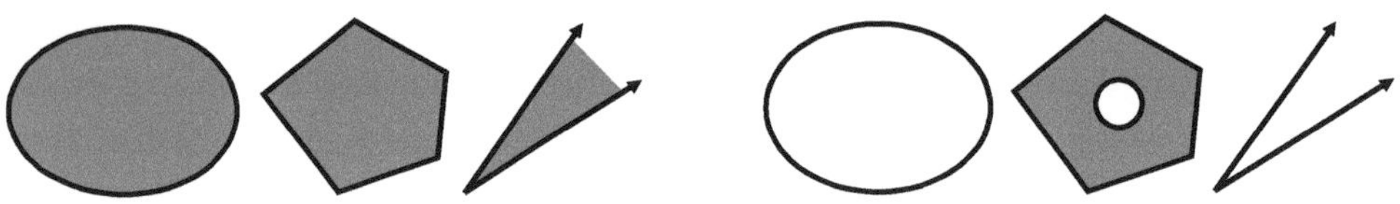

Fig. 1.15: Some convex (left) and non-convex (right) sets in 2-dimensions ($\mathbb{R}^2$).

Balls, cubes, lines, quadrants, planes, and the entire n-dimensional space ($\mathbb{R}^n$) are examples of convex sets because a line segment between any two points in such a set will be fully in the set. Sets with dents or holes aren't convex; cf. Figure 1.15.

We next consider functions, the mathematical construct we'll use throughout to quantify how cost, loss, risk, and other performance metrics vary across the possible decisions.

Functions. A function takes an n-dimensional vector as input and produces a scalar output according to a formula or rule. Often, we denote the function by f and its input by x so that the output becomes $f(x)$ but notation can vary. While $f(x)$ is typically a real number, it can sometimes be ∞ and $-\infty$ to model situations in which certain decisions need to be barred and some formulae aren't valid in the traditional sense. For example, when $f(x) = 1/x$ for $x \neq 0$, one might assign $f(0) = \infty$ and thereby achieve a function that returns a value for all x.

When minimizing a function f, an argument with value ∞ would correspond to the worst conceivable situation; think of $f(x)$ as the cost associated with decision x. In the pathological situation when $f(x) = \infty$ for every x, then we define $\operatorname{argmin} f = \emptyset$. Thus, there are no minimizers of this function. All decision alternatives are equally intolerable. This leads us to define the *set of minimizers* as

$$\operatorname{argmin} f = \{x^\star \mid f(x^\star) \le f(x)\ \forall x \text{ and } f(x^\star) < \infty\}.$$

While we're used to draw the graph of a function f, it turns out that its *epigraph*, written

$$\operatorname{epi} f = \{(x, \alpha) \mid \alpha \ge f(x)\},$$

is more informative. The epigraph is the subset of points in $n+1$ dimensions ($\mathbb{R}^{n+1}$) on and above the graph of f; see Figure 1.16.

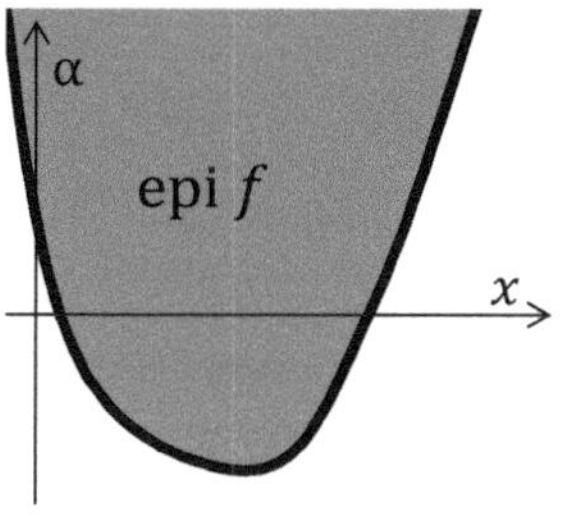

Fig. 1.16: Epigraph of a univariate function f.

Definition 1.5 (convex function). *A function f in n variables is* convex *if its epigraph is a convex set in the $(n+1)$-dimensional space $\mathbb{R}^{n+1}$.* □

A convex function is either linear, *affine* (i.e., linear plus a constant), or has an epigraph with just one "valley" as illustrated in Figure 1.17; the function in Figure 1.16 also turns out to be convex. Moreover, the absolute value function given by $f(x) = |x|$ and the exponential functions given by $f(x) = e^{\alpha x}$ are convex functions regardless of α. The sine function expressed by $f(x) = \sin x$ serves as a typical example of nonconvexity.

It becomes harder to visualize functions in higher dimensions, but Figure 1.18(left) shows the graph when

$$f(x) = \tfrac{3}{2}x_1^2 + 2\left(x_2 - \tfrac{1}{2}\right)^2.$$

Figure 1.18(right) shows the function $g = -f$, which is the upside-down version of f. While f is convex, g isn't because of the inward bulge in its epigraph; recall that the epigraph consists of all the points on and *above* the graph.

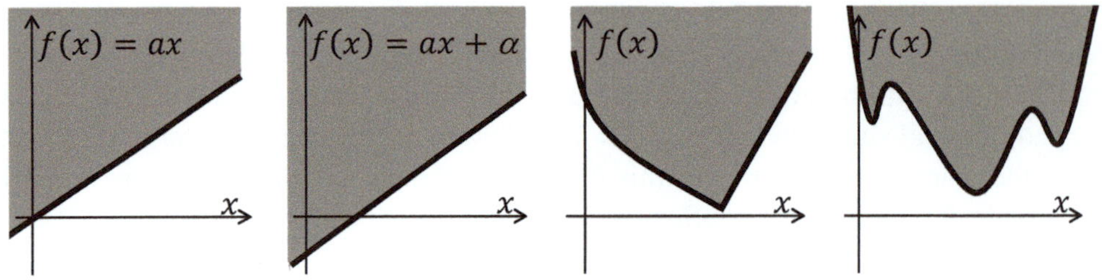

Fig. 1.17: Epigraphs of convex functions: linear (left), affine (middle left), and nonlinear (middle right). Epigraph of nonconvex function (right).

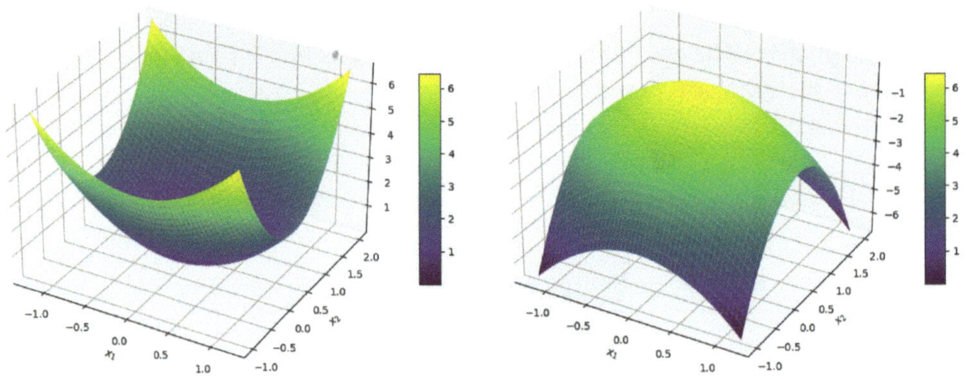

Fig. 1.18: Graphs for $f(x) = (3/2)x_1^2 + 2(x_2 - 1/2)^2$ (left) and $g = -f$ (right), with f being convex but g is nonconvex.

A main consequence of convexity is that the Fermat rule specifies a condition that's necessary *and* sufficient for a point to be a minimizer. Thus, algorithms can concentrate on finding a point with zero gradient. Figure 1.18 shows the essential difference between convex and nonconvex functions. The function graphed on the left has gradient

$$\nabla f(x) = \begin{bmatrix} 3x_1 \\ 4(x_2 - 1/2) \end{bmatrix},$$

which vanishes at $x = (0, 1/2)$. Thus, the Fermat rule correctly identifies the unique minimizer of f. The situation is different for the nonconvex function $g = -f$ in the right portion of the figure. While

$$\nabla g(x) = -\nabla f(x) = \begin{bmatrix} -3x_1 \\ -4(x_2 - 1/2) \end{bmatrix}$$

also vanishes at $(0, 1/2)$, that point isn't a minimizer of g. In fact, it's a maximizer!

Formally, we've the following fundamental consequence of convexity. We refer to [30, Section 1.D] for additional facts and proofs.

Theorem 1.6 (optimality condition). *For a smooth convex function f, one has*

$$x^\star \in \operatorname{argmin} f \quad \textit{if and only if} \quad \nabla f(x^\star) = 0.$$

□

For a smooth convex function f, Theorem 1.6 formally establishes that solving the vector equation $\nabla f(x) = 0$ is equivalent to finding a minimizer of f. This holds regardless of the dimension of x. Thus, the seemingly hopeless process of checking different x-points across an n-dimensional space in search for a minimizer can be replaced by more tractable equation solving.

A main goal is therefore to determine whether a given function is convex, and more generally, whether a given set is convex. We summarize a few key rules for checking convexity.

Proposition 1.7 (convex sets). *The following sets are convex:*

(a) *An affine set $\{x \mid Ax = b\}$, for any $m \times n$-matrix A and m-dimensional vector b.*
(b) *A level-set $\{x \mid f(x) \le \alpha\}$, for any convex function f and scalar α.*
(c) *An intersection $\cap_{\alpha \in \mathbb{A}} C_\alpha$, for any collection $\{C_\alpha,\ \alpha \in \mathbb{A}\}$ of convex sets.*

□

A special case of an affine set in Proposition 1.7(a) has

$$x = \begin{bmatrix} x_1 \\ x_2 \end{bmatrix}, \quad A = \begin{bmatrix} 1 & 1 \\ 0 & 1 \end{bmatrix}, \quad b = \begin{bmatrix} 2 \\ 3 \end{bmatrix}.$$

Then $Ax = b$ is a system of two equations in two variables. The affine set $\{x \mid Ax = b\}$ is the set of solutions to that system. Writing out the expression, we obtain

$$\begin{bmatrix} 1 & 1 \\ 0 & 1 \end{bmatrix} \begin{bmatrix} x_1 \\ x_2 \end{bmatrix} = \begin{bmatrix} 2 \\ 3 \end{bmatrix}, \quad \text{which simplifies to} \quad x_1 + x_2 = 2,\ \ x_2 = 3.$$

Thus $\{x \mid Ax = b\} = \{(-1, 3)\}$, a single point. Definition 1.4 already confirms that any single point is a convex set. Proposition 1.7(a) allows us to reach the same conclusion without having to solve the system of equations.

Figure 1.19 illustrates the level-set rule (left portion of the figure) and the intersection rule (right portion) in Proposition 1.7(b,c), respectively. A level-set consists of points satisfying a certain inequality. While the figure on the right only shows the intersection of two sets, the intersection of any number of convex sets is convex.

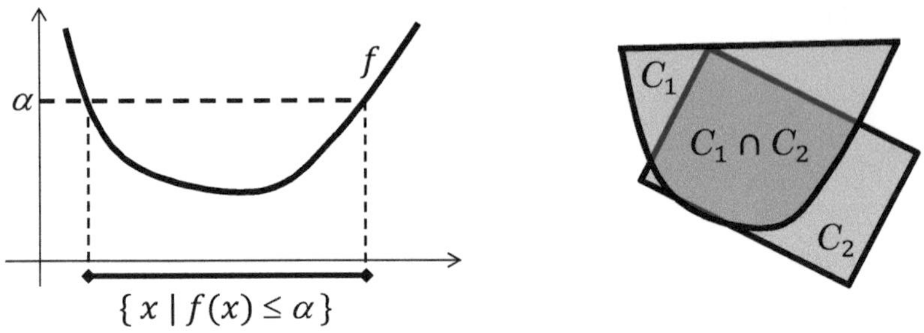

Fig. 1.19: Level-set rule (left) and intersection rule (right) establish convex sets.

Proposition 1.8 (convex functions). *The following functions are convex:*

(a) $f(x) = \sum_{i=1}^{m} f_i(x)$ *for convex functions* f_i (sum rule).
(b) $f(x) = \alpha g(x)$ *for convex function* g *and* $\alpha \geq 0$ (multiplication rule).
(c) $f(x) = \max_{i=1,\ldots,m} f_i(x)$ *for convex functions* f_i (max rule).
(d) $f(x) = g(Ax - b)$ *for convex function* g*,* $m \times n$*-matrix* A*, and* m*-dimensional vector* b (composition rule).

□

Proposition 1.8 is our main workhorse for checking convexity of a function defined in terms of other functions that we already know are convex. For example, the epigraph of the function given by $g(z) = \exp(z)$ is convex. This means that g is convex and also $f(x) = \exp(2x_1 + 3x_2 - 1)$ by the composition rule in Proposition 1.8(d). We note that g is a function in one variable z, but f has two variables $x = (x_1, x_2)$. This is unproblematic as f and g are functions of n and m variables, respectively, in Proposition 1.8(d).

We also use second-order derivatives to check whether a function is convex. Since the derivative of a univariate function f is a function in its own right, we can repeat the process of derivation and compute the derivative of the derivative at a point $\bar{x}$, which we denote by $f''(\bar{x})$. For example, one has

$$f(x) = 2x^2 + 3x - 1, \quad f'(x) = 4x + 3, \quad f''(x) = 4.$$

The concept extends to multivariate functions. For a function f in n dimensions and a point $\bar{x}$, we recall that $f'_{x_i}(\bar{x})$ is the partial derivative of f with respect to the ith variable x_i at $\bar{x}$. This partial derivative is in of itself a multivariate function and we may compute its partial derivative with respect to x_j at $\bar{x}$, which we denote by $f''_{x_i x_j}(\bar{x})$. Typically, we put together all these second-order partial derivatives at $\bar{x}$ in the $n \times n$ *Hessian* matrix

$$\nabla^2 f(\bar{x}) = \begin{bmatrix} f''_{x_1 x_1}(\bar{x}) & \cdots & f''_{x_1 x_n}(\bar{x}) \\ \vdots & \ddots & \vdots \\ f''_{x_n x_1}(\bar{x}) & \cdots & f''_{x_n x_n}(\bar{x}) \end{bmatrix}.$$

The order of differentiation is immaterial so $\nabla^2 f(\bar{x})$ is a symmetric matrix. (Generally, a matrix A is *symmetric* if $A = A^\top$.) Since both partial derivatives and second-order partial derivatives rely on certain limits to exist (recall the definition in (1.2)), a given function may not have well-defined gradients and Hessians. We say that f is *twice smooth* if $\nabla^2 f(x)$ is defined for all x and it varies continuously. For twice smooth functions, the question of convexity is addressed by linear algebra.

Semidefinite Matrices. An $n \times n$-matrix Q is *positive semidefinite* if $y^\top Q y \geq 0$ for all y. When Q is symmetric, it's positive semidefinite if and only if all its eigenvalues are nonnegative. The 2×2-matrix

$$Q = \begin{bmatrix} a & b \\ b & c \end{bmatrix}$$

is positive semidefinite if and only if $a \geq 0$, $c \geq 0$, and $ac - b^2 \geq 0$.

Proposition 1.9 (Hessian rule for convexity). *For a twice smooth function f, one has*

$$f \text{ is convex } \quad \text{if and only if} \quad \nabla^2 f(x) \text{ is positive semidefinite for all } x.$$

□

The proposition translates the question of convexity into one from linear algebra by using differentiation. Calculating Hessian matrices can be tedious and even computationally demanding for complicated functions, but there are exceptions as seen in the next example.

Example 1.10 (quadratic functions). A function f is *quadratic* if it's expressible as

$$f(x) = \tfrac{1}{2} x^\top Q x + c^\top x + \alpha$$

for some symmetric $n \times n$-matrix Q, n-dimensional vector c, and scalar α. Then,

$$\nabla f(x) = Qx + c \quad \text{and} \quad \nabla^2 f(x) = Q,$$

so f is convex if and only if Q is positive semidefinite by Proposition 1.9. These formulas are convenient and they apply for any number of variables n.

Detail. If $\alpha = 1$, $c = (3, 6)$, and

$$Q = \begin{bmatrix} 5 & 2 \\ 2 & 3 \end{bmatrix},$$

then the function f has the specific formula

$$f(x) = \frac{1}{2}[x_1 \ \ x_2]\begin{bmatrix} 5 & 2 \\ 2 & 3 \end{bmatrix}\begin{bmatrix} x_1 \\ x_2 \end{bmatrix} + [3 \ \ 6]\begin{bmatrix} x_1 \\ x_2 \end{bmatrix} + 1.$$

This expression simplifies to

$$f(x) = \tfrac{5}{2}x_1^2 + \tfrac{3}{2}x_2^2 + 2x_1x_2 + 3x_1 + 6x_2 + 1.$$

Its partial derivatives become

$$f'_{x_1}(x) = 5x_1 + 2x_2 + 3, \quad f'_{x_2}(x) = 2x_1 + 3x_2 + 6,$$

which then coincide with the result from the claimed formula $\nabla f(x) = Qx + c$. For the second-order partial derivatives, we find

$$f'_{x_1x_1}(x) = 5, \quad f'_{x_1x_2}(x) = 2, \quad f'_{x_2x_1}(x) = 2, \quad f'_{x_2x_2}(x) = 3,$$

which results in $\nabla^2 f(x) = Q$. □

Length of Vectors. For any $x = (x_1, \ldots, x_n)$, we compute its length as

$$\|x\|_2 = \sqrt{x_1^2 + \cdots + x_n^2} = \sqrt{x^\top x}.$$

This is the usual *Euclidean distance*, but one can also compute the *Manhattan distance*

$$\|x\|_1 = |x_1| + \cdots + |x_n|.$$

Generally, functions computing lengths of vectors in various ways are called *norms* and they're always convex.

The benefits of convexity aren't contingent on smoothness. (The statement in Theorem 1.6 is preliminary with extensions in Chapter 2.) The main consequence of convexity is that global properties of a function such as its minimizers can be ascertained from a local analysis and this fact is unrelated to smoothness.

We say that $x^\star$ is a *local minimizer* of f if $f(x^\star) < \infty$ and there's $\varepsilon > 0$ such that

$$f(x^\star) \le f(x) \quad \text{when } \|x - x^\star\|_2 \le \varepsilon.$$

A minimizer is sometimes called a *global minimizer*, for emphasis. Every minimizer is a local minimizer because then ε can be taken to be any positive number. Figure 1.20 illustrates that every local minimizer of a *convex* function is actually a minimizer. However, a general function might have a graph with many "valleys" and thus many local minimizers too.

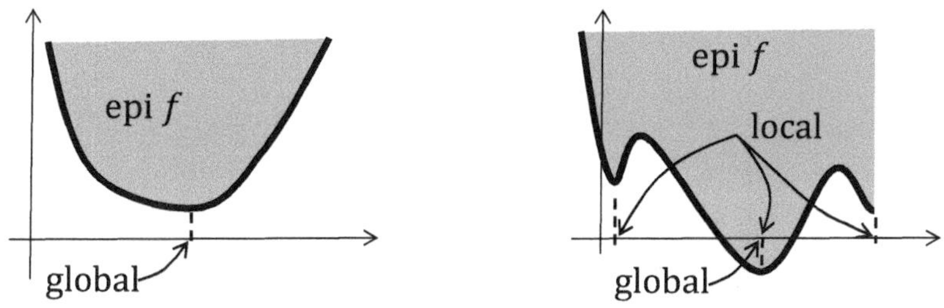

Fig. 1.20: Local and global minimizers for convex (left) and nonconvex (right) functions. The function on the right has value infinity to the right of the right-most local minimizer.

Theorem 1.11 (local implies global minimizers). *Every local minimizer of a convex function is a (global) minimizer.* □

Throughout, we put a firm emphasis on optimization problems for which the goal is to *minimize*, not maximize, a function, i.e., *minimization problems*. This one-sided focus has implications for definitions and concepts. For example, convexity dominates *concavity* (A function f is concave if $-f$ is convex.) We'll occasionally deal with the parallel universe of *maximization problems*, but a comprehensive treatment is largely redundant as most concepts are obtained through a sign change. In particular, maximizing a function f is equivalent to minimizing the function $-f$. Thus, a *maximizer* of f is simply a minimizer of $-f$.

1.E Estimation and Classification

A common task in data analytics, learning, and statistics is to predict a quantity of interest from available data. A prime example is the problem of estimating the density function of a random variable on the basis of observations of its values. Often, we seek a density function that "best" fits the data in some sense and this lands us squarely in optimization territory.

Suppose that we observe m times the outcome of a random phenomenon such as the height of this year's graduating students. Let $\{\xi_1, \dots, \xi_m\}$ be the observed values. We view the data as the outcome of m random variables $\{\boldsymbol{\xi}_1, \dots, \boldsymbol{\xi}_m\}$. How can we estimate the density function of these random variables and thereby obtain a *model of uncertainty* about future values? A common approach is to adopt a collection of density functions $\mathcal{P}$ and then determine $p \in \mathcal{P}$ that maximizes the *likelihood*

$$p(\xi_1)p(\xi_2) \cdot \ \dots \ \cdot p(\xi_m) = \prod\nolimits_{i=1}^{m} p(\xi_i)$$

of the observations.

Figure 1.21 shows that we seek a density function p with high (i.e., likely) values at the observations. The resulting p is a *maximum likelihood* estimate. (We're here making

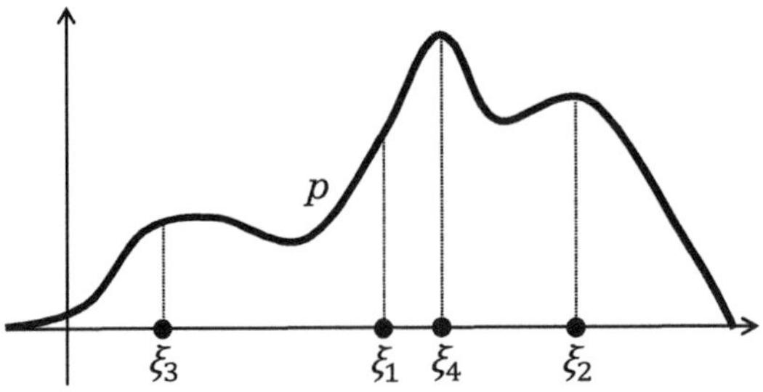

Fig. 1.21: Likelihood of observing $\xi_1, \ldots, \xi_4$ under density function p is the product of the "heights" at these points.

a nontrivial assumption that $\boldsymbol{\xi}_1, \ldots, \boldsymbol{\xi}_m$ are statistically independent and have the same distribution, which should be checked carefully in practice.) This leads to the optimization model

$$\underset{p \in \mathcal{P}}{\text{maximize}} \prod\nolimits_{i=1}^{m} p(\xi_i),$$

but let's convert it into a minimization problem and also introduce a logarithm. Since the logarithmic function is strictly increasing, we can replace the objective function by its logarithm:

$$\underset{p \in \mathcal{P}}{\text{maximize}} \ \ln \prod\nolimits_{i=1}^{m} p(\xi_i) = \sum\nolimits_{i=1}^{m} \ln p(\xi_i).$$

As the final step, the shift to minimization yields

$$\underset{p \in \mathcal{P}}{\text{minimize}} -\frac{1}{m} \sum\nolimits_{i=1}^{m} \ln p(\xi_i), \tag{1.10}$$

where the factor $1/m$ furnishes a normalization without changing the set of minimizers. A minimizer of the objective function in (1.10) is a density function that's most likely to generate the observed data $\{\xi_1, \ldots, \xi_m\}$. For example in the context of Section 1.C, with the data representing demand for newspapers in the past, the density function is a model of uncertainty about future demand. The difficulty of solving the problem depends on $\mathcal{P}$, but there are many tractable cases; see the next example and Section 3.C.

Example 1.12 (estimating a normal distribution). A *normal* (Gaussian) random variable $\boldsymbol{\xi}$ has a density function of the form

$$p(\xi) = \frac{1}{\sqrt{2\pi}\sigma} \exp\left(-\frac{(\xi - \mu)^2}{2\sigma^2} \right),$$

where

$$\mu = \mathbb{E}[\boldsymbol{\xi}] \qquad \textit{(mean, average, expected value)}$$

$$\sigma^2 = \text{var}(\boldsymbol{\xi}) = \mathbb{E}\big[(\boldsymbol{\xi} - \mu)^2 \big] = \mathbb{E}[\boldsymbol{\xi}^2] - \mu^2 \qquad \textit{(variance)}.$$

It's *standard normal* if $\mu = 0$ and $\sigma^2 = 1$.

Given outcomes $\{\xi_1, \dots, \xi_m\}$ from normally distributed random variables with common mean μ and variance σ^2, we seek to determine the maximum likelihood estimate of μ under the assumption that the variance $\sigma^2 \in (0, \infty)$ is known.

Detail. Since we assume that $\mathcal{P}$ in the maximum likelihood problem contains only normal density functions with a known variance, (1.10) reduces to a minimization problem over μ. For such density functions,

$$-\ln p(\xi_i) = f_i(\mu) = \frac{(\mu - \xi_i)^2}{2\sigma^2} + \ln \sigma + \ln \sqrt{2\pi}.$$

The function f_i, defined by the right-hand side above, is convex by Example 1.10 and the same holds for the whole objective function by Proposition 1.8(a,b). The Fermat rule results in the equation

$$0 = \frac{1}{m} \sum\nolimits_{i=1}^{m} f_i'(\mu) = \frac{1}{m} \sum\nolimits_{i=1}^{m} \frac{\mu - \xi_i}{\sigma^2} = \frac{1}{\sigma^2 m}\Big(m\mu - \sum\nolimits_{i=1}^{m} \xi_i\Big),$$

which has solution

$$\mu^\star = \frac{1}{m} \sum\nolimits_{i=1}^{m} \xi_i.$$

By Theorem 1.6, $\mu^\star$ is a minimizer and, in fact, there are no other minimizers. The maximum likelihood estimate is therefore a normal density function with mean $\mu^\star$ and variance σ^2. □

Example 1.13 (classification and logistic regression). Should an email message be classified as spam on the basis of its characteristics (wording, layout, length, signature, etc.) and those of past email messages that have been identified as spam or not? This is an example of a (statistical) *classification problem* which consists of assigning an observation to a category based on its features (characteristics) and a data set of observations with known features and category-membership. The problem, also referred to as a *learning problem*, can be formulated using an optimization model.

Detail. We're given the observations of m feature vectors $\{x^i \in \mathbb{R}^n, i = 1, \dots, m\}$ and their corresponding labels $\{y_i = \pm 1, i = 1, \dots, m\}$, which specify the category for each observation. In the email context, each i corresponds to an email, the vector x^i describes the features of the email, and y_i specifies the two categories: label 1 (spam) or -1 (non-spam). For example, the ith email in the data set might be

> Urgent! call 800 651 8721 immediately. Your complimentary 4* Las Vegas Holiday or $10,000 cash await collection

with label $y_i = 1$. The vector $x^i = (x_1^i, \dots, x_n^i)$ could describe which words from a dictionary of n words are present: $x_j^i = 1$ if the jth word in the dictionary is present and $x_j^i = 0$ otherwise. There are much more sophisticated representations of text in terms of vectors, for example those underpinning modern chat bots as we see in Example 3.9. Regardless, n tends to be large.

A maximum likelihood approach to classification, referred to as *logistic regression*, determines a function h with the following interpretation: $h(x)$ is the (estimated) probability that feature vector x corresponds to label -1 (non-spam). The probability to observe the data would then be

$$\prod\nolimits_{i \mid y_i=-1} h(x^i) \prod\nolimits_{i \mid y_i=1} \left(1 - h(x^i)\right)$$

and the maximum likelihood estimate is a function h that maximizes this quantity. The situation resembles that of (1.10) and again we need to settle on a collection of functions to optimize over. It's common to assume that

$$h(x) = \frac{1}{1 + \exp(a^\top x + \alpha)}, \quad \text{where } a \text{ is an } n\text{-dimensional vector and } \alpha \text{ is a scalar.}$$

As suitable for a probability, $h(x) \in (0,1)$ and can be any such number depending on the choice of a and α. The problem then reduces to optimizing the values of a and α in the expression

$$\begin{aligned}
&\prod\nolimits_{i \mid y_i=-1} h(x^i) \prod\nolimits_{i \mid y_i=1} \left(1 - h(x^i)\right) \\
&= \prod\nolimits_{i \mid y_i=-1} \frac{1}{1 + \exp(-y_i(a^\top x^i + \alpha))} \prod\nolimits_{i \mid y_i=1} \left(1 - \frac{1}{1 + \exp(y_i(a^\top x^i + \alpha))}\right).
\end{aligned}$$

Simplifying this expression, bringing in the logarithm, normalizing with $1/m$, and switching to minimization, we obtain the model

$$\mathop{\text{minimize}}_{(a,\alpha) \in \mathbb{R}^{n+1}} -\frac{1}{m} \ln \prod\nolimits_{i=1}^{m} \frac{1}{1 + \exp(-y_i(a^\top x^i + \alpha))} = \frac{1}{m} \sum\nolimits_{i=1}^{m} \ln\left(1 + e^{-y_i(a^\top x^i + \alpha)}\right).$$

A solution $(a^\star, \alpha^\star)$ of this problem defines a *classifier* $h^\star$, with

$$h^\star(x) = \frac{1}{1 + \exp(a^{\star\top} x + \alpha^\star)} \tag{1.11}$$

furnishing the probability that a feature vector x corresponds to label -1. When each component of x describes whether a particular word is included, $a^\star$ typically has positive values for the components corresponding to words often used in spam emails and negative values for words often used in non-spam. Thus, if x indeed corresponds to a spam email, then $a^{\star\top} x$ tends to be a high number and consequently $h^\star(x)$ is low: the classifier indeed predicts that x is most likely spam.

In view of Proposition 1.8(a,b,d), the objective function in the model is convex provided that the function given by $g(z) = \ln(1 + e^z)$ is convex. This is indeed the case since $g''(z) = e^z/(1+e^z)^2$ and one can invoke the Hessian rule from Proposition 1.9. In contrast to the previous example, it's not possible to solve this estimation problem analytically; the variables a and α must be determined by an algorithm. □

y_i	x_1^i (income)	x_2^i (height)	x_3^i (west coast)
1	12.1	70	1
1	13.5	69	1
1	23.4	70	0
1	8.1	72	1
1	7.3	67	1
−1	22.9	70	0
−1	2.1	67	0
−1	10.0	68	0
−1	3.7	70	1
−1	9.8	73	0

Table 1.1: Data for classifying a person as a Dodgers fan ($y_i = 1$) or not ($y_i = -1$), with income in tens of thousands of dollars, height in inches, and west coast home indicated by 1.

Example 1.14 (classifying LA Dodgers fans). We seek to classify whether a person is a fan of the Los Angeles Dodgers based on their annual income, height, and the location of their home. Using the label 1 for Dodgers fan and −1 for not Dodgers fan, the available data is given in Table 1.1, where each row represents a person. The income is given in tens of thousands of dollars and height in inches. The last column specifies a west coast home with 1, and anywhere else by 0.

Adopting the logistic regression approach of Example 1.13 with $m = 10$ data points and $n = 3$, we obtain a minimization problem over the decision variables $a = (a_1, a_2, a_3)$ and α that's solved using a Newton-type algorithm (see Section 1.G below for details). Specifically, we obtain

$$
\begin{aligned}
a_1^\star &= 62.4 && \text{(income)}\\
a_2^\star &= -10.1 && \text{(height)}\\
a_3^\star &= 981.2 && \text{(west coast)}\\
\alpha^\star &= -736.3.
\end{aligned}
$$

The probability that a feature vector $x = (x_1, x_2, x_3)$ corresponds to label −1 (not Dodgers fan) is then given by (1.11) with $a^\star = (a_1^\star, a_2^\star, a_3^\star)$. Since $a_1^\star = 62.4 > 0$, a larger value of x_1 (income) implies a smaller value of $h^\star(x)$ and thus higher chance of being a Dodgers fan. Likewise, living on the west coast increases the chances of being a Dodgers fan. The effect is opposite for height; being tall reduces the chance of being a Dodgers fan.

Detail. We test the predictions from $h^\star$ on the data set in Table 1.2 using the convention: If $h^\star(x) < 0.5$, then we predict that x corresponds to label 1 (Dodgers fan); otherwise we predict label -1 (not Dodgers fan).

The second-to-last column in Table 1.2 gives the probabilities $h^\star(x)$ that x isn't corresponding to a Dodgers fan. Thus, we make the correct prediction in each case; the first column (truth) and last column (prediction) in the table match. However, there's some "doubt" in the second-to-last row. The model is only 73% sure about a non-Dodgers fan. The source of confusion is the person's west coast home, but the low income provides sufficient evidence to correctly tilt the prediction toward a non-Dodgers fan. □

y (label)	x_1 (income)	x_2 (height)	x_3 (west coast)	$h^\star(x)$	prediction
1	19.2	71	1	0.00	Dodgers fan
1	8.6	69	1	0.00	Dodgers fan
-1	7.4	70	1	0.73	not Dodgers fan
-1	20.1	70	0	1.00	not Dodgers fan

Table 1.2: Test data for classifying Dodger fans with the last column giving the prediction.

1.F Gradient Descent Method

When we seek to minimize a function f in n dimensions, there are two main approaches. We can immediately leverage the Fermat rule and solve the vector equation $\nabla f(x) = 0$ or we can iteratively move from x-point to x-point with the goal of each time obtaining a lower value $f(x)$. This section adopts the second approach, while the first perspective is exploited in Section 1.G. From both approaches, which turn out to be intimately connected, we obtain far-reaching algorithms for solving optimization problems.

We say that an n-dimensional vector w is a *descent direction* for f at $\bar{x}$ if there's $\bar{\lambda} > 0$ such that

$$f(\bar{x} + \lambda w) < f(\bar{x}) \quad \forall \lambda \in (0, \bar{\lambda}).$$

Thus, an algorithm for minimizing a function f can be based on repeating the following simple but fruitful steps: From a given x, identify a descent direction w, determine $\lambda > 0$ such that

$$f(x + \lambda w) < f(x),$$

and replace x by $x + \lambda w$. One might hope that after many iterations of this kind, the algorithm reaches a point x where there are no decent directions. An algorithm that iteratively reduces the value of f in this manner is a *descent method.*

Proposition 1.15 (descent direction). *For a smooth function f, every vector w with*

$$\nabla f(\bar{x})^\top w < 0$$

is a descent direction for f at $\bar{x}$. □

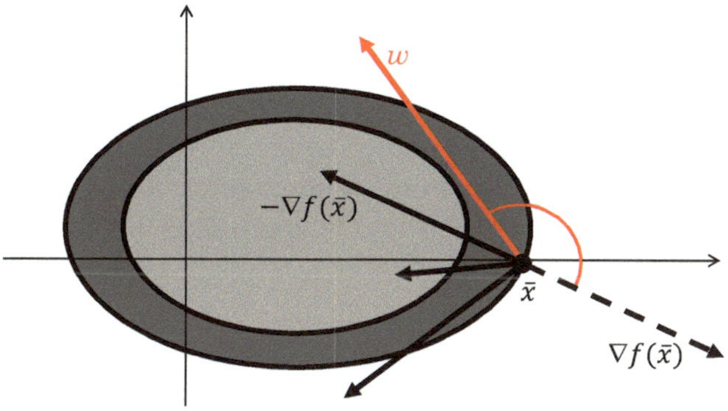

Fig. 1.22: Contours of f and four descent directions at $\bar{x}$, drawn as solid arrows, which each has an angle between 90 and 270 degrees with the gradient $\nabla f(\bar{x})$ (dashed arrow).

One can visualize the condition $\nabla f(\bar{x})^\top w < 0$ as requiring the vector w to form an angle with the vector $\nabla f(\bar{x})$ of between 90 and 270 degrees; see Figure 1.22. The latter vector points in a direction of "greatest steepness" of f at $\bar{x}$. In particular, $-\nabla f(\bar{x})$ is a descent direction for f at $\bar{x}$ provided that $\nabla f(\bar{x}) \neq 0$.

These ideas about descent directions lead to the following fundamental algorithm, where we count iterations using superscript ν (the Greek letter "nu").

Gradient Descent Method.

Data. $x^0 \in \mathbb{R}^n$.
Step 0. Set $\nu = 0$.
Step 1. Stop if $\nabla f(x^\nu) = 0$, otherwise compute a direction

$$w^\nu = -\nabla f(x^\nu).$$

Step 2. Select a positive step size λ^ν.
Step 3. Set $x^{\nu+1} = x^\nu + \lambda^\nu w^\nu$, replace ν by $\nu + 1$, and go to Step 1.

If the algorithm stops in Step 1, then we've obtained a point that satisfies the requirements of the Fermat rule. The challenge is to determine what happens if it keeps on iterating and constructs a sequence $\{x^1, x^2, x^3, \dots\}$ or more concisely $\{x^\nu\}_{\nu=1}^\infty$. The exact performance depends on the rule in Step 2 for determine the *step size* λ^ν.

The simplest step-size rule is to set $\lambda^\nu = \lambda$ for all iterations, with λ sufficiently small. This works in the following sense. If f is smooth and convex, $\operatorname{argmin} f \neq \emptyset$, and, for some $\kappa \in (0, \infty)$, f also satisfies the *Lipschitz condition*

$$\left\|\nabla f(x) - \nabla f(\bar{x})\right\|_2 \leq \kappa \|x - \bar{x}\|_2 \quad \forall x, \bar{x},$$

then

$$f(x^\nu) - \min f \leq \frac{\beta}{2\lambda\nu} \quad \text{provided that } \lambda \in (0, 1/\kappa]. \tag{1.12}$$

Here, β is any constant no smaller than $\|x^0 - x^\star\|_2^2$ for some $x^\star \in \operatorname{argmin} f$. This means that the error bound after ν iterations, specified by the right-hand side in (1.12), is larger if we start the algorithm at a point x^0 far from all minimizers. Still, the conclusion in (1.12) (see, e.g., [12] for a proof) confirms that $f(x^\nu)$ approaches the minimum value of f as the number of iterations ν tends to infinity.

For a complicated function, it's often difficult to determine κ in the Lipschitz condition, at least without being overly conservative, which in turn would force a small step size in (1.12). When f is twice smooth, a valid κ is the largest eigenvalue of the Hessian matrix $\nabla^2 f(x)$ across all x but this can still be hard to compute unless f is quadratic.

Example 1.16 (gradient descent). For $f(x) = x^2$, we've $f'(x) = 2x$ and the Lipschitz condition simplifies to

$$\left\|\nabla f(x) - \nabla f(\bar{x})\right\|_2 = \left|2x - 2\bar{x}\right| = 2\|x - \bar{x}\|_2 \quad \forall x, \bar{x}.$$

Thus, $\kappa = 2$, which also follows from $f''(x) = 2$. Set $\lambda^\nu = 1/3$ for all iterations; any value no greater than $1/2$ is admissible by (1.12). The gradient descent method iterates from $x^0 = 5$ as follows:

$$\begin{aligned} x^1 &= x^0 - \tfrac{1}{3} f'(x^0) = 5 - \tfrac{1}{3} 10 = \tfrac{5}{3} \\ x^2 &= x^1 - \tfrac{1}{3} f'(x^1) = \tfrac{5}{3} - \tfrac{1}{3}\tfrac{10}{3} = \tfrac{5}{9} \\ x^3 &= x^2 - \tfrac{1}{3} f'(x^2) = \tfrac{5}{9} - \tfrac{1}{3}\tfrac{10}{9} = \tfrac{5}{27}. \end{aligned}$$

The corresponding objective function values are $f(x^0) = 25$, $f(x^1) \approx 2.78$, $f(x^2) \approx 0.31$, and $f(x^3) \approx 0.03$, which indeed are converging to the minimum value of zero.

Detail. In this case, $\beta = \|x^0 - x^\star\|_2^2 = 5^2$ so that the right-hand size in (1.12) becomes $\beta/(2\lambda\nu) = 37.5/\nu$. We see that the progress of the algorithm is much faster than the theoretically guaranteed progress. This is expected because (1.12) applies to a broad class of functions. □

The gradient descent method isn't limited to convex functions or to fixed step sizes (as assumed in the derivations underpinning (1.12)). As long as f is smooth, the method applies and can be paired with a variety of step-size rules. In the absence of convexity, however, convergence can typically only be guaranteed to a point x satisfying $\nabla f(x) = 0$; see [30, Section 1.F] and [12].

A remarkable thing about the right-hand side in (1.12) is the absence of the dimension n. The computational cost per iteration is also low for the gradient descent method. Thus, the method is highly scalable and, together with its extensions (see our discussion in Section 3.D), it has become the central tool for large-scale learning problems such as training of neural networks.

A hurdle for the gradient descent method is the step size λ^ν because κ is typically unknown and one might estimate its value too conservatively. Practically, this often causes λ^ν to be chosen too small, making progress slow for the method. Other approaches would be to adaptively select λ^ν (see, e.g., [30, Section 1.F]) or tune it by trail-and-error for a particular application. The gradient descent method may also require many iterations; the right-hand side in (1.12) vanishes but somewhat slowly as $\nu \to \infty$.

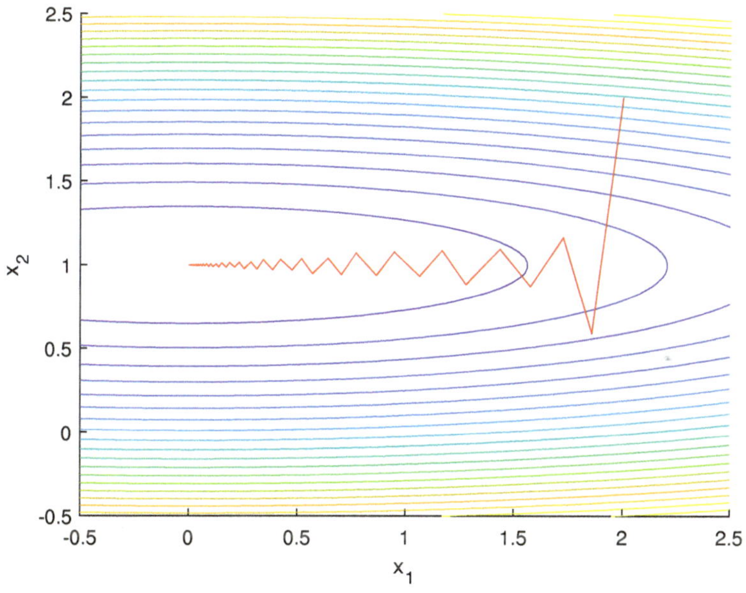

Fig. 1.23: For $\sigma = 20$ in Example 1.17, the gradient descent method zigzags as it starts at the point (2,2) and converges to the point (0,1).

Example 1.17 (zigzagging in gradient descent). Consider

$$f(x) = x_1^2 + \sigma(x_2 - 1)^2,$$

with σ being a positive scalar. The gradient descent method with unfortunate step sizes may exhibit zigzagging behavior when started from $x^0 = (2, 2)$ and this worsens as σ grows despite the fact that $\operatorname{argmin} f = \{(0, 1)\}$ and $\min f = 0$ for any positive σ. Table 1.3 reports the number of iterations to reach $\|\nabla f(x^\nu)\|_2 \le 10^{-7}$ for different σ.

Detail. The results in Table 1.3 leverage the Armijo step-size rule (see, e.g., [30, Section 1.F]), which wisely ensures that $f(x^{\nu+1}) < f(x^\nu)$ in every iteration. However, this doesn't prevent the zigzagging trajectory in Figure 1.23, which worsens as σ grows. Recall that the direction $-\nabla f(x^\nu)$ is always perpendicular to the contour line at x^ν in the figure. Thus, the elongated contours of f cause the gradient descent method to prescribe poor directions. □

σ	2	20	200	2000	20000
iterations	17	180	1785	17376	176835

Table 1.3: The number of iterations of the gradient descent method grows rapidly for increasing σ in Example 1.17.

1.G Newton's Method

The iterations of the gradient descent method can be interpreted as follows: At a current point x^ν, the function f to be minimized is approximated by

$$f^\nu(x) = f(x^\nu) + \nabla f(x^\nu)^\top (x - x^\nu) + \tfrac{1}{2\lambda^\nu}\|x - x^\nu\|_2^2, \tag{1.13}$$

where λ^ν is a positive parameter to be determined. The next point $x^{\nu+1}$ is obtained by minimizing f^ν, which is convex even if f is nonconvex; see Figure 1.24.

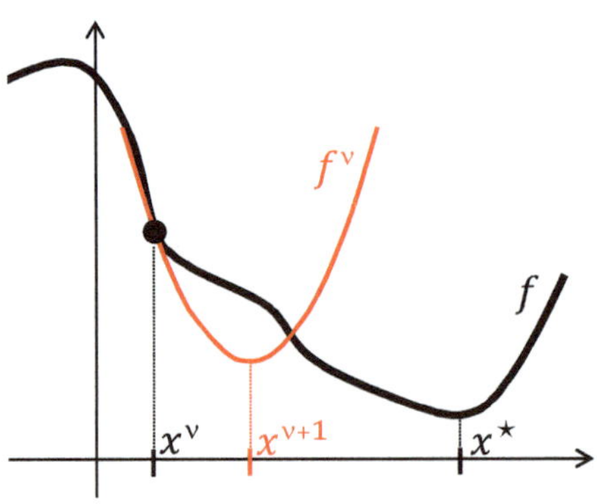

Fig. 1.24: A function f is approximated by f^ν at a point x^ν, with argmin f^ν producing a new point $x^{\nu+1}$.

Thus, by the optimality condition in Theorem 1.6 and the formula for gradients from Example 1.10, the following three conditions are equivalent:

$$x^{\nu+1} \in \operatorname{argmin} f^\nu, \quad \nabla f^\nu(x^{\nu+1}) = 0, \quad x^{\nu+1} = x^\nu - \lambda^\nu \nabla f(x^\nu).$$

We now see that the parameter λ^ν is indeed the step size in the gradient descent method. If f^ν turns out to be a good approximation of f, then we expect $x^{\nu+1}$ to be near argmin f. However, it's obvious that this can only be the case for a limited class of functions regardless of how skillfully we select λ^ν.

Under the additional assumption that f is twice smooth, a more trustworthy approximation of f at x^ν is provided by

$$f^\nu(x) = f(x^\nu) + \nabla f(x^\nu)^\top(x - x^\nu) + \tfrac{1}{2\lambda^\nu}(x - x^\nu)^\top\nabla^2 f(x^\nu)(x - x^\nu), \tag{1.14}$$

again with λ^ν being a positive parameter to be selected appropriately. When $\lambda^\nu = 1$, the approximation f^ν is the multivariate Taylor expansion of f at x^ν up to the second order. Again bringing in Theorem 1.6 and Example 1.10, we find that

$$x^{\nu+1} \in \operatorname{argmin} f^\nu \quad \text{if and only if} \quad \lambda^\nu\nabla f(x^\nu) + \nabla^2 f(x^\nu)(x^{\nu+1} - x^\nu) = 0.$$

If the Hessian $\nabla^2 f(x^\nu)$ is invertible, then the above equation has the unique solution

$$x^{\nu+1} = x^\nu - \lambda^\nu\big(\nabla^2 f(x^\nu)\big)^{-1}\nabla f(x^\nu).$$

The suggested direction,

$$w^\nu = -\big(\nabla^2 f(x^\nu)\big)^{-1}\nabla f(x^\nu),$$

is known as *Newton's direction*. We can then follow the pattern of the gradient descent method, but utilizing Newton's direction instead of the gradient. Newton's direction turns out to also be a descent direction when $\nabla f(x^\nu) \neq 0$. In summary, this leads to an algorithm for minimizing a twice smooth function with invertible Hessians.

Newton's Method.

Data. $x^0 \in \mathbb{R}^n$.
Step 0. Set $\nu = 0$.
Step 1. Stop if $\nabla f(x^\nu) = 0$, otherwise compute the direction

$$w^\nu = -\big(\nabla^2 f(x^\nu)\big)^{-1}\nabla f(x^\nu).$$

Step 2. Select a positive step size λ^ν.
Step 3. Set $x^{\nu+1} = x^\nu + \lambda^\nu w^\nu$, replace ν by $\nu + 1$, and go to Step 1.

A convergence statement for Newton's method resembles those for the gradient descent method, but requires the assumptions that f is twice smooth and the matrices $\nabla^2 f(x^\nu)$ are invertible for all ν; see, e.g., [32, Theorem 5.12]. In implementations, one would again replace Step 2 by a specific step-size rule.

Newton's method has desirable *local* convergence properties. If the initial point x^0 is sufficiently close to a minimizer $x^\star$ with an invertible Hessian matrix $\nabla^2 f(x^\star)$, then

x^ν converges "fast" to $x^\star$ when the step size $\lambda^\nu = 1$. As seen in [30, Section 1.G], the given *rate of convergence* of Newton's method is *quadratic*, which indeed is "fast." The reference gives that

$$\|x^\nu - x^\star\|_2 \leq \left(\alpha\|x^0 - x^\star\|_2\right)^{2^\nu - 1}\|x^0 - x^\star\|_2,$$

where α is a positive constant. For example, if $\alpha\|x^0 - x^\star\|_2 = 0.5$, then the initial distance to $x^\star$ is reduced with a factor of 10^{-1}, 10^{-2}, 10^{-5}, and 10^{-10} after 2, 3, 4, and 5 iterations, respectively. The number of correct digits in the approximating solution x^ν is roughly doubled in each iteration. In Example 1.17, which causes difficulties for the gradient descent method, Newton's method requires only a single iteration regardless of x^0. We realize this by recalling that in each iteration, Newton's method finds the minimizer of a convex quadratic approximation. Since in Example 1.17 the function is already convex and quadratic, the approximation turns out to be exact.

The improved performance of Newton's method relative to the gradient descent method could be expected as it leverages second-order information (Hessians) about the function to be minimized and not only first-order information (gradients). However, the advantage diminishes and might vanish if the algorithms start relatively far away from a minimizer; the analysis of the rate of convergence of Newton's method is local and only relies on properties of f near $x^\star$. Newton's method also breaks down when $\nabla^2 f(x^\nu)$ isn't invertible and some alternative strategy needs to be devised at such points. In Figure 1.24, this would be the case if x^ν had been further to the left.

Newton's method might require few iterations, but each one of them is much more computationally costly than those of the gradient descent method. The most efficient way of computing w^ν in Step 1 isn't to determine the inverse $(\nabla^2 f(x^\nu))^{-1}$ but rather to solve the system of equations

$$\nabla^2 f(x^\nu) w^\nu = -\nabla f(x^\nu).$$

Typically, this is achieved by Cholesky decomposition, which still requires order n^3 operations (number of variables cubed) unless the Hessian matrix is sparse, i.e., contains many zeros. For large-scale problems without much sparsity, Newton's method becomes less appealing.

Our strategy thus far has been to mimic the gradient descent method, while introducing, hopefully, a better direction w^ν. As alluded to at the beginning of Section 1.F, another possibility is to directly solve the system of equations $\nabla f(x) = 0$, which by Theorem 1.6 is equivalent to minimizing a convex f. This perspective also brings us to Newton's method.

Given smooth functions $g_1, \dots, g_n$ in n dimensions, the system of equations

$$g_1(x) = 0, \dots, g_n(x) = 0$$

can be written concisely as the vector equation $G(x) = 0$, where G is a mapping with

$$G(x) = \left(g_1(x), \dots, g_n(x)\right).$$

The *Jacobian* of G at x is the $n \times n$-matrix $\nabla G(x)$ with gradients $\nabla g_j(x)^\top$, $j = 1, \dots, n$ as rows. A first-order Taylor expansion of G at x^ν gives the approximation

$$G^\nu(x) = G(x^\nu) + \nabla G(x^\nu)(x - x^\nu).$$

If $\nabla G(x^\nu)$ is invertible, then the solution of the (approximating) vector equation $G^\nu(x) = 0$ is

$$x^{\nu+1} = x^\nu - \big(\nabla G(x^\nu)\big)^{-1} G(x^\nu).$$

This provides the foundation for an iterative method for solving the vector equation $G(x) = 0$.

Newton-Raphson Algorithm.

Data. $x^0 \in \mathbb{R}^n$.
Step 0. Set $\nu = 0$.
Step 1. Stop if $G(x^\nu) = 0$, otherwise set

$$w^\nu = -\big(\nabla G(x^\nu)\big)^{-1} G(x^\nu).$$

Step 2. Set $x^{\nu+1} = x^\nu + w^\nu$, replace ν by $\nu + 1$, and go to Step 1.

We realize that if $G(x) = \nabla f(x)$ for some function f to be minimized, then the Newton-Raphson algorithm coincides with Newton's method under step size $\lambda^\nu = 1$. Consequently, after making the appropriate change of notation, one realizes that the Newton-Raphson algorithm has the same rate of convergence as Newton's method. The insight that Newton's method can be viewed as a procedure for computing a point satisfying the optimality condition $\nabla f(x) = 0$ reveals vividly the important connections between optimization problems and equation solving.

1.H Summary and Study Guide

In this chapter, we learned how a multivariate function can model a real-world problem and how its minimizers provide recommended decisions.

- A function f takes an n-dimensional vector x as input and produces a scalar output $f(x)$ according to some formula or rule. We often seek to find an x-vector (also viewed as a point in n-dimensional space) with the lowest value $f(x)$. Any such x is a minimizer of f, with the resulting $f(x)$ being the minimum value denoted by $\min f$. The set of all minimizers is written as $\operatorname{argmin} f$.
- *Modeling* is about deriving a reasonable function f that represents a real-world problem. *Solving* is about finding a minimizer of f using an algorithm.

- The Fermat rule says that every minimizer $x^\star$ of a smooth function satisfies $\nabla f(x^\star) = 0$.
- A convex function tends to be easier to minimize than nonconvex ones because solving $\nabla f(x) = 0$ is guaranteed to produce a minimizer of f.
- A convex set doesn't have any "holes" or a boundary with inward cusps.
- Affine sets, level-sets of convex functions, and intersections of convex sets are all convex sets.
- A function is convex if its epigraph is a convex set.
- The sum rule, multiplication rule, max rule, composition rule, and Hessian rule help us determine whether a function is convex.
- The gradient descent method and Newton's method aim to minimize a function f by iterating $x^{\nu+1} = x^\nu + \lambda^\nu w^\nu$, where ν is the iteration counter, $x^{\nu+1}$ is the next point, x^ν is the current point, λ^ν is the step size, and w^ν is the direction given by

$$w^\nu = -\nabla f(x^\nu) \qquad \text{(gradient descent method)}$$

$$w^\nu = -\big(\nabla^2 f(x^\nu)\big)^{-1}\nabla f(x^\nu) \qquad \text{(Newton's method)}.$$

  The gradient descent method tends to involve many computationally cheap iterations, while Newton's method requires few but costly iterations.

1.I Exercises

Exercise 1.1 (function in one variable). Consider the cubic function given by $f(x) = \frac{1}{3}x^3 - x + 1$. Determine its derivatives. Plot the graphs of f and f'. Use the Fermat rule to identify points that can't be minimizers of f.

Exercise 1.2 (function in one variable). Consider the cubic function given by $f(x) = 3x^2 + x^3$. Determine its derivatives. Use the Fermat rule to identify points that can't be minimizers of f.

Exercise 1.3 (function in two variables). Consider the quadratic function given by $f(x) = 2x_1^2 + (x_2 - 1)^2$ for $x = (x_1, x_2)$. Determine its gradients. Plot the graphs of f, f'_{x_1}, and f'_{x_2}. Use the Fermat rule to identify points that can't be minimizers of f.

Exercise 1.4 (function in two variables). Consider the quadratic function given by $f(x) = 3x_1^2 + x_2 + x_1x_2$. Determine its gradients. Use the Fermat rule to identify points that can't be minimizers of f.

Exercise 1.5 (data smoothing). Using $x = (0, 0.1, 0.2, \ldots, 0.9, 1)$ and $y = (0.85, 0.61, 0.89, 0.46, 0.82, 0.62, 0.72, 0.18, 0.54, 0.41, 0.26)$, fit polynomials of degree $n = 1, 2, \ldots, 10$. For each n, visualize the fit by plotting the data as well as the graph of the resulting polynomial and also report the minimum value. Discuss which n gives the "best" result.

Exercise 1.6 (newsvendor problem). With $o = 3$, $s = 7$, and ξ uniformly distributed on $[10, 20]$, write an expression for the expected loss, graph the distribution function P, and find the minimizing order quantity. Show that the expected demand 15 isn't a minimizer.

Guide. Set $p(\xi) = 0.1$ for $\xi \in [10, 20]$ and zero otherwise.

Exercise 1.7 (convexity of quadratic function). For the quadratic function in Exercise 1.3, check whether it's convex. Determine all its minimizers.

Exercise 1.8 (convexity of quadratic function). For the quadratic function given by $f(x) = \sum_{i=1}^{n}(x_i - 3)^2$, with $x = (x_1, \dots, x_n)$, check whether it's convex.

Exercise 1.9 (convexity of set). Show that the set $\{(x_1, x_2) \mid 2x_1^2 + 3x_2^2 - 1 \le 0,\ x_1 + x_2 = 0.1\}$ is convex.

Exercise 1.10 (convexity of composite function). Show that the function given by

$$f(x) = \max_{k=1,\dots,q} \sum\nolimits_{i=1}^{m} \alpha_{ik} \exp\left(c_{ik}^\top x + \beta_{ik}\right)$$

is convex, where $\alpha_{ik} \ge 0$, β_{ik} is a scalar, and c_{ik} is an n-dimensional vector.

Exercise 1.11 (convexity of composite function). For positive scalars $\alpha_1, \dots, \alpha_m$, n-dimensional vectors $a_1, \dots, a_m$, and

$$f(x) = \sum\nolimits_{i=1}^{m} \alpha_i \exp(a_i^\top x),$$

show that (a) f is a convex function and (b) $C = \{x \mid f(x) \le 10\}$ is a convex set.

Exercise 1.12 (convexity in search problem). In planning the search for floating debris from an aircraft lost over the ocean, one would divide the area of interest into n squares and then determine how much time and resources should be put towards searching each square. For simplicity, suppose that the search will be carried out by one aircraft flying at s knots, with a sensor that effectively can reach out $w/2$ nautical miles, and that the debris is located in one square and is stationary. Prior knowledge stipulates the probability that the debris is in square j as $p_j > 0$, with $\sum_{j=1}^{n} p_j = 1$. Let the size of the squares be $a_1, \dots, a_n$ square nautical miles and the total amount of search time available be t. The search problem of determining the time allocation $x = (x_1, \dots, x_n)$ that minimizes the probability of not finding the debris can be formulated as (see, e.g., [33])

$$\operatorname*{minimize}_{x \in C} \sum\nolimits_{j=1}^{n} p_j \exp(-swx_j/a_j),$$

with $C = \{x \mid \sum_{j=1}^{n} x_j = t,\ x_j \ge 0, j = 1, \dots, n\}$. Show that f and C are convex.

Exercise 1.13 (descent directions). Given $f(x) = (x_1 - 1)^2 + 2x_2^2$ and the point $\bar{x} = (1, 2)$, identify some descent directions for f at $\bar{x}$.

Guide. It might be helpful to draw a contour plot of f as well as $\nabla f(\bar{x})$ to verify your answer.

Exercise 1.14 (step size in gradient descent). Consider the function given by $f(x) = x_1^2 + x_2^2$, with $x = (x_1, x_2)$, and the gradient descent method. Explain why a step size of 1/2 is reasonable.

Exercise 1.15 (gradient descent). For f in Example 1.17 and $\sigma = 2$, carry out three iterations of the gradient descent method starting at $x^0 = (2, 2)$ and using a permissible step size according to the derivations leading up to (1.12). Compute the values of $f(x^\nu)$ for iterations $\nu = 0, 1, 2, 3$, and compare them with the guaranteed progress according to (1.12). Visualize the progress of the method.

Exercise 1.16 (Newton's method). For f in Example 1.17 and $\sigma = 2$, carry out one iteration of Newton's method starting at $x^0 = (2, 2)$ and using step size $\lambda^0 = 1$.

Exercise 1.17 (implementation of Newton's). Implement Newton's method with fixed step size ($\lambda^\nu = 1$) and apply it to the Rosenbrock function from (1.3); see Figures 1.7 and 1.8. Starting at $x^0 = (-1.2, 1)$, determine how many iterations it takes to obtain x^ν with $\|\nabla f(x^\nu)\|_2 \leq 10^{-7}$.

Guide. The Rosenbrock function has a unique minimizer at $(1, 1)$.

Chapter 2
Hitting a Wall: Constraints and Nonsmoothness

Optimization problems are often specified in terms of an objective function to be minimized and a set that defines permissible decisions. The newsvendor in Section 1.C may not be able to order more than α newspapers and would need to select an order quantity from the interval $[0, \alpha]$. A maximum likelihood problem (Section 1.E) permits only certain types of estimates from a set of density functions. This chapter initiates our treatment of such optimization problems with *constraints*. As we'll see, a problem of this general type can be formulated in terms of a modified function that incorporates the effect of both the set of permissible decisions and the original objective function. This perspective will allow us to achieve far-reaching extensions of the principles laid out in Chapter 1 and address nonsmooth problems broadly. Since it's most accessible, we'll concentrate on the convex case.

2.A Formulations

In applications, we might be given an objective function f_0 in n dimensions representing cost, loss, error, or some other quantity that we would like to minimize as well as a set C; any decision not in C is deemed unacceptable. This leads to the minimization problem

$$\underset{x \in \mathbb{R}^n}{\text{minimize}}\, f_0(x) \ \text{ subject to } \ x \in C \qquad \text{or, more concisely,} \qquad \underset{x \in C}{\text{minimize}}\, f_0(x).$$

Any $x \in C$ that also satisfies $f_0(x) < \infty$ is a *feasible point (decision)* for the problem. The *feasible set* is the collection of all such feasible points. The problem is *feasible* if there's at least one feasible point and is *infeasible* otherwise.

Constraints are conditions that make the feasible set a strict subset of $\mathbb{R}^n$, which then make the problem *constrained*. Otherwise, the problem is *unconstrained*.

Example 2.1 (shortest path). A *directed graph* $G = (V, E)$ consists of a vertex set V and an edge set E. Figure 2.1 shows a specific case with $V = \{1, \ldots, 8\}$ and

J. O. Royset, Optimization Concepts, *Springer Series in Operations Research and Financial Engineering*,
https://doi.org/10.1007/978-3-032-36819-5_2

$$E = \{(1,2), (1,3), (1,5), (2,5), (2,7), (3,4), (3,5), (3,6), (4,6), (5,7), (6,8), (7,8)\}.$$

Each (directed) *edge* $(i, j) \in E$ connects distinct *vertices* $i \in V$ and $j \in V$, and it possesses *length* $c_{ij} \in [0, \infty)$. We refer to a graph equipped with data as a *network*. A directed s-t path is an ordered set of edges of the form $\{(s, i_1), (i_1, i_2), \ldots, (i_{\nu-1}, t)\}$ for some natural number ν. Given two distinct vertices $s, t \in V$, the *shortest-path problem* seeks to determine a directed s-t path such that the sum of the edge lengths along the path is minimized.

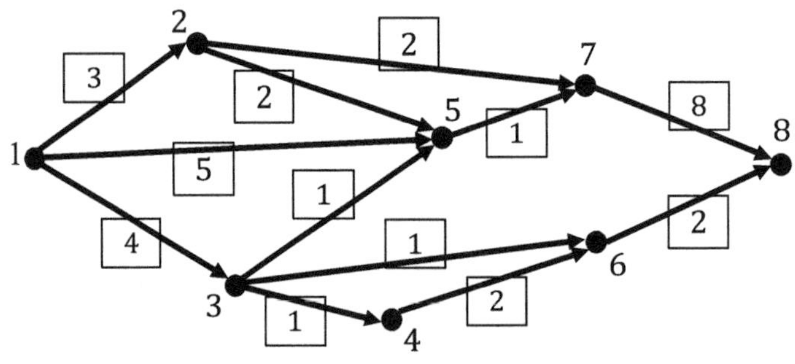

Fig. 2.1: A directed graph with vertex set $V = \{1, \ldots, 8\}$ and 12 edges with lengths indicated by boxed numbers.

With start vertex $s = 1$ and terminal vertex $t = 8$, the directed graph in Figure 2.1 has six s-t paths:

1-2-7-8	with length: 13
1-2-5-7-8	with length: 14
1-5-7-8	with length: 14
1-3-5-7-8	with length: 14
1-3-6-8	with length: 7
1-3-4-6-8	with length: 9

and it's clear that 1-3-6-8 is the shortest one. While this simple approach based on path enumeration is always possible in principle, the number of paths in a graph tends to grow exponentially in the number of vertices and edges. Thus, there's a need for a better solution approach to the shortest-path problem. An optimization model with constraints furnishes an alternative approach.

Detail. Suppose that m is the number of vertices in V and n is the number of edges in E. Let A denote the $m \times n$-vertex-edge incidence matrix such that if edge $e = (i, j)$, then

$$A_{ie} = 1, \quad A_{je} = -1, \quad A_{i'e} = 0 \quad \text{for any } i' \neq i, j.$$

Also, let $b_s = 1$, $b_t = -1$, and $b_i = 0$ for $i \in V \backslash \{s, t\}$ and collect them in the vector b. With c being the column vector with components c_{ij}, $(i, j) \in E$, the shortest-path problem may then be formulated as

$$\underset{x \in \mathbb{R}^n}{\text{minimize}}\ c^\top x \quad \text{subject to } Ax = b,\ x \in [0, 1]^n$$

where x is the column vector with components x_{ij}, $(i, j) \in E$. The side conditions $Ax = b$ and $x \in [0, 1]^n$ are constraints that define a feasible set. We can also write the model in terms of the feasible set $C = \{x \in [0, 1]^n \mid Ax = b\}$:

$$\underset{x \in C}{\text{minimize}}\ c^\top x.$$

To make the constraints $Ax = b$ clearer, let's write them out explicitly in the setting of Figure 2.1:

$$\begin{aligned}
x_{12} + x_{13} + x_{15} &= 1 && \text{(balance at vertex 1)}\\
-x_{12} + x_{25} + x_{27} &= 0 && \text{(balance at vertex 2)}\\
-x_{13} + x_{34} + x_{35} + x_{36} &= 0 && \text{(balance at vertex 3)}\\
-x_{34} + x_{46} &= 0 && \text{(balance at vertex 4)}\\
-x_{15} - x_{25} - x_{35} + x_{57} &= 0 && \text{(balance at vertex 5)}\\
-x_{36} - x_{46} + x_{68} &= 0 && \text{(balance at vertex 6)}\\
-x_{27} - x_{57} + x_{78} &= 0 && \text{(balance at vertex 7)}\\
-x_{68} - x_{78} &= -1 && \text{(balance at vertex 8).}
\end{aligned}$$

We see that x_{ij} may be interpreted as the amount of *flow* on edge (i, j) and the constraints $Ax = b$ impose *flow balance* at each vertex. For instance at vertex 1, one unit of flow must be distributed among edges $(1, 2)$, $(1, 3)$, and $(1, 5)$. At vertex 2, the flow coming in through edge $(1, 2)$ must match the flowing going out on edges $(2, 5)$ and $(2, 7)$. The situation is similar at vertices $3, 4, \ldots, 7$. At the terminal vertex 8, one unit of flow must come in. Thus, $Ax = b$ forces one unit of flow to move through the directed graph. There are typically many such feasible flows; Figure 2.1 has six paths across which the flow can be distributed. The objective function

$$c^\top x = \sum\nolimits_{(i,j) \in E} c_{ij} x_{ij}$$

determines which one is better.

While it isn't necessary to formulate the shortest-path problem in this manner—we give a specialized algorithm in Section 4.E—there are advantages to such an approach. First, it enables the use of well-developed general-purpose optimization algorithms as seen in the next sections. Second, it becomes straightforward to add constraints to the model and thus adapt to related but different situations; see, e.g., Example 5.10. □

Example 2.2 (constrained data fitting). As seen in Section 1.B, we can fit a polynomial p_n to given data by minimizing

$$f_0(c) = (Dc - y)^\top (Dc - y) = \|Dc - y\|_2^2,$$

where $c = (c_0, c_1, \dots, c_n)$ are the coefficients of the polynomial. Suppose that additional information dictates that $p_n(0) \in [\alpha, \beta]$ and $p_n(1) = \gamma$ for some given scalars $\alpha \le \beta$ and γ. These requirements lead to the feasible set

$$C = \{c \in \mathbb{R}^{1+n} \mid c_0 \ge \alpha,\ c_0 \le \beta,\ c_n + \cdots + c_1 + c_0 = \gamma\}$$

and the optimization model

$$\underset{c \in \mathbb{R}^{1+n}}{\text{minimize}}\ f_0(c)\ \text{ subject to }\ c_0 \ge \alpha,\ c_0 \le \beta,\ c_n + \cdots + c_1 + c_0 = \gamma.$$

Detail. The need for data fitting arises especially in time-series analysis, which examines how a phenomenon evolves over time. For example, we might be interested in the level of activity in the housing market, the evolution of consumer preferences, the spread of an epidemic, and so on. An autoregressive (AR) model describes, or predicts, by means of a (linear) transfer function, the random state $\boldsymbol{\xi}_t$ of a system at time t in terms of the states at times $t-1, t-2, \dots, t-n$:

$$\boldsymbol{\xi}_t = c_0 + c_1 \boldsymbol{\xi}_{t-1} + \cdots + c_n \boldsymbol{\xi}_{t-n} + \boldsymbol{\varepsilon}_t,$$

where $\boldsymbol{\varepsilon}_t$ is a random variable that accounts for disturbances not captured by the transfer function. This leads to an optimization model similar to the one in Section 1.B.

Based on observations $\xi_{1-n}, \dots, \xi_m$ of past states at times $1-n, \dots, m$, respectively, we're looking for estimates of the coefficients $c = (c_0, c_1, \dots, c_n)$ that minimize the role played by the disturbances. A least-squares approach leads to the model

$$\underset{c \in \mathbb{R}^{1+n}}{\text{minimize}}\ \frac{1}{m} \sum_{i=1}^{m} (c_0 + c_1 \xi_{i-1} + \cdots + c_n \xi_{i-n} - \xi_i)^2.$$

Suppose that on the basis of physical laws, economic considerations, or simply common sense, we would like the coefficients to satisfy $c_1 \ge \cdots \ge c_n \ge 0$, i.e., the influence on a state by a prior state tends be less if it's more in the past. Figure 2.2 illustrates the effect of including such constraints, where an AR model with $n = 20$ is fitted to $m = 137$ data points about annual average global temperatures (given in terms of deviation from the average during 1901-2000) up to 2016. The figure shows an upward trend for the last 57 years of record. The solid and dashed lines give the model's predictions with and without constraints, respectively. The constraints dampen undesirable oscillations in the predictions. □

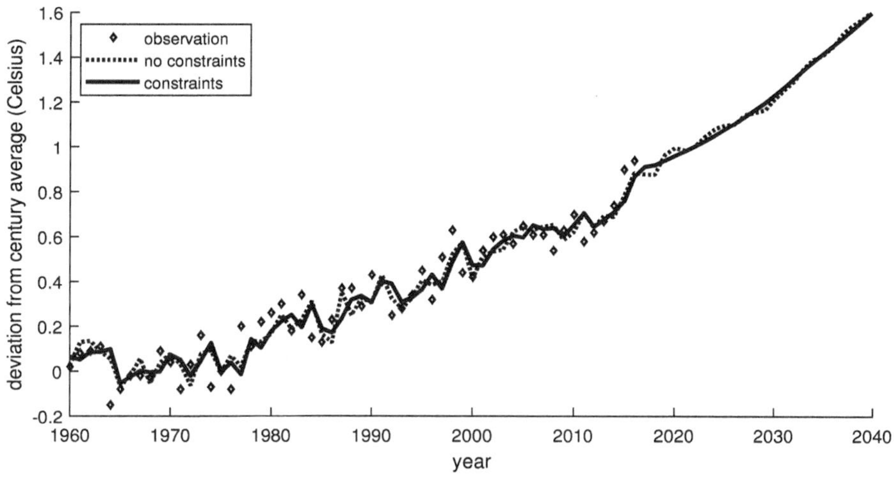

Fig. 2.2: Predictions of rise in global temperatures using AR model.

Constraints that are listed in a model formulation are called *explicit constraint*; the examples above have constraints of this type. However, an objective function f_0 might bring its own *induced constraints* that make $f_0(x) = \infty$ for some x. There are no induced constraints from f_0 in Example 2.2 because it's real-valued. However, Section 1.E describes an objective function with terms of the form $-\ln \alpha$, which implies the induced constraint $\alpha > 0$. Expressions involving expected values, integrals, and maximization might have ∞ as value and likewise cause induced constraints.

Since we permit functions to take the value ∞, a minimization problem involving a function f_0 and a set C can just as well be stated in terms of another function that incorporates the effect of both. In addition to the simplified notation resulting from this perspective, we'll be able to naturally extend the Fermat rule from Chapter 1 and generally address multiple theoretical challenges. Specifically, given a function f_0 in n dimensions and a set $C \subset \mathbb{R}^n$, we can always define a new function f by setting

$$f(x) = \begin{cases} f_0(x) & \text{if } x \in C \\ \infty & \text{otherwise.} \end{cases}$$

Thus, minimizing $f_0(x)$ subject to $x \in C$ is equivalent to minimizing $f(x)$ subject to $x \in \mathbb{R}^n$. The (common) minimizers and minimum value of these problems are denoted by

$$\operatorname{argmin}_C f_0 = \operatorname{argmin} f \quad \text{and} \quad \min_C f_0 = \min f = f_0(x^\star) \text{ for } x^\star \in \operatorname{argmin}_C f_0.$$

While we call a point in these argmin-sets a "minimizer," they're also commonly known as *optimal solutions* or simply "solutions." However, we prefer to use "solution" more broadly encompassing feasible points and/or those satisfying the Fermat rule. Depending on the difficulty of an optimization problem, various notions might serve as satisfactory "solutions" of the problem.

We say that a minimization problem is a *convex problem* when it can be expressed as minimizing a convex function f.

The reformulation in terms of a function f highlights the inherent lack of smoothness in the presence of constraints. It triggers a jump in f from $f(x) < \infty$ to $f(x) = \infty$ as x changes from being feasible to being infeasible and thus makes f neither continuous nor smooth.

Example 2.3 (online retail). An online retailer seeks to optimize the prices of n items under an *elastic demand* model so that its cost is minimized and the revenue is at least r. Suppose that the demand for item j under price vector $x = (x_1, \dots, x_n)$ is

$$d_j(x) = \beta_j - \alpha_j(x_j - p_j), \quad j = 1, \dots, n,$$

where β_j is a baseline demand, $\alpha_j > 0$ reflects the fact that demand decreases as the price x_j increases, and p_j is a reference price for item j. Let $d(x)$ be the column vector with components $d_j(x)$, $j = 1, \dots, n$. The revenue is obtained by summing up "price times demand," i.e., $x^\top d(x)$, and the cost by $(c - x)^\top d(x)$, where $c = (c_1, \dots, c_n)$ lists the per-item expenses incurred by the retailer so that $x_j - c_j$ is the margin for item j. (A negative cost corresponds to making money.) Under the bounds $a_j \le x_j \le b_j$, this leads to the model

$$\underset{x \in \mathbb{R}^n}{\text{minimize}}\ (c - x)^\top d(x) \ \text{ subject to } \ x^\top d(x) \ge r, \ \ a \le x \le b,$$

where $a = (a_1, \dots, a_n)$ and $b = (b_1, \dots, b_n)$. Here, and throughout, we interpret an inequality between two vectors componentwise, i.e., $x \le b$ means that $x_j \le b_j$ for all $j = 1, \dots, n$.

An equivalently model is to minimize the function given by

$$f(x) = \begin{cases} (c - x)^\top d(x) & \text{if } x^\top d(x) \ge r, \ \ a \le x \le b \\ \infty & \text{otherwise.} \end{cases}$$

Regardless, the feasible set is $\{x \mid x^\top d(x) \ge r, \ a \le x \le b\}$. □

Equivalent Problems. The above discussion illustrates that a particular situation always has multiple formulations. We can use this to our advantage and seek those that are computationally and/or theoretically more tractable. Two problems are *equivalent* if we can obtain minimizers of one from those of the other with only minor effort and vice versa. What constitutes "minor effort" depends on the circumstances, but it usually means just some arithmetic operations or possibly the solution of linear equations.

For example, a problem in the decision variable x with the constraint

$$g(x) \leq 0$$

is equivalent to a problem in the decision variables x and y with the constraints

$$g(x) + y = 0 \quad \text{and} \quad y \geq 0.$$

The x-component of a minimizer of the latter problem furnishes a minimizer of the original problem. A minimizer $\bar{x}$ of the original problem specifies $(\bar{x}, -g(\bar{x}))$ as a minimizer of the other problem. Generally, a problem *reformulation* implies that we've obtained an equivalent problem, without introducing any approximation.

The three examples above result in convex problems. We establish this by looking at a broader class involving the *indicator function* ι_C of a set $C \subset \mathbb{R}^n$ with values

$$\iota_C(x) = \begin{cases} 0 & \text{if } x \in C \\ \infty & \text{otherwise.} \end{cases}$$

Figure 2.3 illustrates its epigraph.

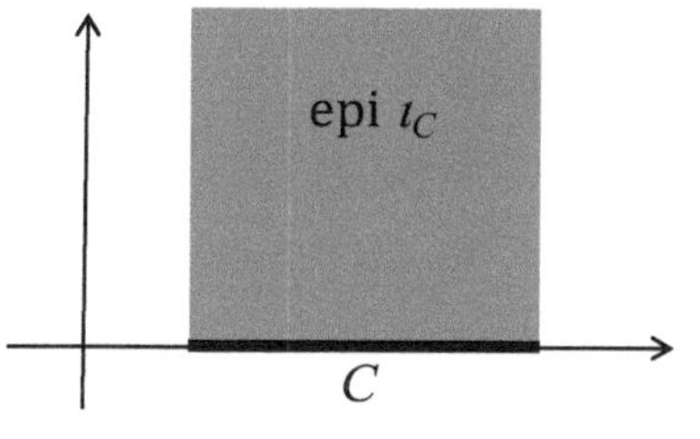

Fig. 2.3: The epigraph of an indicator function ι_C of a set C.

Indicator functions allow us to reformulate a minimization problem given by an objective function f_0 and a set C into one with a single function through *addition*:

$$f = f_0 + \iota_C.$$

The resulting function has

$$f(x) = f_0(x) + \iota_C(x) = \infty \quad \text{whenever} \quad x \notin C,$$

which then rules out such points as minimizers. This holds even if $f_0(x) = -\infty$ because we universally adopt the convention $-\infty + \infty = \infty$. In theoretical development, we can then simply consider the function f. For example, convexity of f, and thus of the whole problem, is settled easily as seen next (cf. [30, Proposition 2.3] for a proof).

Proposition 2.4 (convexity of indicator function). *For a set $C \subset \mathbb{R}^n$, the indicator function ι_C is convex if and only if C is convex. Moreover, if the n-dimensional function f_0 is convex and the set C is convex, then $f_0 + \iota_C$ is convex.* □

In view of this proposition, we can establish the convexity of problems by examining the defining functions and sets separately. Let's consider some common constraints that include those from Examples 2.1-2.3.

The n-dimensional row vectors A_i, with components $A_{i1}, \dots, A_{in}$, and scalars b_i, $i = 1, \dots, m$, give rise to the collection of *linear equations*

$$\sum\nolimits_{j=1}^{n} A_{ij}x_j = b_i, \quad i = 1, \dots, m.$$

A visual interpretation of $2x_1 + x_2 = 1$ appears in Figure 2.4 as a line.

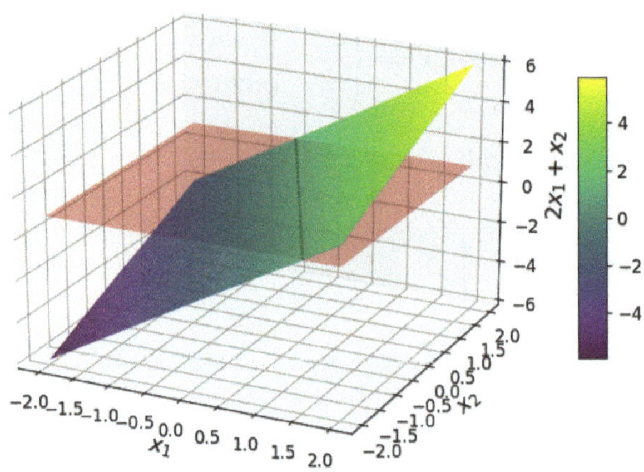

Fig. 2.4: The values of $2x_1 + x_2$ and the target value of 1 illustrated as planes.

We express the m linear equations equivalently by writing

$$A_i x = b_i, \quad i = 1, \dots, m \qquad \text{or} \qquad Ax = b,$$

where A is the $m \times n$-matrix with rows $A_1, \dots, A_m$ and $b = (b_1, \dots, b_m)$. Example 2.1 provides a concrete instance of such linear equations.

The set of solutions $\{x \mid Ax = b\}$ is convex by Proposition 1.7(a). If A_i isn't the zero vector, then $\{x \mid A_i x = b_i\}$ is called a *hyperplane* because it extends the notion of a plane in three-dimensional space; see Figure 2.4 for an example of the simpler case of two dimensions.

Companions to linear equations are *linear inequalities* obtained by replacing "=" by "≤" above. If A_i isn't the zero vector, then $\{x \mid A_i x \le b_i\}$ is called a *halfspace* because it specifies the "half" of $\mathbb{R}^n$ on one side of the hyperplane $\{x \mid A_i x = b_i\}$; see Figure 2.5. Since the function given by $f(x) = A_i x$ is linear and thus convex, $\{x \mid A_i x \leq b_i\}$ is convex by virtue of being a level-set of a convex function; cf. Proposition 1.7(b). Also,

$$\{x \mid Ax \leq b\} = \bigcap_{i=1}^{m} \{x \mid A_i x \leq b_i\}$$

and is thus convex by the intersection rule in Proposition 1.7(c).

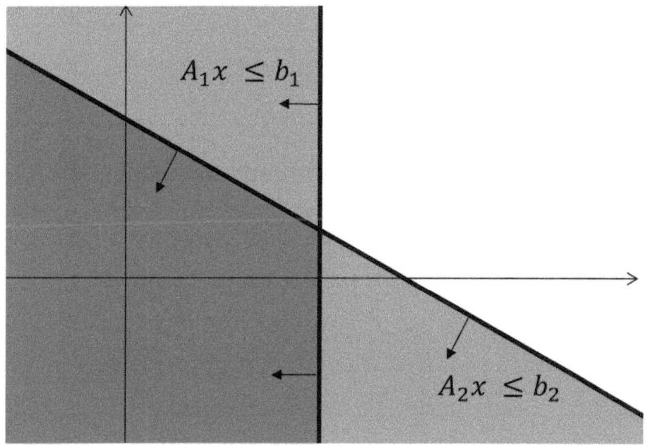

Fig. 2.5: Linear inequalities define two halfspaces and their intersection.

A set $C \subset \mathbb{R}^n$ defined by those points satisfying a finite number of linear equalities and linear inequalities is called a *polyhedron* or a *polyhedral set*; see Figure 2.6 for an example in three dimensions. In particular, $C = \emptyset$ and $C = \mathbb{R}^n$ are polyhedral sets.

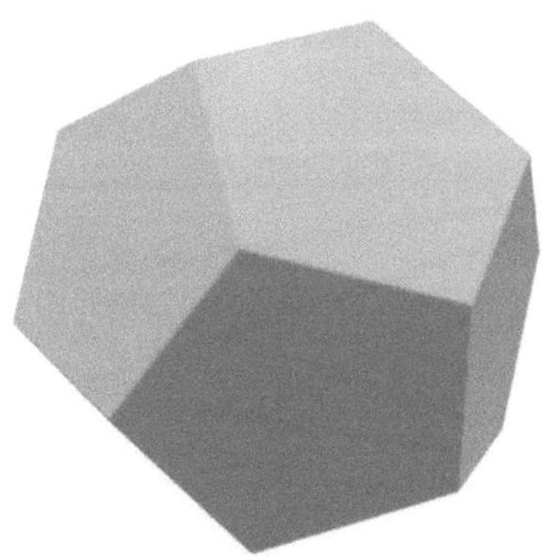

Fig. 2.6: Polyhedral set in three dimensions.

Since a linear equality $a^\top x = \alpha$ can be written as the two inequalities $a^\top x \leq \alpha$ and $a^\top x \geq \alpha$, a polyhedron can always be expressed in terms of a finite number of inequalities.

Although sometimes convenient theoretically, a reformulation of linear equalities in this manner is rarely beneficial computationally.

A polyhedral set is an intersection of convex sets and thus convex by Proposition 1.7(c); the feasible sets in Examples 2.1 and 2.2 furnish concrete instances. In Example 2.3, the feasible set can be expressed as the intersection of a polyhedron and the convex set $\{x \mid x^\top d(x) \geq r\}$, and must then be convex again due to the intersection rule in Proposition 1.7(c).

A *quadratic optimization problem* takes the form

$$\underset{x\in\mathbb{R}^n}{\text{minimize}}\ \tfrac{1}{2}x^\top Qx + c^\top x \ \text{ subject to }\ Ax = b,\ Dx \leq d$$

where Q is a symmetric $n \times n$-matrix. The number of rows in the matrix A determines the number of linear equalities and, likewise, the number of rows in D gives the number of inequalities. A quadratic optimization problem is convex if and only if Q is positive semidefinite; cf. Example 1.10. The model in Example 2.2 is a convex quadratic optimization problem. If Q is the positive semidefinite matrix with all its elements being zero, then the problem reduces to a *linear optimization problem*:

$$\underset{x\in\mathbb{R}^n}{\text{minimize}}\ c^\top x \ \text{ subject to }\ Ax = b,\ Dx \leq d$$

The shortest-path problem in Example 2.1 gives rise to a linear optimization problem.

Quadratic and linear optimization problems are sometimes referred to as quadratic and linear *programs*, despite having little to do with computer programs. The reason is that "programming" was used in the meaning of "planning" among Pentagon staff in the middle of the last century, and much pioneering optimization research occurred in that context.

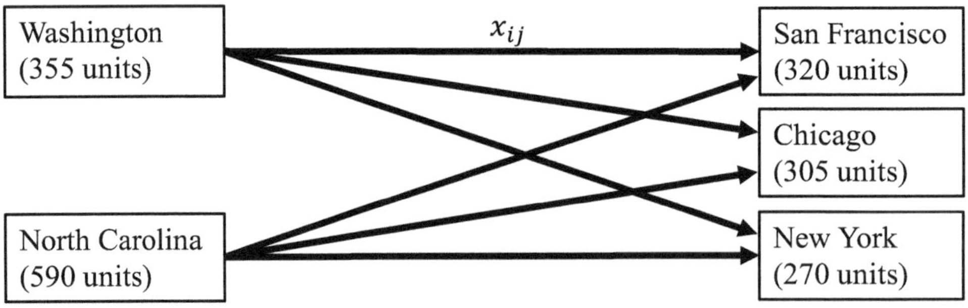

Fig. 2.7: Warehouses (left) and demand locations (right) in a transportation problem.

Example 2.5 (transportation problem). An online retailer has one warehouse in Washington State (WA), with 355 units of a particular product in stock, and one in North Carolina (NC), with 590 units. Customers in San Francisco, Chicago, and New York demand 320, 305, and 270 units of the product, respectively; see Figure 2.7. The per-unit cost of shipping from WA to San Francisco, Chicago, and New York is \$1.20, \$1.70, and \$2.80, respectively. The corresponding numbers from NC are \$1.90, \$2.90, and \$1.20. We would like to decide the number of units to ship from each warehouse to each location such that the total shipping cost is minimized. This is a *transportation problem*.

Detail. Let x_{ij} be the number of products to be shipped from warehouse i to location j. Here, i is WA or NC, but in general we may have m warehouse, i.e., $i = 1, 2, \ldots, m$. Similarly, j is San Francisco, Chicago, or New York, but in general, $j = 1, 2, \ldots, n$. We denote by c_{ij} the per-unit transportation cost from warehouse i to location j, by w_i the stock level in warehouse i, and by b_j the demand at location j. With x being the column vector with components x_{ij}, $i = 1, \ldots, m$, $j = 1, \ldots, n$, we obtain the model

$$\begin{aligned}\underset{x\in\mathbb{R}^{mn}}{\text{minimize}} \sum_{i=1}^{m}\sum_{j=1}^{n} c_{ij}x_{ij} \text{ subject to } & \sum_{j=1}^{n} x_{ij} \leq w_i, \quad i = 1, \ldots, m \\ & \sum_{i=1}^{m} x_{ij} = b_j, \quad j = 1, \ldots, n \\ & x_{ij} \geq 0, \quad i = 1, \ldots, m; \; j = 1, \ldots, n.\end{aligned}$$

The first m linear inequalities ensure that no warehouse exceeds its inventory and the n linear equalities guarantee that all demand is being satisfied. The feasible set is polyhedral and can be described concisely by

$$Ax = b \quad \text{and} \quad Dx \leq d,$$

where A is an $n \times mn$-matrix consisting of 0 and 1, b is an n-vector with elements b_j, D is an $(m + mn) \times mn$-matrix with 1, -1, and 0, and d is an $(m + mn)$-vector with w_i and zero as elements. Thus, the model is a linear optimization problem. Table 2.1 specifies a minimizer, with minimum value \$1415.50, computed by an algorithm from Section 2.D or 2.E. The minimizer provides the insight that it's better to avoid the intuitive solution of supplying San Francisco entirely from WA and New York from NC. Although WA and NC are the most cost efficient warehouses for San Francisco and New York, respectively, such an intuitive solution (also given in Table 2.1) results in expensive shipping to Chicago from NC with a total cost of \$1550.50, which is 9% higher than the minimum value. □

	Demand locations			Cost
Warehouses	San Francisco	Chicago	New York	
Washington State	50	305	0	\$1415.50
North Carolina	270	0	270	
Washington State	320	35	0	\$1550.50
North Carolina	0	270	270	

Table 2.1: Shipping quantities for two different decisions: a minimizer (top) and an intuitive but expensive one (bottom).

Constraints introduce nonsmoothness by assigning certain decisions the value ∞, but smoothness also fails when the graph of a real-valued function has kinks, as would be the case when it represents a physical phenomenon with abrupt changes. Several commonly used functions are nonsmooth as well.

Example 2.6 (lasso and regularization). In the data fitting of Section 1.B, we may prefer a "simple" polynomial that has only a few nonzero coefficients. This is often desirable because it would highlight the important terms and avoid an unnecessarily intricate fit. One way to encourage sparsity of this kind is to augment the model in Section 1.B and consider the *lasso regression problem*

$$\underset{c\in\mathbb{R}^{1+n}}{\text{minimize}}\ \|Dc - y\|_2^2 + \theta\|c\|_1 \tag{2.3}$$

where θ is a positive parameter and $\|c\|_1$ is the Manhattan distance of c defined by

$$\|c\|_1 = \sum\nolimits_{j=0}^{n} |c_j|.$$

We can think of the added term as a penalty for having a nonzero c. Thus, the optimization problem aims to find a balance between a good fit with a low value of $\|Dc - y\|_2^2$ and a sparse fit with many zeros in c so that $\theta\|c\|_1$ is also low. This general approach is often used in data analytics and learning since it tends to perform well from a statistical point of view. However, the objective function is now nonsmooth.

Detail. For $g(c) = \|c\|_1 = \sum_{j=0}^{n} |c_j|$, the partial derivative

$$g'_{c_j}(c) = \begin{cases} 1 & \text{if } c_j > 0 \\ -1 & \text{if } c_j < 0 \\ \text{undefined} & \text{if } c_j = 0. \end{cases}$$

Thus, g fails to have a gradient at any point c with a zero component. Although nonsmooth, the objective function in (2.3) is convex by virtue of being a sum of two convex functions; cf. Proposition 1.8(a). The first function is convex by Example 1.10 and the second one by properties of norms and Proposition 1.8(b). □

2.B Subgradients

Since optimization models often involve nonsmooth functions, it becomes essential to extend the definitions and rules of differential calculus. We need to replace the notions of derivatives and gradients by appropriate substitutes. This leads us to *subgradients*.

Even for convex functions, nonsmoothness materializes in several ways. At the point x_1 in Figure 2.8, f has no derivative due to a kink in its graph. At the point x_2, it becomes

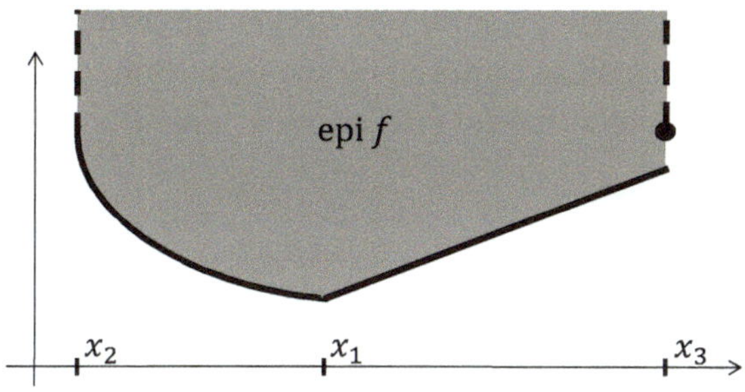

Fig. 2.8: The epigraph of a nonsmooth convex function.

meaningless to talk about a derivative as the function isn't even continuous at x_2; its value jumps to ∞ to the left of x_2. The function isn't continuous at x_3 either. In fact, regardless of the side from which we approach x_3, function values don't tend to $f(x_3)$. Although smoothness of a function may hold at "most" points, it often breaks down exactly at points of greatest interest: the minimizers of the function; see x_1 in Figure 2.8. So we can't simply ignore the issue of nonsmoothness.

The key insight is that the gradient $\nabla f(\bar{x})$ at a point $\bar{x}$ of smoothness furnishes the "slopes" of f along the various axes. When smoothness breaks down, it means that there might be *many* possible slopes describing the local behavior of the function near $\bar{x}$; see Figure 2.9 for an example with $f(x) = -x$ if $x < 0$ and $f(x) = x^2$ if $x \geq 0$. At $\bar{x} = 0$, the function has a kink and there are many lines representing the "slope" of f at the point.

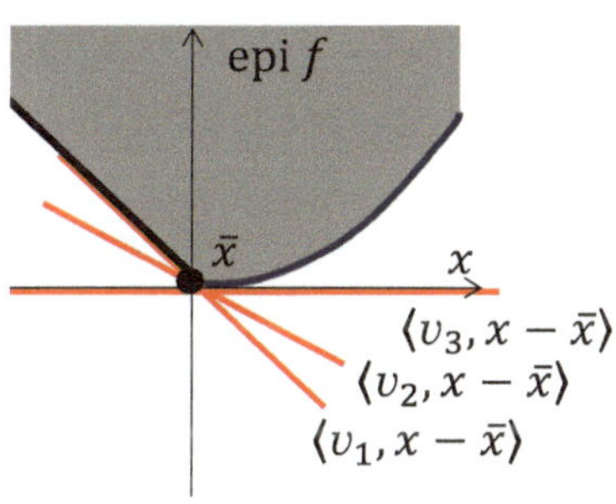

Fig. 2.9: Subgradients v_1, v_2, v_3 gives the slopes of supporting lines (hyperplanes in higher dimensions) of epi f.

Definition 2.7 (subgradient). *For an n-dimensional convex function f and a point $\bar{x}$ at which $f(\bar{x})$ is finite, an n-dimensional vector v is a* subgradient *of f at $\bar{x}$ when*

$$f(x) \geq f(\bar{x}) + v^\top (x - \bar{x}) \;\; \forall x.$$

The set of all subgradients at $\bar{x}$ is denoted by $\partial f(\bar{x})$. □

In Figure 2.9, v_1, v_2, and v_3 are examples of subgradients of f at 0 because they define lower-bounding affine functions as required by the definition. Generally, it's clear that if f is actually smooth at a point $\bar{x}$, then $\partial f(\bar{x})$ contains only the single vector $\nabla f(\bar{x})$.

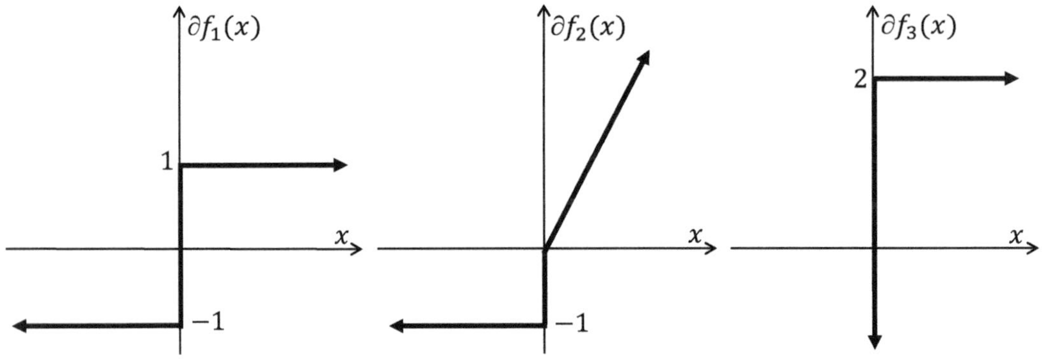

Fig. 2.10: Graphs of ∂f_1, ∂f_2, and ∂f_3 from Example 2.8.

Example 2.8 (subgradients). Consider the three functions given by

$$f_1(x) = |x|, \qquad f_2(x) = \begin{cases} -x & \text{if } x < 0 \\ x^2 & \text{otherwise,} \end{cases} \qquad f_3(x) = \begin{cases} 2x & \text{if } x \geq 0 \\ \infty & \text{otherwise,} \end{cases}$$

with Figure 2.9 providing an illustration for f_2. The subgradients for f_1 and f_2 are

$$\partial f_1(\bar{x}) = \begin{cases} \{-1\} & \text{if } \bar{x} < 0 \\ [-1, 1] & \text{if } \bar{x} = 0 \\ \{1\} & \text{otherwise,} \end{cases} \qquad \partial f_2(\bar{x}) = \begin{cases} \{-1\} & \text{if } \bar{x} < 0 \\ [-1, 0] & \text{if } \bar{x} = 0 \\ \{2x\} & \text{otherwise.} \end{cases}$$

The curly brackets indicate that $\partial f_1(\bar{x})$ and $\partial f_2(\bar{x})$ are always *sets* even though they may contain a single element sometimes.

The third function f_3 is finite only for $x \geq 0$ and has subgradients only at such points:

$$\partial f_3(\bar{x}) = \begin{cases} (-\infty, 2] & \text{if } \bar{x} = 0 \\ \{2\} & \text{if } \bar{x} > 0. \end{cases}$$

Figure 2.10 illustrates the subgradients of f_1, f_2, and f_3.

Detail. The subgradients of f_1, f_2, and f_3 follow from Definition 2.7. For instance, $\partial f_2(0)$ consists of v such that

$$x \geq vx \text{ if } x < 0 \quad \text{and} \quad x^2 \geq vx \text{ if } x > 0.$$

Any $v < -1$ invalidates the first expression and any positive v would violate the second expression. Thus, we're left with $v \in [-1, 0]$. □

In addition to algorithmic possibilities based on affine approximations of convex functions as explored in Exercise 2.12, subgradients lead to an extension of the Fermat rule for nonsmooth convex functions (see [30, Theorem 2.19] for a proof).

Theorem 2.9 (the Fermat rule; extended). *For an n-dimensional convex function f and a point $x^\star$ at which f is finite, one has*

$$x^\star \in \operatorname{argmin} f \quad \textit{if and only if} \quad 0 \in \partial f(x^\star).$$

□

The theorem extends Theorem 1.6 to the nonsmooth case by replacing gradients by subgradients, with the earlier result now becoming a special case. While in the smooth case solving the vector equation $0 = \nabla f(x)$ is a viable approach to minimizing f, we're now presented with the possibility of solving the *generalized equation* $0 \in \partial f(x)$. Thus, we shift from finding x such that $\nabla f(x)$ is a zero-vector to finding x such that the zero-vector is a subgradient of f at x.

The functions in Example 2.8 are convex and satisfy the optimality condition in Theorem 2.9 at $x = 0$ because Figure 2.10 shows that zero is a subgradient at that point for all three functions. Consequently,

$$\operatorname{argmin} f_1 = \operatorname{argmin} f_2 = \operatorname{argmin} f_3 = \{0\}.$$

The most expedient way to obtain expressions for subgradients is to leverage rules in the same manner as we compute gradients using rules from differential calculus. Chapters 2, 4, and 6 of [30] provide many rules. Here, we limit the discussion to a sum rule.

The gradient of the sum of two smooth functions is simply the sum of gradients of the individual functions; recall the discussion in Section 1.A. Specifically, for $f(x) = f_1(x) + f_2(x)$, with both f_1 and f_2 being smooth, one has

$$\nabla f(x) = \nabla f_1(x) + \nabla f_2(x).$$

We extend this rule to subgradients.

A main motivation for our focus on sums of two convex functions is that many minimization problems can be formulated in terms of a function $f = f_0 + \iota_C$, where f_0 is a smooth convex objective function and C is a convex feasible set; see Section 2.A for examples. We start by examining indicator functions more closely.

Example 2.10 (subgradients of indicator functions). The indicator function of a convex set C at a point $\bar{x} \in C$ has subgradients given by

$$\partial \iota_C(\bar{x}) = \big\{ v \;\big|\; v^\top (x - \bar{x}) \leq 0 \;\; \forall x \in C \big\}.$$

The expression follows by Definition 2.7 because $\iota_C(\bar{x}) = 0$ and $\iota_C(x) = \infty$ for $x \notin C$; see Figure 2.11. □

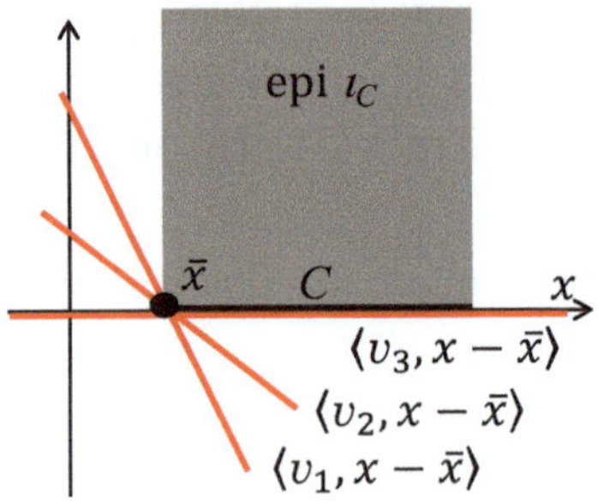

Fig. 2.11: At the point $\bar{x}$, the one-dimensional vectors v_1, v_2, and v_3 are examples of subgradients of the indicator function ι_C. In fact, $\partial\iota_C(\bar{x}) = (-\infty, 0]$.

The set of subgradients of an indicator function has a geometric interpretation, which motivates the following terminology.

Definition 2.11 (normal cones of convex sets). *For a nonempty convex set C, a vector v is a* normal *to C at $\bar{x} \in C$ when $v \in \partial\iota_C(\bar{x})$. The set of such v, denoted by $N_C(\bar{x})$, is the* normal cone *to C at $\bar{x}$, i.e.,*

$$N_C(\bar{x}) = \partial\iota_C(\bar{x}) = \{v \mid v^\top(x - \bar{x}) \le 0 \ \forall x \in C\}.$$

□

If C is a hyperplane, then $N_C(\bar{x})$ is simply the normal space from linear algebra consisting of vectors perpendicular to the hyperplane. Specifically, $v \in N_C(\bar{x})$ if and only if $v^\top(x - \bar{x}) = 0$ for all $x \in C$. Figure 2.12 illustrates the situation for more general convex sets using the geometric interpretation that the vectors $x - \bar{x}$ need to be between 90 and 270 degrees away from a normal vector v for any choice of $x \in C$ in Definition 2.11. This means that if C has a "smooth" boundary at $\bar{x}$ as in Figure 2.12(left), then all $v \in N_C(\bar{x})$ must point in the same direction. If C has a "corner" at $\bar{x}$ as in Figure 2.12(right), then $N_C(\bar{x})$ fans out in different directions.

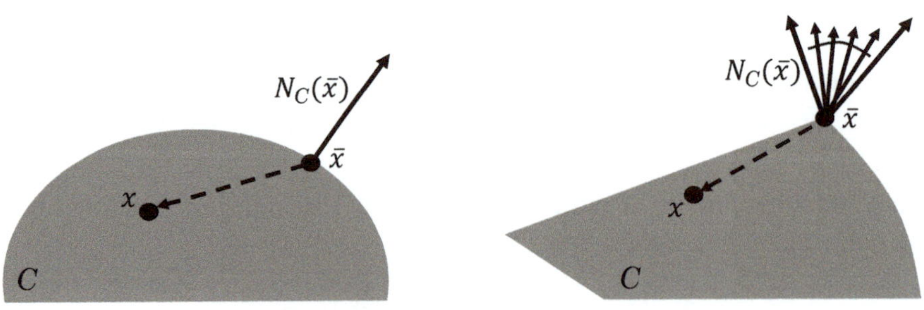

Fig. 2.12: Two examples of a normal cone $N_C(\bar{x})$ to a convex set C at a point $\bar{x}$.

In general, if $N_C(\bar{x})$ contains non-zero vectors, then it has vectors of any length:

$$\text{If } v \in N_C(\bar{x}) \text{ and } \lambda > 0, \text{ then } \lambda v \in N_C(\bar{x}).$$

Thus, one can scale a normal vector arbitrary to produce another normal vector.

Theorem 2.12 (sum rule for subgradients). *For a smooth convex function f_1 and a convex function f_2, the sum-function $f(x) = f_1(x) + f_2(x)$ has subgradients given by*

$$\partial f(\bar{x}) = \nabla f_1(\bar{x}) + \partial f_2(\bar{x})$$

at any point $\bar{x}$ at which f_2 is finite. □

For example, if $f(x) = (x-1)^2 + |x|$, then the subgradients of f at 0 are

$$\partial f(0) = 2(0-1) + [-1, 1] = [-3, -1],$$

where we use Example 2.8 to compute the subgradients of the absolute value function.

It's also clear from Theorem 2.12 that if f_0 is a smooth convex function and C is a convex set, then the function given by $f(x) = f_0(x) + \iota_C(x)$ has

$$\partial f(x) = \nabla f_0(x) + N_C(x)$$

at any point $x \in C$. In conjunction with Theorem 2.9, this insight leads to the following optimality condition.

Theorem 2.13 (optimality for constrained convex problems). *For a smooth convex function f in n dimensions and a convex set $C \subset \mathbb{R}^n$, one has*

$$x^\star \in \operatorname{argmin}_C f \quad \textit{if and only if} \quad -\nabla f(x^\star) \in N_C(x^\star).$$

□

Example 2.14 (optimality for convex functions). Consider the function f and the set C given by

$$f(x) = x_1^2 + x_2^2, \qquad C = \{x \mid x_1 + x_2 \geq 1\}.$$

Theorem 2.13 allows us to compute all the minimizers in $\operatorname{argmin}_C f$.

Detail. Convexity of f follows by Proposition 1.9. The set C is a polyhedron and thus convex. We compute $\nabla f(x) = (2x_1, 2x_2)$. Figure 2.13 indicates the nature of the normal cone of C at a point on the boundary of C. It consists of vectors v pointing in the southwest direction. These vectors are of the form $v = -(y, y)$ for some nonnegative y. For feasible x *not* on the boundary of C, i.e., $x_1 + x_2 > 1$, one has the normal cone $N_C(x) = \{(0, 0)\}$ as seen from the definition of a normal cone.

Thus, we've two cases: (i) x isn't on the boundary of C. Then, we need to find x such that $-\nabla f(x) = (0,0)$, but that isn't possible for any feasible x. (ii) x is on the boundary of C. Then, we need to find x such that

$$-\nabla f(x) = -\begin{bmatrix} 2x_1 \\ 2x_2 \end{bmatrix} = -\begin{bmatrix} y \\ y \end{bmatrix}$$

for some nonnegative y. These equations hold at the feasible point $x = (1/2, 1/2)$, which is on the boundary of C, and $y = 1$. It doesn't hold at $\bar{x}$ in Figure 2.13 as the two arrows don't point in opposite directions. We conclude that $\operatorname{argmin}_C f = \{(1/2, 1/2)\}$. □

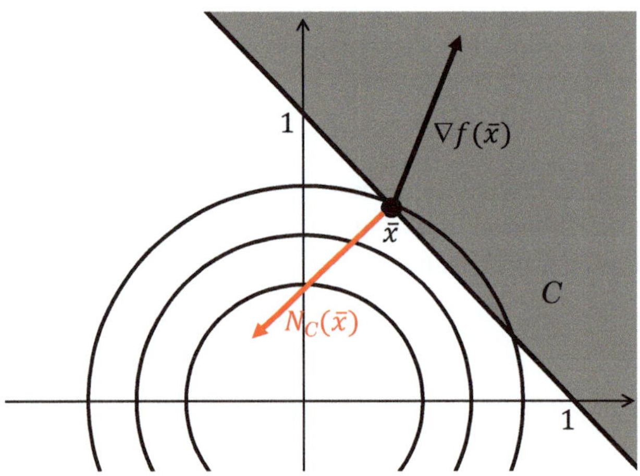

Fig. 2.13: Gradient of $f(x) = x_1^2 + x_2^2$ and normal cone of $C = \{x \mid x_1 + x_2 \geq 1\}$ at a point $\bar{x}$; see Example 2.14.

Example 2.15 (optimality in lasso regression). The lasso regression problem from Example 2.6 has a nonsmooth objective function and its gradients aren't defined at all points. Still, the extended Fermat rule $0 \in \partial f(c)$ from Theorem 2.9 applies. The objective function has

$$f(c) = \|Dc - y\|_2^2 + \theta\|c\|_1,$$

with θ being positive. An expression for the subgradients follows by Theorem 2.12.

Detail. The smooth function $f_0(c) = \|Dc - y\|_2^2$ has gradient $\nabla f_0(c) = 2D^\top(Dc - y)$; see (1.7). The subgradients of the Manhattan distance, scaled by θ, are given by

$$C_0 \times C_1 \times \cdots \times C_n, \text{ where } C_j = \begin{cases} \{-\theta\} & \text{if } c_j < 0 \\ [-\theta, \theta] & \text{if } c_j = 0 \\ \{\theta\} & \text{otherwise.} \end{cases}$$

Here, $\times$ denotes a *Cartesian product* of single points and line segments making $C_0 \times C_1 \times \cdots \times C_n$ a $(1+n)$-dimensional box, with (potentially) some dimensions collapsed.

By the extended Fermat rule in Theorem 2.9, we then conclude that $c^\star$ is a minimizer of f if and only if

$$\forall j = 0, 1, \ldots, n: \qquad 2(D^\top y - D^\top Dc^\star)_j \in \begin{cases} \{-\theta\} & \text{if } c_j^\star < 0 \\ [-\theta, \theta] & \text{if } c_j^\star = 0 \\ \{\theta\} & \text{otherwise,} \end{cases}$$

where $(u)_j$ is the jth component of a vector u. Solving for $c^\star$ in these conditions requires some trial and error. One would need to guess the sign of each component of $c^\star$ or set a component to zero. This produces equations that can be used to pin down the exact value of $c^\star$, which in turn needs to be consistent with the initial guess. However, trial and error isn't needed if we merely seek to verify optimality of a given vector $c^\star$ because then we already know the signs of the components of the vector. □

2.C Linear Constraints: Normal Cones and Feasibility

Feasible sets that are polyhedral allow us to compute normal cones and other quantities of interest relatively easily. Since they're specified by linear equalities and inequalities, properties and procedures from linear algebra become central. In this section, we develop formulas for normal cones that can be paired with the optimality condition in Theorem 2.13 to produce the celebrated Karush-Kuhn-Tucker condition for optimality in the next section. We also discuss how to check if a problem is feasible.

We start with a fact from [30, Proposition 2.39] about normal cones to affine sets.

Proposition 2.16 (normal cone of affine set). *For an $m \times n$-matrix A and an m-dimensional column vector b, the normal cone to the set*

$$C = \{x \mid Ax = b\}$$

at a point in C is the range of the matrix $A^\top$, i.e., regardless of $x \in C$, one has

$$N_C(x) = \{A^\top y \mid y \in \mathbb{R}^m\}.$$

□

Figure 2.14 illustrates the situation in the proposition for the special case with a single linear equality, i.e., $C = \{x \mid a^\top x = b\}$. Then, the normal cone comprises the vectors ya produced by any scalar y.

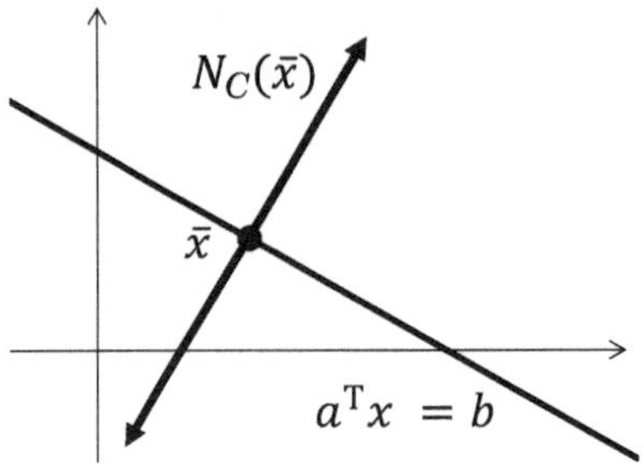

Fig. 2.14: Normal cone of an affine set $\{x \mid a^\top x = b\}$.

Example 2.17 (minimum norm problems). When there are multiple solutions to $Ax = b$, we may want to identify the "best" one in some sense. For a symmetric positive semidefinite $n \times n$-matrix Q, this might lead to the quadratic optimization problem

$$\underset{x \in \mathbb{R}^n}{\text{minimize}} \ \tfrac{1}{2} x^\top Q x \ \text{ subject to } \ Ax = b.$$

If $Q = I$, the identity matrix, then the objective function takes the form $\frac{1}{2}\|x\|_2^2$ and the problem identifies the smallest solution to $Ax = b$. The optimality condition in Theorem 2.13 reduces in this case to a system of linear equations involving an auxiliary vector that needs to be determined together with x.

Detail. In view of Proposition 2.16 and Example 1.10, the optimality condition in Theorem 2.13 ensures that x is a minimizer if and only if $-Qx = A^\top y$ for some m-dimensional vector y. This equation together with the feasibility condition $Ax = b$ lead to the linear system of equations

$$\begin{bmatrix} Q & A^\top \\ A & 0 \end{bmatrix} \begin{bmatrix} x \\ y \end{bmatrix} = \begin{bmatrix} 0 \\ b \end{bmatrix}.$$

If $Q = I$, then the system of equations simplifies to

$$x + A^\top y = 0, \quad Ax = b.$$

After eliminating x, we obtain

$$-AA^\top y = b.$$

If the rank of A is m, then $AA^\top$ is invertible and $y = -(AA^\top)^{-1}b$. This results in the minimizer

$$x = -A^\top y = A^\top (AA^\top)^{-1} b$$

of the minimum norm problem. □

The intermediate vector y plays a seemingly minor role in Example 2.17, but can trace its history back to the eminent mathematician and astronomer J.-L. Lagrange (1736-1813). In fact, the components of such vectors are called *Lagrange multipliers* or simply multipliers. Lagrange used such vectors to develop optimality conditions for the more general problem

$$\underset{x\in\mathbb{R}^n}{\text{minimize}}\ f_0(x) \ \text{ subject to } \ f_1(x)=0,\ldots,f_m(x)=0,$$

with all the functions being smooth. His fundamental insight was that optimization subject to equality constraints reduces (at least in the convex case) to solving a system of equations, linear or nonlinear, that's analogous to the Fermat rule of Chapter 1 for unconstrained optimization. However, the system of equations now involves one additional variable per constraint, the multipliers, that also need to be determined. It will become clear that multipliers are central to the development of broad classes of optimality conditions. Still, we'll have to wait to Chapter 5 before we see their full significance. There, multipliers emerge as equal partners with the decision variables and, in fact, become solutions to corresponding dual optimization problems.

We might be hopeful that one can extend the successful treatment of linear equality constraints in Example 2.17 to also cover inequality constraints. However, this doesn't come without complications.

Example 2.18 (optimization over a box). Consider the problem of minimizing a smooth convex function f over a box

$$C=\{x \mid \alpha_j \le x_j \le \beta_j,\ j=1,\ldots,n\},$$

where $-\infty \le \alpha_j \le \beta_j \le \infty$. We can combine expressions for normal cones of a box with the optimality condition in Theorem 2.13 to obtain:

$$x\in \operatorname{argmin}_C f \ \text{ if and only if } \ \forall j=1,\ldots,n:\ -f'_{x_j}(x)\in\begin{cases}(-\infty,\infty) & \text{if } \alpha_j=x_j=\beta_j\\ \{0\} & \text{if } \alpha_j<x_j<\beta_j\\ [0,\infty) & \text{if } \alpha_j<x_j=\beta_j\\ (-\infty,0] & \text{if } \alpha_j=x_j<\beta_j.\end{cases}$$

In contrast to in Example 2.17, these conditions aren't equations and resemble more the cases encountered in Example 2.15. If f is quadratic, then the partial derivatives become linear, but the issue lingers as the right-hand side remains conditional on x.

Detail. Directly from their definition, the normal cones of a line segment $C_j=\{x_j \mid \alpha_j \le x_j \le \beta_j\}$ are

$$N_{C_j}(x_j)=\begin{cases}(-\infty,\infty) & \text{if } \alpha_j=x_j=\beta_j\\ \{0\} & \text{if } \alpha_j<x_j<\beta_j\\ [0,\infty) & \text{if } \alpha_j<x_j=\beta_j\\ (-\infty,0] & \text{if } \alpha_j=x_j<\beta_j;\end{cases}$$

see Figure 2.11 for a related normal cone. These formulas lead to the asserted conditions. The difficulty is, however, that we don't know ahead of time which of the four cases apply for each j; we need to know x_j to make this determination. Let's ignore the trivial case with $\alpha_j = \beta_j$, which we can identify from the problem data, and suppose that $\alpha_j < \beta_j$, with both α_j and β_j being finite for all $j = 1, \ldots, n$. Then, we're down to three cases for each j, a total of 3^n possibilities. One such possibility amounts to an assumption about which case each component of a solution x satisfies. Suppose that a solution x satisfies

$$\begin{aligned} \alpha_j < x_j < \beta_j &\text{ for } j \in J_2 \\ x_j = \beta_j &\text{ for } j \in J_3 \\ x_j = \alpha_j &\text{ for } j \in J_4, \end{aligned}$$

with $J_2 \cup J_3 \cup J_4 = \{1, 2, \ldots, n\}$. Then, we only need to set $x_j = \beta_j$ for $j \in J_3$ and $x_j = \alpha_j$ for $j \in J_4$ and solve the equations

$$-f'_{x_j}(x) = 0 \quad \forall j \in J_2$$

to determine $\{x_j, j \in J_2\}$, which all together produce an n-dimensional vector x. Finally, we verify that $\alpha_j < x_j < \beta_j$ for $j \in J_2$ and check that

$$f'_{x_j}(x) \leq 0 \quad \forall j \in J_3 \quad \text{and} \quad f'_{x_j}(x) \geq 0 \quad \forall j \in J_4.$$

If this goes well, the obtained x satisfies the optimality condition. However, we may reach a contradiction: the assumption isn't compatible with the checks. Thus, in the worst case, one needs to cycle through 3^n possible assumptions and this quickly becomes intractable for large n. □

The challenge seen in the example is caused by inequality constraints. While a normal cone to an affine set has an explicit formula by Proposition 2.16, the situation for halfspaces is conditioned on the point under consideration.

Proposition 2.19 (normal cone of halfspace). *For an n-dimensional column vector a and the scalar* α, *let*

$$C = \{x \mid a^\top x \leq \alpha\}.$$

Then, for any $\bar{x} \in C$, *the normal cone*

$$N_C(\bar{x}) = \{az \mid z \in [0, \infty) \ \textit{if}\ a^\top \bar{x} = \alpha, \ \ z = 0 \ \textit{otherwise}\}.$$

□

In contrast to the situation for affine sets, only nonnegative scalars z need to be considered in the formula for $N_C(\bar{x})$ as negative ones would have produced vectors pointing "into" C; see Figure 2.15. This proposition follows largely from Definition 2.11 of normal cones. If $a^\top \bar{x} < \alpha$, then for all x near $\bar{x}$ one has $a^\top x \leq \alpha$. Thus, $v^\top(x - \bar{x}) \leq 0$ for all $x \in C$ only if $v = 0$. This means that $N_C(\bar{x}) = \{0\}$ when $\bar{x}$ is "inside" the halfspace.

When $a^\top \bar{x} = \alpha$ and $a \neq 0$, then a normal vector v needs to be perpendicular to the hyperplane specified by that equation, i.e., of the form az for some scalar z. This is the case shown in Figure 2.15. Moreover, to satisfy $v^\top(x - \bar{x}) \leq 0$ for all $x \in C$, z can't be negative. If $a = 0$, then $\alpha \geq 0$ because otherwise C would have been empty. Thus, $C = \mathbb{R}^n$ and the zero-vector is the only normal vector. This justifies the formula in Proposition 2.19.

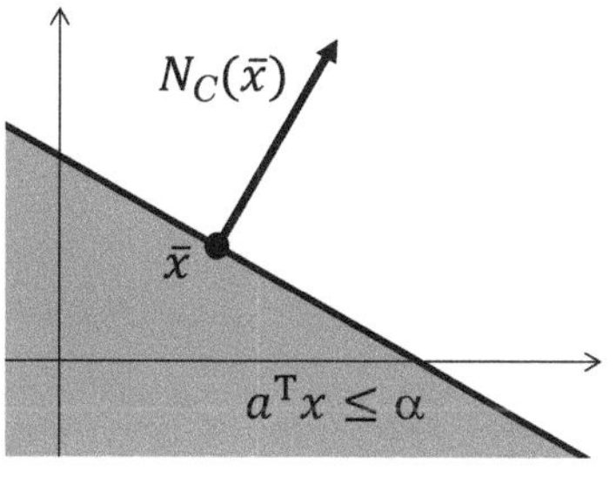

Fig. 2.15: Normal cone at the boundary of a halfspace.

We end the section with a discussion of feasibility, which can't be taken for granted in an optimization problem. If x isn't even feasible, then it can't be a minimizer.

Suppose that a feasible set is described by the system of equations

$$Ax = b,$$

where A is an $m \times n$-matrix. Is an optimization problem involving such a set feasible? A fundamental result of linear algebra is that there's a solution to $Ax = b$ if and only if the rank of $[A \;\; b]$ is the same as the rank of A, where $[A \;\; b]$ is the $m \times (n + 1)$-matrix obtained by augment A with the additional column b. Since the rank of a matrix can be computed by standard techniques, this check is usually easily carried out. If a solution exists, then the set of *all* solutions is

$$\{x \mid x = \bar{x} + y,\ y \in \operatorname{null} A\},$$

where $\bar{x}$ is any solution satisfying $A\bar{x} = b$ and $\operatorname{null} A = \{x \mid Ax = 0\}$, the *nullspace* of A. Thus, a solution is unique if and only if the nullspace of A is simply $\{0\}$.

Example 2.20 (polynomial interpolation). The problem of *interpolating* a given set of points $\{(x_i, y_i) \in \mathbb{R}^2, i = 1, 2, \ldots, m\}$ by a polynomial of degree n amounts to finding a solution to a system of linear equations. This problem is generally different than finding the closest curve as examined in Section 1.B.

Detail. A polynomial of degree n takes the form

$$p_n(x) = c_n x^n + \cdots + c_2 x^2 + c_1 x + c_0,$$

where $c = (c_0, c_1, \ldots, c_n)$ are coefficients to be determined such that

$$p_n(x_i) = y_i, \quad i = 1, \ldots, m.$$

Thus, we need to solve the equations

$$\begin{aligned}
c_0 + c_1 x_1 + c_2 x_1^2 + \cdots + c_n x_1^n &= y_1 \\
c_0 + c_1 x_2 + c_2 x_2^2 + \cdots + c_n x_2^n &= y_2 \\
\vdots \qquad\qquad \ddots \qquad &\quad \vdots \\
c_0 + c_1 x_m + c_2 x_m^2 + \cdots + c_n x_m^n &= y_m,
\end{aligned}$$

or concisely expressed as $Dc = y$ in the notation of Section 1.B. In particular, interpolation of the points $(0, 1), (1, 0), (2, 1)$ by a polynomial of degree 2 requires us to solve

$$\begin{bmatrix} 1 & 0 & 0 \\ 1 & 1 & 1 \\ 1 & 2 & 4 \end{bmatrix} \begin{bmatrix} c_0 \\ c_1 \\ c_2 \end{bmatrix} = \begin{bmatrix} 1 \\ 0 \\ 1 \end{bmatrix}.$$

The rank of the matrix is 3 in this case, and the rank remains 3 after augmenting the matrix with the right-hand side vector. Thus, a solution exists and in fact it's unique because null $D = \{0\}$. If the degree n were one, then there would be no solution and no interpolating polynomial. □

Example 2.21 (least-squares problems). When A is an $m \times n$-matrix that isn't necessarily invertible, finding a solution to $Ax = b$ or verify that none exists can be achieved by solving the *least-squares problem*

$$\boxed{\operatorname*{minimize}_{x \in \mathbb{R}^n} \|Ax - b\|_2^2}$$

The optimality condition in Theorem 1.6 (or Theorem 2.9) allows us to reduce this problem to one of linear algebra.

Detail. If the minimum value of the problem turns out to be zero, then any minimizer furnishes a solution to $Ax = b$. If the minimum value is positive, then there's no solution to the system of equations. The objective function is convex because it can be written as

$$f(x) = g(Ax - b), \quad \text{where } g(z) = \|z\|_2^2 = \sum_{i=1}^m z_i^2$$

and then Proposition 1.8(d) applies; g is convex by the Hessian rule in Proposition 1.9. Since

$$f(x) = (Ax - b)^\top (Ax - b) = x^\top A^\top Ax - 2b^\top Ax + \|b\|_2^2,$$

Example 1.10 shows that

$$\nabla f(x) = 2A^\top Ax - 2A^\top b.$$

The optimality condition in Theorem 1.6 therefore reduces to $A^\top Ax^\star = A^\top b$. This new system of equations has a solution always. When A has rank n, then $A^\top A$ is invertible and, e.g., Cholesky decomposition can provide the unique solution. □

When equalities are supplemented by inequalities, linear algebra doesn't fully suffice to determine whether there's a feasible point. However, the issue can be settled by solving a linear optimization problem that in fact generates a feasible point if there's one. Specifically, consider the polyhedral set

$$C = \{x \mid Ax = b, Dx \le d\},$$

where A and D are $m \times n$ and $q \times n$ matrices, respectively. We would like to find $x \in C$ or verify that none exists. We achieve this by formulating the model

$$\begin{aligned} \underset{(x,u,v,w)\in\mathbb{R}^{n+2m+q}}{\text{minimize}} \quad & \sum\nolimits_{i=1}^{m} (u_i + v_i) + \sum\nolimits_{i=1}^{q} w_i \\ \text{subject to} \quad & Ax = b + u - v \\ & Dx \le d + w \\ & u, v, w \ge 0, \end{aligned} \tag{2.4}$$

which is a linear optimization problem. This problem has a nonempty feasible set as confirmed by the point:

$$\begin{aligned} & x = 0, \;\; u_i = \max\{0, -b_i\}, \;\; v_i = \max\{0, b_i\}, \;\; i = 1, \dots, m \\ & w_i = \max\{0, -d_i\}, \;\; i = 1, \dots, q. \end{aligned}$$

If the minimum value is zero, then the x-component of a minimizer is a point in C. Otherwise, C is empty. As we discuss in Sections 2.D and 2.E, linear optimization problems are easily solved unless exceptionally large.

2.D Karush-Kuhn-Tucker Condition

If a feasible set is a polyhedron, then the optimality condition in Theorem 2.13 takes an explicit form that leads to computational procedures. The key challenge is to determine the normal cones of the feasible set. Propositions 2.16 and 2.19 provide the foundations but need to be combined to handle multiple constraints. It turns out that a normal vector for an intersection of sets is simply the sum of normal vectors for the individual sets. This produces the following formula; we refer to [30, Section 4.H] for a proof.

Proposition 2.22 (normal cone of polyhedral set). *Suppose that*

$$C = \{x \mid Ax = b,\ Dx \le d\},$$

where A and D are $m \times n$ and $q \times n$ matrices, respectively. The normal cone to C at one its points $\bar{x}$ is given by the formula

$$N_C(\bar{x}) = \left\{A^\top y + D^\top z \,\middle|\, y \in \mathbb{R}^m;\ \ z_i \ge 0 \ \text{if}\ i \in \mathbb{A}(\bar{x}),\ \ z_i = 0 \ \text{otherwise}\right\},$$

where $\mathbb{A}(\bar{x}) = \{i \mid D_i\bar{x} = d_i\}$, with D_i being the ith row of D and d_i being the ith component of d. □

The set $\mathbb{A}(\bar{x})$ specifies the inequality constraints that are *active* at $\bar{x}$. Recall that $Dx \le d$ is equivalently stated as the q inequality constraints $D_i x \le d_i$, $i = 1, \ldots, q$. So the active constraints $\mathbb{A}(\bar{x})$ are those for which there's no slack at the point $\bar{x}$. Inequalities not in $\mathbb{A}(\bar{x})$ are *inactive* and don't influence $N_C(\bar{x})$ because the corresponding $z_i = 0$. This is meaningful because inactive constraints at $\bar{x}$ have no impact on the local geometry of C near $\bar{x}$. The situation arises already for a single inequality constraints as discussed after Proposition 2.19; see also Figure 2.15.

We can now state the celebrated *Karush-Kuhn-Tucker* (KKT) condition for optimality of a smooth convex function under linear equality and inequality constraints; its validity is immediate from Proposition 2.22 and Theorem 2.13.

Theorem 2.23 (KKT condition for polyhedral sets). *Suppose that f is a smooth, convex, and n-dimensional function and*

$$C = \{x \mid Ax = b,\ Dx \le d\},$$

where A and D are $m \times n$ and $q \times n$ matrices, respectively. Then,

$$\begin{aligned} x^\star \in \operatorname{argmin}_C f \quad \text{if and only if} \quad & x^\star \in C \text{ and there are} \\ & y \in \mathbb{R}^m \text{ and } z_i \ge 0,\ i \in \mathbb{A}(x^\star),\ \ z_i = 0,\ i \notin \mathbb{A}(x^\star) \\ & \text{such that } -\nabla f(x^\star) = A^\top y + D^\top z. \end{aligned}$$

Here, the set $\mathbb{A}(x^\star) = \{i \mid D_i x^\star = d_i\}$, D_i is the ith row of D, and d_i is the ith component of d. □

The components of y and z in the theorem are called *multipliers*; there's one per constraint. The requirement that $z_i = 0$ if $D_i x^\star < d_i$ is a *complementary slackness condition*. It's the main source of difficulty when solving the KKT condition.

Example 2.24 (solving using KKT). Consider the quadratic optimization problem

$$\underset{x\in\mathbb{R}^2}{\text{minimize}}\ x_1 + 2x_2^2 \quad \text{subject to} \quad 2x_1 + x_2 \le 1,\ x_1 \ge 0.$$

Since $f(x) = x_1 + 2x_2^2$ defines a smooth convex function, the KKT condition in Theorem 2.23 applies with $x = (x_1, x_2)$, $d = (1, 0)$, and

$$D = \begin{bmatrix} 2 & 1 \\ -1 & 0 \end{bmatrix}.$$

The problem has no equality constraints so we can ignore A and b in the theorem.

Detail. The gradient $\nabla f(x) = (1, 4x_2)$ so that $-\nabla f(x) = A^\top y + D^\top z$ becomes

$$-\begin{bmatrix} 1 \\ 4x_2 \end{bmatrix} = \begin{bmatrix} 2 & -1 \\ 1 & 0 \end{bmatrix} \begin{bmatrix} z_1 \\ z_2 \end{bmatrix}.$$

This simplifies to $-1 = 2z_1 - z_2$ and $-4x_2 = z_1$. We seek x_1, x_2, z_1, z_2 satisfying these two equations as well as the feasibility condition $Dx \le d$. Moreover, the multipliers z_1 and z_2 need to be nonnegative and actually zero if the corresponding inequality constraint is inactive by the complementary slackness condition. We don't know which constraints are active and consider all four possibilities:

(i) Suppose that both constraints are active, i.e., $\mathbb{A}(x) = \{1, 2\}$. Then, $2x_1 + x_2 = 1$ and $x_1 = 0$. This implies that $x_2 = 1$. However, the resulting z_1 is negative because $-4x_2 = z_1$ so there's no valid solution.

(ii) Suppose that neither constraint is active, i.e., $\mathbb{A}(x) = \emptyset$. Then, $z_1 = z_2 = 0$. This isn't valid either because $-1 = 2z_1 - z_2$ can't hold.

(iii) Suppose that the first constraint is active and the second one is inactive, i.e., $\mathbb{A}(x) = \{1\}$. Then, $2x_1 + x_2 = 1$ and $z_2 = 0$. This isn't valid either because z_1 is negative due to $-1 = 2z_1 - z_2$.

(iv) Suppose that the first constraint is inactive and the second one is active, i.e., $\mathbb{A}(x) = \{2\}$. Then, $x_1 = 0$ and $z_1 = 0$. From $-4x_2 = z_1$ we conclude that $x_2 = 0$. The equation $-1 = 2z_1 - z_2$ yields $z_2 = 1$. Since $x = (0, 0)$ satisfies $Dx \le d$, we conclude that this point, with $z = (0, 1)$, satisfies the KKT condition and $x = (0, 0)$ is the unique minimizer of the quadratic problem by Theorem 2.23. □

Example 2.25 (projection on a polyhedron). For n-dimensional vector $\bar{x}$, $m \times n$-matrix A, $q \times n$-matrix D, and a nonempty polyhedral set

$$C = \{x \mid Ax = b,\ Dx \le d\},$$

the projection of $\bar{x}$ onto C is the point given by the unique minimizer of the problem

$$\underset{x\in C}{\text{minimize}}\ \tfrac{1}{2}\|x - \bar{x}\|_2^2.$$

It's characterized by the KKT condition in Theorem 2.23.

Detail. By Theorem 2.23, the projection point $x^\star$ satisfies

$$\bar{x} - x^\star = A^\top y + D^\top z, \text{ with } y \in \mathbb{R}^m \text{ and } z_i \geq 0,\ i \in \mathbb{A}(x^\star),\ \ z_i = 0,\ i \notin \mathbb{A}(x^\star).$$

In contrast to the situation in Example 2.17, this isn't a system of equations because it also involves inequalities and a complementary slackness condition. As in Examples 2.18 and 2.24, we could *assume* the active constraints at a minimizer, say $\mathbb{A}_0 \subset \{1, \ldots, q\}$, set $z_i = 0$ for $i \notin \mathbb{A}_0$, and solve

$$\begin{bmatrix} I & A^\top & \hat{D}^\top \\ A & 0 & 0 \\ \hat{D} & 0 & 0 \end{bmatrix} \begin{bmatrix} x \\ y \\ \hat{z} \end{bmatrix} = \begin{bmatrix} \bar{x} \\ b \\ \hat{d} \end{bmatrix},$$

where $\hat{D}$ consists of the rows $\{D_i, i \in \mathbb{A}_0\}$ of D and $\hat{d}$ consists of the elements $\{d_i, i \in \mathbb{A}_0\}$ from d. If there's a solution $(x^\star, y^\star, \hat{z}^\star)$ of this system of equations with $\hat{z}^\star \geq 0$ and the set of active inequalities at $x^\star$ coincides with the assumed $\mathbb{A}_0$, then $x^\star$ satisfies the KKT condition and is a minimizer. However, if there's no such $(x^\star, y^\star, \hat{z}^\star)$, then the process needs to be repeated with another assumption about $\mathbb{A}_0$; there are 2^q possibilities as each inequality constraint can be either active or inactive. We find the (unique) minimizer in this manner, but at a substantial computational cost. This highlights the combinatorial aspect caused by inequality constraints.

Of course, if we simply want to assess whether a given $x^\star$ is the projection of $\bar{x}$ on C, then it's easy to determine the corresponding active set $\mathbb{A}(x^\star)$, the assumption about $\mathbb{A}_0$ can be made correctly, and there's no need to cycle through a potentially exponentially large number of possibilities. □

Examples 2.24 and 2.25 point to the possibilities and challenges associated with using the KKT condition in Theorem 2.23 to determine a minimizer of a convex quadratic optimization problem. The prospect of cycling through all possible combinations of active constraints seems unattractive and *interior-point methods* seek to avoid it by cleverly adjusting the KKT condition into *nonlinear* equations and then using the Newton-Raphson method of Section 1.G. This results in algorithms with worst-case run times that grow only polynomially in the number of variables and constraints, and not exponentially as one would face when cycling through the active constraints. We refer to [30, Section 2.G] for details.

2.E Simplex Method

Solution of the KKT condition in Theorem 2.23 by cycling through the possible active constraints leads to the celebrated *simplex method* for solving linear optimization problems. Even though the method has to examine many possible sets of active constraints, with the number of possibilities being exponentially large in the number of constraints, it can be remarkably efficient in practice, in part due to convenient shortcuts available in the linear case. There are several versions of the method.

In this section, we describe the *revised simplex method* as applied to a *standard linear optimization problem:*

$$\boxed{\operatorname*{minimize}_{x\in\mathbb{R}^n}\ c^\top x \ \text{ subject to } \ Ax = b,\ x \geq 0}$$

A standard linear optimization problem is a special case of the (general) case from Section 2.A, which also includes the inequalities $Dx \leq d$. We get to the standard form by setting the matrix $D = -I$, the negative identity matrix, and $d = 0$. Then, the inequality constraints $Dx \leq d$ simplifies to $x \geq 0$. We can also go the other way. Every general linear optimization problem can be converted into one in standard form, albeit with more variables. Starting with the general problem of minimizing $c^\top x$ subject to the constraints $Ax = b$ and $Dx \leq d$, we replace x by the difference of two n-dimensional nonnegative vectors u and v to obtain

$$\operatorname*{minimize}_{u,v\in\mathbb{R}^n}\ c^\top(u-v) \ \text{ subject to } \ Au - Av = b,\ Du - Dv \leq d,\ u, v \geq 0.$$

Next, convert the inequalities $Du - Dv \leq d$ into equalities by introducing a q-dimensional nonnegative slack vector w, where q is the number of rows in the matrix D, to achieve

$$\operatorname*{minimize}_{u,v\in\mathbb{R}^n, w\in\mathbb{R}^q}\ c^\top(u-v) \ \text{ subject to } \ Au - Av = b,\ Du - Dv = d - w,\ \ u, v, w \geq 0,$$

which is in standard form with $(2n + q)$-dimensional decision vector (u, v, w), objective function coefficients $(c, -c, 0)$, right-hand side (b, d), and matrix

$$\begin{bmatrix} A & -A & 0 \\ D & -D & I \end{bmatrix}.$$

The equivalence is clear because a minimizer of the general linear optimization problem can be used to construct a minimizer of the one in standard form, and vice versa.

Hence, there's no loss of generality in focusing on standard linear optimization problems. Implementations of the simplex method would allow a user to input a linear optimization problem in any form, and the solver will carry out a transformation to the most computationally convenient form.

Conceptually, the revised simplex method for standard linear optimization problems resembles the gradient descent method of Section 1.F by computing in each iteration a direction of descent w, a step size $\lambda > 0$, and a new iterate $x + \lambda w$ from the current iterate x. However, now we need to account for the constraints and aim to ensure that all iterates are feasible. Throughout the section we assume that the $m \times n$-matrix A has $m < n$ and the rank of A is m. If m had been equal or larger than n, then either there are many redundant equality constraints, the problem is infeasible, and/or there's a unique solution to $Ax = b$, which are better treated separately. Moreover, we denote the feasible set by C, i.e.,

$$C = \{x \mid Ax = b, x \geq 0\}.$$

We use concepts from linear algebra to express the feasible set in a manner suitable for computations. We recall from Section 1.B that vectors $v^1, \ldots, v^m$ are linearly independent if $\alpha_1 v^1 + \ldots \alpha_m v^m = 0$ implies that $\alpha_1 = 0, \ldots, \alpha_m = 0$.

Definition 2.26 (basic feasible point). *A point $x \in C$ is* basic feasible *for C if its components can be divided into an m-dimensional vector x_B and an $(n-m)$-dimensional vector x_N of all zeros such that the m columns of A corresponding to x_B are linearly independent.*

The $m \times m$-matrix B formed by these columns is the basis matrix *associated with x. The remaining columns of A form the associated $m \times (n-m)$* nonbasic matrix *N. The components of x_B are* basic variables *and those of x_N are* nonbasic variables.

We denote by I_B and I_N the indices corresponding to x_B and x_N, respectively. Thus, $x_B = (x_j, j \in I_B)$ and $x_N = (x_j, j \in I_N)$. □

As an example, consider three constraints and five variables $x = (x_1, \ldots, x_5)$:

$$3x_1 + 2x_2 + 1x_3 + 0x_4 + 0x_5 = 120$$
$$5x_1 + 0x_2 + 0x_3 + 1x_4 + 0x_5 = 100$$
$$0x_1 + 1x_2 + 0x_3 + 0x_4 + 1x_5 = 40.$$

The constraints can be written concise as $Ax = b$, with $b = (120, 100, 40)$ and

$$A = \begin{bmatrix} 3 & 2 & 1 & 0 & 0 \\ 5 & 0 & 0 & 1 & 0 \\ 0 & 1 & 0 & 0 & 1 \end{bmatrix}.$$

In this case, the feasible set C has five basic feasible points. One basic feasible point has basic variables (x_3, x_4, x_5) and nonbasic variable (x_1, x_2), i.e., $I_B = \{3, 4, 5\}$ and $I_N = \{1, 2\}$. The associated basis and nonbasic matrices are

$$B = \begin{bmatrix} 1 & 0 & 0 \\ 0 & 1 & 0 \\ 0 & 0 & 1 \end{bmatrix}, \quad N = \begin{bmatrix} 3 & 2 \\ 5 & 0 \\ 0 & 1 \end{bmatrix}.$$

Given B and N, the constraint $Ax = b$ can be written as

$$Bx_B + Nx_N = \begin{bmatrix} 1 & 0 & 0 \\ 0 & 1 & 0 \\ 0 & 0 & 1 \end{bmatrix} \begin{bmatrix} x_3 \\ x_4 \\ x_5 \end{bmatrix} + \begin{bmatrix} 3 & 2 \\ 5 & 0 \\ 0 & 1 \end{bmatrix} \begin{bmatrix} x_1 \\ x_2 \end{bmatrix} = b = \begin{bmatrix} 120 \\ 100 \\ 40 \end{bmatrix}.$$

If x is a basic feasible point, then $x_N = 0$ by definition. Moreover, because the columns in B are linearly independent, B is invertible and $x_B = B^{-1}b$. This produces $x_B =$

(120, 100, 40), which indeed is a point in C. We can analysis the other four basic feasible points of C in a similar manner. They're defined by the nonbasic variables (x_2, x_4), (x_3, x_4), (x_3, x_5), and (x_1, x_5), respectively.

The following facts underpin the revised simplex method and show that we can largely focus on basic feasible points; see [7, Chapter 2] for proofs.

Theorem 2.27 (linear optimization problems in standard form). *For a linear optimization problem in standard form with $m < n$ and the rank of A being m, the following hold:*

(a) If the feasible set C is nonempty, then there exists at least one basic feasible point.
(b) If the problem has a minimizer, then at least one minimizer is a basic feasible point.
(c) If the feasible set C is bounded, then the problem has a minimizer.

□

While basic feasible points are defined using concepts from linear algebra, they're fully characterized by a geometric property.

Definition 2.28 (extreme point). *An* extreme point *of a convex set X is a point $\bar{x}$ in the set that can't be expressed as $\bar{x} = \lambda x + (1 - \lambda)y$ by distinct points x and y in the set and $\lambda \in (0, 1)$.* □

There's a one-to-one relationship between extreme points and basic feasible points for the feasible set C, which is also clear from Figure 2.16.

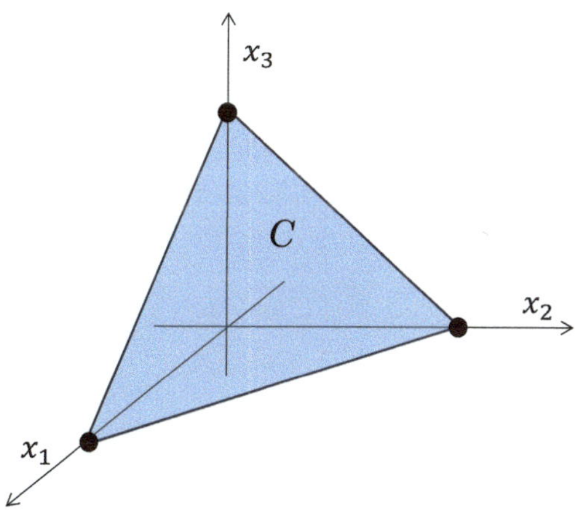

Fig. 2.16: Feasible set $C = \{x \mid Ax = b, x \geq 0\}$ with $A = [1\ 1\ 1]$ and $b = 1$, and its three extreme points $(1, 0, 0)$, $(0, 1, 0)$, $(0, 0, 1)$, which are also basic feasible points by Proposition 2.29.

Proposition 2.29 (extreme points and basic feasible points). *A point $\bar{x}$ is basic feasible for C if and only if $\bar{x}$ is an extreme point of C.* □

These concepts allow us to define and efficiently compute the specific direction w and step size λ for the revised simplex method. It results in an algorithm that moves from extreme point to extreme point until reaching a minimizer. We start by defining the direction.

Suppose x is a basic feasible point with basis matrix B, nonbasic matrix N, basic variables x_B, and nonbasic variables x_N. A direction w is *feasible* at x if $x + \lambda w \in C$ for sufficiently small $\lambda > 0$. This requires that $A(x + \lambda w) = b$ for sufficiently small $\lambda > 0$, which takes place when $Aw = 0$ because we know that $Ax = b$.

A feasible set C is *nondegenerate*, if all its basic feasible points have basic variables x_B that are positive. We proceed under this assumption, but return to the issue at the end of the section.

To ensure that a direction w is feasible we also need that $x+\lambda w \geq 0$ for sufficiently small $\lambda > 0$. Let w_B and w_N be the vectors formed by the components of w corresponding to x_B and x_N, respectively. Thus, w_B is m-dimensional and w_N is $(n-m)$-dimensional. Since x_B contains only strictly positive components by the nondegeneracy assumption, any w_B would be acceptable because $x_B + \lambda w_B \geq 0$ for sufficiently small λ. However, $x_N = 0$ and therefore $w_N \geq 0$ is required. In summary, under the nondegeneracy assumption, a direction w is feasible if

$$Aw = 0 \quad \text{and} \quad w_N \geq 0.$$

Among the collection of feasible directions, the revised simplex method makes a particular choice.

A *simplex direction* is a feasible direction of the form:

$$\begin{aligned} &w_N \text{ has one element set to 1 and the others to 0} \\ &w_B = -B^{-1}A^j, \end{aligned}$$

where A^j is the column of A corresponding to the nonbasic variable x_j for which a component in w_N is set to 1. The resulting direction is indeed feasible because

$$Aw = Bw_B + Nw_N = B(-B^{-1}A^j) + A^j = 0$$

and $w_N \geq 0$ by construction.

Since we've $n - m$ nonbasic variables, there's the same number of possible elements in w_N to set to 1 and thus $n - m$ simplex directions. Among these directions we seek one that's a descent direction. We recall from Section 1.F that w is a descent direction at x for a smooth objective function f if

$$\nabla f(x)^\top w < 0.$$

We now have $f(x) = c^\top x$ and $\nabla f(x) = c$. Thus, the requirement for a descent direction is $c^\top w < 0$.

We let c_B and c_N be the components of the vector c that correspond to the basic variables x_B and nonbasic variable x_N, respectively. Moreover, let c_j be the coefficient associated with nonbasic variable x_j, where j is the index singled out when settling on a simplex direction. With the adopted form of w, the descent condition translates into having

$$0 > c^\top w = c_B^\top w_B + c_N^\top w_N = c_j - c_B^\top B^{-1} A^j.$$

The quantity on the right-hand side is the *reduced cost* of nonbasic variable x_j and we denote it by $\bar{c}_j$, i.e.,

$$\bar{c}_j = c_j - c_B^\top B^{-1} A^j.$$

Putting these pieces together, we obtain the following algorithm.

Revised Simplex Method.

Data. Basic feasible point x, with basis matrix B and nonbasic matrix N.
Step 1. For each $j \in I_N$, compute the reduced cost $\bar{c}_j = c_j - c_B^\top B^{-1} A^j$.
Step 2. If $\bar{c}_j \geq 0$ for all $j \in I_N$, then stop.
Step 3. Select $i \in I_N$ with $\bar{c}_i < 0$.
Step 4. Set $w_N = (w_j, j \in I_N)$, with $w_i = 1$ and $w_j = 0$, $j \in I_N, j \neq i$.
Step 5. Compute $w_B = (w_j, j \in I_B) = -B^{-1} A^i$.
Step 6. Set $\lambda = -x_{j^\star} / w_{j^\star}$, where

$$j^\star \in \operatorname{argmin}_{j \in I_B} \{-x_j / w_j \mid w_j < 0\}.$$

Step 7. Update B by replacing the column corresponding to index $j^\star$ by the one corresponding to index i. Update N by replacing the column corresponding to index i by the one corresponding to index $j^\star$.
Step 8. With $w = (w_1, \ldots, w_n)$, replace x by $x + \lambda w$, and go to Step 1.

Step 1 is called *pricing* and computes the reduced costs. If there's no descent direction, the algorithm stops in Step 2. The stopping condition can also be interpreted from the point of view of the KKT condition in Theorem 2.23. In the present setting of a standard linear optimization problem, the KKT condition specializes to

$$-c = A^\top y - z,$$

with y being m-dimensional, z being n-dimensional, and the components $z_j = 0$ for any index with $x_j > 0$ but otherwise $z_j \geq 0$. An informed guess leads us to $y = -(B^{-1})^\top c_B$. Let's check this choice by plugging it into the KKT condition:

$$0 = c + A^\top y - z = c - A^\top (B^{-1})^\top c_B - z.$$

We write out this vector equation and obtain for each $j = 1, \ldots, n$:

$$0 = c_j - (A^j)^\top (B^{-1})^\top c_B - z_j = c_j - c_B^\top B^{-1} A^j - z_j.$$

Under the nondegeneracy assumption $x_j > 0$ for $j \in I_B$ and, by definition, $x_j = 0$ for $j \in I_N$. Thus, $z_j = 0$ for $j \in I_B$ and $z_j \geq 0$ for $j \in I_N$. We conclude that the KKT condition amounts to

$$c_j - c_B^\top B^{-1} A^j \geq 0, \quad j \in I_N$$
$$c_j - c_B^\top B^{-1} A^j = 0, \quad j \in I_B.$$

The second row always holds because one can combine the expressions for the different $j \in I_B$ and recall that B is the matrix with columns $A^j, j \in I_B$. This yields

$$c_B^\top - c_B^\top B^{-1} B = 0.$$

The first row is the check in Step 2 of the revised simplex method. Consequently, the method checks whether the KKT condition holds in each iteration.

If the condition doesn't hold, then the algorithm selects a nonbasic variable that's associated with a negative reduced cost in Step 3. This can be done in several ways; *Dantzig's rule* selects $i \in I_N$ with the lowest $\bar{c}_i$. Steps 4 and 5 compute a simplex direction. Step 6 identifies the step size, which is made exactly so long that a basic variable reduces its value to zero. If multiple basic variables reach zero for the same step size an arbitrary choice can be made. Step 7 *pivots*, i.e., the columns of A that make up B and N are updated. Step 8 computes the numerical value of the basic feasible point corresponding to the new basis matrix. We refer to [7, Chapter 3] for a theoretical justification of the simplex method, which includes the following fact.

Theorem 2.30 (convergence of revised simplex method). *If the feasible set C is nondegenerate and bounded, then the revised simplex method terminates in a finite number of steps with a basic feasible point that's a minimizer.* □

A linear optimization problem is *unbounded* if there exists a sequence $\{x^\nu\}_{\nu=1}^\infty$ in the feasible set C such that $c^\top x^\nu \to -\infty$ as $\nu \to \infty$. The revised simplex method can identify this situation in Step 6. If $w_j \geq 0$ for all $j \in I_B$, then $x + \lambda w$ is feasible for all step sizes $\lambda \geq 0$. Since $\bar{c}_i < 0$, this means that we can improve the objective function value at $x + \lambda w$ arbitrarily by increasing λ and the problem is unbounded.

The revised simplex method also identifies whether there are multiple minimizers. If the method terminates in Step 2 with $\bar{c}_j > 0$ for all $j \in I_N$, then the minimizer x is unique. If there are indices $j \in I_N$ with $\bar{c}_j = 0$ at termination, then multiple minimizers are present and they can be characterized by examining the direction w and step size λ in the manner laid out in Steps 4-6.

The revised simplex method requires a basic feasible point for initialization. Although finding such point manually could be hard, it's achieved by solving a feasibility problem in the form of a linear optimization problem, similar to (2.4), for which a basic feasible point is easily identified.

The assumption of nondegeneracy in Theorem 2.30 is *not* superfluous. The revised simplex method as stated may fail to obtain a minimizer for degenerate feasible sets because then the possibility of zero step size can't be ruled out for some choices of basis. However, this is easily addressed by making sure that the algorithm examines candidate bases without cycling back and revisiting a basis that's already identified as resulting in a zero step size. Among several such anti-cycling rules, *Bland's rule* is often used in practice and with this modification the revised simplex method is guaranteed to obtain a minimizer (when one exists) in a finite number of iterations even under degeneracy.

Example 2.31 (revised simplex method). Consider the standard linear optimization problem with $c = (-10, -5, 0, 0, 0)$, $b = (120, 100, 40)$, and

$$A = \begin{bmatrix} 3 & 2 & 1 & 0 & 0 \\ 5 & 0 & 0 & 1 & 0 \\ 0 & 1 & 0 & 0 & 1 \end{bmatrix}.$$

We use the revised simplex method to find a minimizer.

Detail. Among the five decision variables, let x_1 and x_2 be nonbasic, and then x_3, x_4, and x_5 must be basic variables, i.e., $I_N = \{1, 2\}$ and $I_B = \{3, 4, 5\}$. This means that $c_B = (0, 0, 0)$,

$$B = \begin{bmatrix} 1 & 0 & 0 \\ 0 & 1 & 0 \\ 0 & 0 & 1 \end{bmatrix}, \quad N = \begin{bmatrix} 3 & 2 \\ 5 & 0 \\ 0 & 1 \end{bmatrix}.$$

The initial basic feasible point is obtained from $Ax = Bx_B + Nx_N = b$ by computing $x_B = B^{-1}b = (120, 100, 40)$.

Using the columns $A^1 = [3\ \ 5\ \ 0]^\top$ and $A^2 = [2\ \ 0\ \ 1]^\top$, the pricing in Step 1 gives

$$\bar{c}_1 = c_1 - c_B^\top B^{-1} A^1 = -10 - [0, 0, 0] \begin{bmatrix} 1 & 0 & 0 \\ 0 & 1 & 0 \\ 0 & 0 & 1 \end{bmatrix} \begin{bmatrix} 3 \\ 5 \\ 0 \end{bmatrix} = -10$$

$$\bar{c}_2 = c_2 - c_B^\top A^2 = -5 - [0, 0, 0] \begin{bmatrix} 1 & 0 & 0 \\ 0 & 1 & 0 \\ 0 & 0 & 1 \end{bmatrix} \begin{bmatrix} 2 \\ 0 \\ 1 \end{bmatrix} = -5.$$

The optimality check in Step 2 fails because there's a negative reduced cost corresponding to a nonbasic variable. In fact, in this case, all the reduced costs are negative.

In Step 3, we set $i = 1$ in accordance with Dantzig's rule of selecting the nonbasic variables with the lowest reduced cost.

In Step 4, we set $w_1 = 1$ and $w_2 = 0$, and Step 5 gives $w_B = -(3, 5, 0)$.

Step 6 considers $j \in I_B = \{3, 4, 5\}$ with $w_j < 0$. Since $w_B = (-3, -5, 0)$, we only need to examine $j = 3$ and $j = 4$ and this leads to the step size

$$\lambda = \min\big\{ -120/(-3), -100/(-5)\big\} = 20.$$

The second term (corresponding to $j = 4$) attains the minimum, and thus $j^\star = 4$.

Step 7 updates the basis matrix B by replacing the column A^4 (because $j^\star = 4$) by A^1 (because $i = 1$), with corresponding changes in the nonbasic matrix N, i.e.,

$$B = \begin{bmatrix} 1 & 3 & 0 \\ 0 & 5 & 0 \\ 0 & 0 & 1 \end{bmatrix}, \quad N = \begin{bmatrix} 0 & 2 \\ 1 & 0 \\ 0 & 1 \end{bmatrix}.$$

We now have $I_B = \{3, 1, 5\}$ and $I_N = \{4, 2\}$. The columns in B need to be arranged in a manner corresponding to x_B and c_B, which permits simple matrix multiplication below.

In Step 8, we compute using $w = (1, 0, -3, -5, 0)$, the new iterate

$$x = \begin{bmatrix} 0 \\ 0 \\ 120 \\ 100 \\ 40 \end{bmatrix} + 20 \begin{bmatrix} 1 \\ 0 \\ -3 \\ -5 \\ 0 \end{bmatrix} = \begin{bmatrix} 20 \\ 0 \\ 60 \\ 0 \\ 40 \end{bmatrix}.$$

This ends the first iteration of the revised simplex method. We proceed with the next iteration.

Step 1: We need to make sure c_B is arranged properly to match B. The columns in B correspond to x_3, x_1, and x_5, respectively. Thus, $c_B = (c_3, c_1, c_5) = (0, -10, 0)$. Since

$$B^{-1} = \begin{bmatrix} 1 & 3 & 0 \\ 0 & 5 & 0 \\ 0 & 0 & 1 \end{bmatrix}^{-1} = \begin{bmatrix} 1 & -3/5 & 0 \\ 0 & 1/5 & 0 \\ 0 & 0 & 1 \end{bmatrix},$$

the reduced costs are

$$\bar{c}_2 = c_2 - c_B^\top B^{-1} A^2 = -5 - [0, -10, 0] \begin{bmatrix} 1 & -3/5 & 0 \\ 0 & 1/5 & 0 \\ 0 & 0 & 1 \end{bmatrix} \begin{bmatrix} 2 \\ 0 \\ 1 \end{bmatrix} = -5$$

$$\bar{c}_4 = c_4 - c_B^\top A^4 = 0 - [0, -10, 0] \begin{bmatrix} 1 & -3/5 & 0 \\ 0 & 1/5 & 0 \\ 0 & 0 & 1 \end{bmatrix} \begin{bmatrix} 0 \\ 1 \\ 0 \end{bmatrix} = 2.$$

The optimality check in Step 2 fails because there's a negative reduced cost corresponding to a nonbasic variable. In Step 3, we set $i = 2$ as there's no other choice.

In Step 4, we set $w_2 = 1$ and $w_4 = 0$. Step 5 gives

$$w_B = -B^{-1} A^2 = - \begin{bmatrix} 1 & -3/5 & 0 \\ 0 & 1/5 & 0 \\ 0 & 0 & 1 \end{bmatrix} \begin{bmatrix} 2 \\ 0 \\ 1 \end{bmatrix} = \begin{bmatrix} -2 \\ 0 \\ -1 \end{bmatrix}.$$

Step 6 computes the step size by setting

$$\lambda = \min\left\{ -60/(-2), -40/(-1)\right\} = 30,$$

which is achieved using $j^\star = 3$ because the minimum is attended by the first term and it corresponds to $j = 3$.

Step 7 updates the basis matrix B by replacing the column A^3 (because $j^\star = 3$) by A^2 (because $i = 2$), with corresponding changes in the nonbasic matrix N, i.e.,

$$B = \begin{bmatrix} 2 & 3 & 0 \\ 0 & 5 & 0 \\ 1 & 0 & 1 \end{bmatrix}, \quad N = \begin{bmatrix} 0 & 1 \\ 1 & 0 \\ 0 & 0 \end{bmatrix}.$$

We now have $I_B = \{2, 1, 5\}$ and $I_N = \{4, 3\}$.

In Step 8, we compute the new iterate using $w = (0, 1, -2, 0, -1)$:

$$x = \begin{bmatrix} 20 \\ 0 \\ 60 \\ 0 \\ 40 \end{bmatrix} + 30 \begin{bmatrix} 0 \\ 1 \\ -2 \\ 0 \\ -1 \end{bmatrix} = \begin{bmatrix} 20 \\ 30 \\ 0 \\ 0 \\ 10 \end{bmatrix}.$$

This ends the second iteration.

Step 1 of the next iteration gives $\bar{c}_3 = 5/2$ and $\bar{c}_4 = 1/2$, and we conclude in Step 2 that $x = (20, 30, 0, 0, 10)$ is a minimizer. □

2.F Support Vector Machines

The binary classification problem aims to predict whether an observation (or feature vector) belongs to one category (labeled −1) or another (labeled 1) based on its characteristics and can be addressed using logistic regression; cf. Example 1.13. An alternative formulation of the problem leads to *support vector machines*.

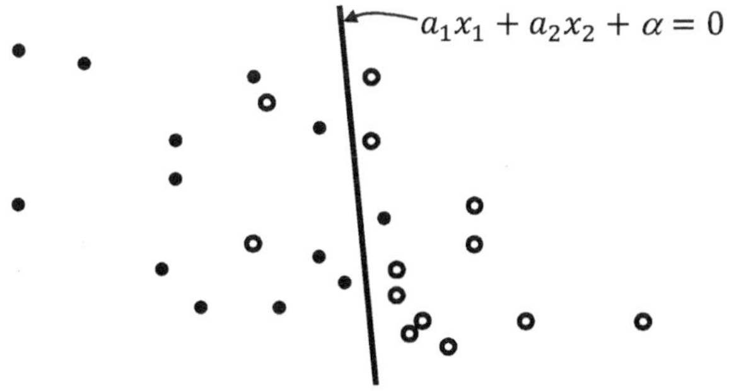

Fig. 2.17: Classification data with label −1 (dots) and label 1 (circles).

Suppose that we've m observations of feature vectors $\{x^i \in \mathbb{R}^n,\ i = 1,\dots,m\}$ and corresponding labels $\{y_i \in \{-1,1\},\ i = 1,\dots,m\}$; see Figure 2.17 for a case with $n = 2$ and $m = 26$. A support vector machine aims to identify a function f, called a classifier, that "best" predicts the unknown label of a new observation $\bar{x} \in \mathbb{R}^n$. We adopt the following prediction rule:

$$\text{predict label 1 if } f(\bar{x}) > 0 \qquad\qquad \text{predict label } -1 \text{ if } f(\bar{x}) < 0$$

and leave it undetermined if $f(\bar{x}) = 0$. Let's consider affine functions with $f(x) = a^\top x + \alpha$ and the problem becomes how to find (a, α).

Figure 2.17 shows a line defined by $a^\top x + \alpha = 0$ that represents the boundary between the observations that we predict to have label 1 and those to have label -1. At an observation $\bar{x}$ to the left of the line, $a^\top \bar{x} + \alpha < 0$ and we predict label -1. To the right, values are positive and the prediction is 1. The vector a and the scalar α defining the line in Figure 2.17 seem reasonable; only three misclassifications are made across the data set. How can we determine the "best" a and α?

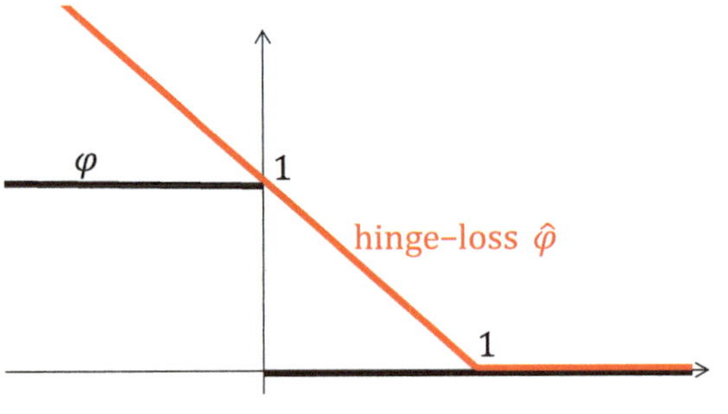

Fig. 2.18: The 0-1 function φ and its convex hinge-loss surrogate $\hat{\varphi}$.

A goal might be to find a and α that minimize the average number of misclassifications across the data as obtained by solving

$$\underset{(a,\alpha)\in\mathbb{R}^{n+1}}{\text{minimize}}\ \frac{1}{m}\sum_{i=1}^{m} \varphi\big(y_i(a^\top x^i + \alpha)\big), \quad \text{where } \varphi(\gamma) = \begin{cases} 1 & \text{if } \gamma \le 0 \\ 0 & \text{otherwise.} \end{cases}$$

Here, the 0-1 function φ returns 1 if its argument is nonpositive. Thus, the above sum computes the number of times $y_i(a^\top x^i + \alpha) \le 0$. The term $y_i(a^\top x^i + \alpha)$ is negative if the sign of y_i doesn't match that of $a^\top x^i + \alpha$ indicating a misclassification. The objective function therefore gives the average number of misclassification, counting also $a^\top x^i + \alpha = 0$ as a mistake regardless of y_i.

The problem is hard to solve at the scale needed in practice because of its nonconvexity; the objective functions isn't even continuous. As a surrogate, we consider the convex hinge-loss function given by $\hat{\varphi}(\gamma) = \max\{0, 1 - \gamma\}$ (see Figure 2.18) and this leads to the *hinge-loss problem*

$$\boxed{\underset{(a,\alpha)\in\mathbb{R}^{n+1}}{\text{minimize}}\ \frac{1}{m}\sum_{i=1}^{m}\max\left\{0, 1 - y_i(a^\top x^i + \alpha)\right\}}$$

with a convex but nonsmooth objective function; cf. Proposition 1.8. Let

$$\mathbb{A}^+ = \{i \mid y_i = 1\} \quad \text{and} \quad \mathbb{A}^- = \{i \mid y_i = -1\}.$$

If the minimum value of the hinge-loss problem is zero with minimizer $(a^\star, \alpha^\star)$, then

$$a^{\star\top} x^i + \alpha^\star \geq 1 \ \ \forall i \in \mathbb{A}^+ \quad \text{and} \quad a^{\star\top} x^i + \alpha^\star \leq -1 \ \ \forall i \in \mathbb{A}^-.$$

This means that

$$y_i(a^{\star\top} x^i + \alpha^\star) \geq 1, \quad i = 1, \dots, m$$

so the average number of misclassifications is zero. Thus, $(a^\star, \alpha^\star)$ is also a minimizer of the problem with the 0-1 objective function. This situation has a geometric interpretation: $(a^\star, \alpha^\star)$ defines a hyperplane that separates $\{x^i, i \in \mathbb{A}^+\}$ from $\{x^i, i \in \mathbb{A}^-\}$. There's even a buffer; no data point has $a^{\star\top} x^i + \alpha^\star \in (-1, 1)$. Figure 2.19 illustrates the situation, which is more clear-cut than in Figure 2.17.

In general, a minimizer of the hinge-loss problem may not minimize the actual one, but empirically we find that the hinge-loss problem often produces few misclassifications across a data set and, most significantly, reliable predictions in practice, especially when combined with the enhancements discussed below.

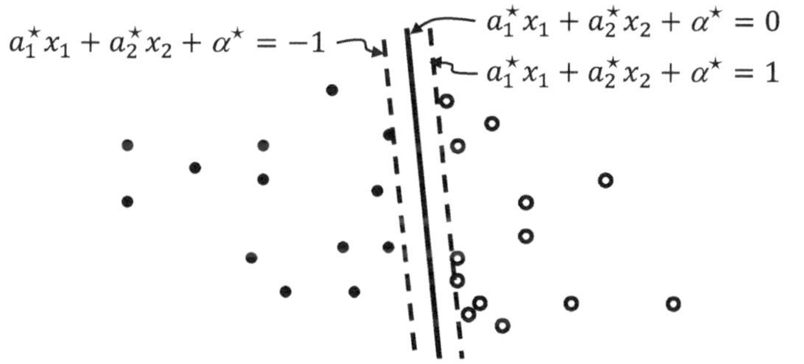

Fig. 2.19: Separation of data with label −1 (dots) and label 1 (circles) by a solid line, even leaving a buffer as indicated by the dashed lines.

Example 2.32 (subgradients of hinge-loss). Given n-dimensional vector x, scalar y, and integer m, let

$$h(a, \alpha) = \tfrac{1}{m}\max\left\{0, 1 - y(a^\top x + \alpha)\right\}.$$

The subgradients of h at (a, α) are given by

$$\partial h(a,\alpha) = \begin{cases} \{0\} & \text{if } 1 - y(a^\top x + \alpha) < 0 \\ \{(-yx/m, -y/m)\} & \text{if } 1 - y(a^\top x + \alpha) > 0 \\ \{\beta(-yx/m, -y/m) \mid \beta \in [0,1]\} & \text{otherwise.} \end{cases}$$

The objective function in the hinge-loss problem is the sum of such h with $x = x^i$ and $y = y_i$. The subgradients of the objective function at (a, α) is simply the sum of subgradients from the corresponding $\partial h(a, \alpha)$ by [30, Theorem 2.26]. □

As in regression and least-squares problems (cf. Examples 2.6), the practical performance of support vector machines improves by adding a regularization term of the form $\theta\|a\|_1$ or $\theta\|a\|_2^2$ for some positive parameter θ. This leads to the *regularized hinge-loss problem*

$$\underset{(a,\alpha)\in\mathbb{R}^{n+1}}{\text{minimize}} \ \frac{1}{m}\sum_{i=1}^{m} \max\left\{0, 1 - y_i(a^\top x^i + \alpha)\right\} + \theta\|a\|_1 \tag{2.5}$$

Although nonsmooth, the regularized hinge-loss problem can be reformulated as the equivalent linear optimization problem

$$\begin{aligned} \underset{(a,\alpha)\in\mathbb{R}^{n+1}, u\in\mathbb{R}^n, v\in\mathbb{R}^m}{\text{minimize}} \quad & \frac{1}{m}\sum_{i=1}^{m} v_i + \theta\sum_{j=1}^{n} u_j \\ \text{subject to} \quad & 1 - y_i(a^\top x^i + \alpha) \le v_i, \ i = 1, \ldots, m \\ & 0 \le v_i, \ i = 1, \ldots, m \\ & -a_j \le u_j, \ a_j \le u_j, \ j = 1, \ldots, n. \end{aligned}$$

If $(a^\star, \alpha^\star, u^\star, v^\star)$ minimizes this linear optimization problem, then $(a^\star, \alpha^\star)$ minimizes the regularized hinge-loss problem. Conversely, if $(a^\star, \alpha^\star)$ minimizes the regularized hinge-loss problem and $u^\star$ and $v^\star$ have components

$$u_j^\star = |a_j^\star| \quad \text{and} \quad v_i^\star = \max\left\{0, 1 - y_i(a^{\star\top} x^i + \alpha^\star)\right\},$$

then $(a^\star, \alpha^\star, u^\star, v^\star)$ minimizes the linear optimization problem. Thus, in principle, the regularized hinge-loss problem can be solved, via this reformulation, by the revised simplex method (Section 2.E) as well as interior-point methods (Section 2.D). However, the reformulation lifts the problem from having $n + 1$ variables and no constraints to having $m + 2n + 1$ variables and $2m + 2n$ constraints, which might be excessive in real-world applications. For large-scale instances, it's better to tackle the regularized hinge-loss problem directly by leveraging the expressions for subgradients in Example 2.32. One concrete possibility is laid out in the next section.

2.G Subgradient Method

The gradient descent method (Section 1.F) minimizes a smooth function f by moving from a current point x^ν in the direction $-\nabla f(x^\nu)$. It works because $-\nabla f(x^\nu)$ is a descent direction unless x^ν already satisfies the requirement of the Fermat rule. Consequently, with an appropriately small step size λ, we've

$$f\big(x^\nu - \lambda \nabla f(x^\nu)\big) < f(x^\nu).$$

An idea for extending the method to nonsmooth functions would be to simply replace $\nabla f(x^\nu)$ by a subgradient of f at x^ν. At least when f is convex, this indeed leads to viable algorithms but not without potential pitfalls.

Figure 2.20 illustrates the fundamental shift taking place when we pass from smooth to nonsmooth functions. On the left, $-\nabla f(\bar{x})$ provides a descent direction at $\bar{x}$ because f is smooth. On the right, we consider the nonsmooth function given by

$$f(x) = \max\left\{\tfrac{1}{2}x_1 + x_2,\ \tfrac{1}{2}x_1 - x_2,\ -\tfrac{1}{2}x_1 + x_2,\ -\tfrac{1}{2}x_1 - x_2\right\} - 1,$$

which defines an upside down pyramid. At $\bar{x} = (2,0)$, this function has subgradients $\partial f(\bar{x}) = \{(w_1, w_2) \mid w_1 = 1/2, w_2 \in [-1,1]\}$ illustrated by a fan of arrows in Figure 2.20(right). The subgradient $w = (1/2, 1)$ gives

$$f(\bar{x} - \lambda w) = \tfrac{3}{4}\lambda > f(\bar{x}) = 0$$

regardless of $\lambda > 0$. Consequently, $-w$ isn't a descent direction for f at $\bar{x}$.

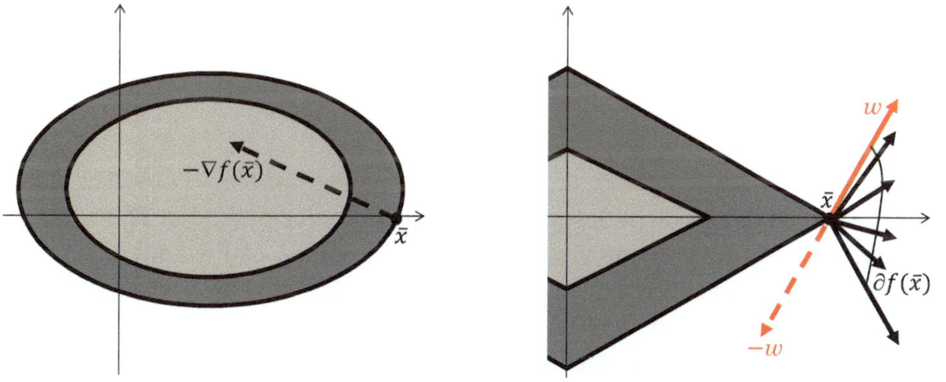

Fig. 2.20: Contours of a smooth (left) and a nonsmooth (right) function. On the left, $-\nabla f(\bar{x})$ provides a descent direction but on the right the subgradient $w = (1/2, 1)$ gives $f(\bar{x} - \lambda w) = 3\lambda/4 > f(\bar{x}) = 0$.

Despite this disappointment, subgradients are still useful. For the function in Figure 2.20(right), the minimizer is $x^\star = (0,0)$. Thus, movement along the direction $-w$ from $\bar{x}$ does get us *closer* to $x^\star$ as long as we don't move too far. This is an example of a general situation for convex functions. One can show that $\bar{x} - \lambda w$ is closer to a minimizer than $\bar{x}$ as long as the step size λ is small enough. Unfortunately, the bound on what constitutes a sufficiently small step size depends on $\min f$, which usually isn't known; see [30, Section 2.I] for details. Nevertheless, we arrive at an algorithm for

$$\operatorname*{minimize}_{x\in\mathbb{R}^n} f(x),$$

with f being convex and real-valued. It's *not* a descent method, but rather a procedure for solving the generalized equation $0 \in \partial f(x)$ by moving steadily closer to such a solution. Thus, it's related to Newton's method with its focus on the Fermat rule. By the optimality condition in Theorem 2.9, the generalized equation is of course equivalent to solving the minimization problem.

Subgradient Method.

Data. $x^0 \in \mathbb{R}^n$ and $\lambda^\nu \in (0, \infty), \nu = 0, 1, 2, \dots$.
Step 0. Set $\nu = 0$.
Step 1. Compute $w^\nu \in \partial f(x^\nu)$.
Step 2. Set

$$x^{\nu+1} \in x^\nu - \lambda^\nu w^\nu.$$

Step 3. Replace ν by $\nu + 1$ and go to Step 1.

Compared to the gradient descent method, which converges with rate $1/\nu$, the subgradient method is slower. For a specific choice of step size, we obtain that

$$\min_{k=0,1,\dots,\nu} f(x^k) - \min f \leq \frac{\beta\|x^0 - x^\star\|_2}{\nu^{1/2}}, \tag{2.6}$$

where $x^\star$ is a minimizer of f. Thus, the rate of convergence is merely $1/\sqrt{\nu}$. The slowdown can be expected as the gradient descent method relies on smoothness. The check across all iterations in (2.6) is reasonably since the "last" iteration of the subgradient method may not furnish the lowest function value; there's no guaranteed descent in each iteration.

The statement in (2.6) also involves $x^\star$, which is an arbitrary minimizer of f. Naturally, the right-hand side becomes smaller if the starting point x^0 is closer to some minimizer. The statement also involves the constant β, which is a bound on the subgradients of f:

$$\|w\|_2 \leq \beta \quad \forall w \in \partial f(x) \text{ and } x. \tag{2.7}$$

The statement in (2.6) relies on using a constant step size (across iterations):

$$\lambda = \frac{\|x^0 - x^\star\|_2}{\beta\nu^{1/2}},$$

where ν is the number of iterations to be carried out. Consequently, one would need to select an iteration limit and then start the calculations with the proposed step size.

The main advantage of the subgradient method is its broad applicability to nonsmooth convex problems. It also requires little memory and has low computational cost per iteration. The rate of convergence is also independent of n, the dimension of the problem, making the method attractive for large-scale problems. In fact, the rate is the best any algorithm can achieve for this broad class of convex functions [21, Section 3.2.3]. Still, the subgradient method tends to be sensitive to the choice of step size and some tuning is needed as the theoretically recommended value might not be known and anyhow is often on the small side in practice.

2.H Summary and Study Guide

In this chapter, we learned how undesirable decisions can be barred from consideration using constraints and how it's still possible to compute minimizers.

- Minimizing $f_0(x)$ over $x \in C$ can equivalently be written as minimizing $f(x)$ over $x \in \mathbb{R}^n$, where $f(x) = f_0(x) + \iota_C(x)$, because the indicator function ι_C has the value ∞ outside the feasible set C and thus $f(x) = \infty$ for all $x \notin C$.
- Quadratic and linear optimization problems commonly appear in applications and involve constraints specified by $Ax = b$ (linear equality constraints) and $Dx \leq d$ (linear inequality constraints).
- Constraints (via the indictor function ι_C) and functions involving $|x|$ and $\|x\|_1$ produce objective functions that are nonsmooth, which means that their gradients aren't defined everywhere or, if defined, the gradients don't vary continuously.
- Subgradients generalize gradients by allowing for many "slopes" at points of nonsmoothness. The set of all subgradients of f at x is denoted by $\partial f(x)$.
- The subgradients of the indicator function ι_C at x are called normal vectors to C at x and define the normal cone $N_C(x)$.
- The Fermat rule $\nabla f(x) = 0$ generalizes to $0 \in \partial f(x)$ when f is nonsmooth. If $f(x) = f_0(x) + \iota_C(x)$ for a convex smooth function f_0 and a convex set C, then the rule specializes to $-\nabla f_0(x) \in N_C(x)$. If C is given by linear equality and inequality constraints, then this specializes further to the KKT condition.
- The simplex method applies to linear optimization problems and seeks to solve the corresponding KKT condition by iteratively moving from extreme point to extreme point of the feasible set.
- The subgradient method minimizes a convex (real-valued) function f even if it's nonsmooth. It mimics the gradient descent method by computing $x^{\nu+1} = x^\nu - \lambda^\nu w^\nu$, with w^ν being a subgradient of f at the current point x^ν.

2.I Exercises

Exercise 2.1 (visualization of feasible set). Consider the linear optimization problem

$$\underset{x \in C}{\text{minimize}}\ 2x_1 + x_2, \text{ where } C = \{(x_1, x_2) \mid -x_1 + x_2 \le 2,\ x_1 \le 4,\ x_1 \ge 0,\ x_2 \ge 0\}.$$

Illustrate the feasible set C with a 2-dimensional drawing. Graphically, identify the minimizers and the minimum value. Give examples of what the coefficients in the objective function need to be for the problem to have a minimizer at $(4, 0)$, at $(4, 6)$, and at all points on the straight line between $(4, 0)$ and $(4, 6)$?

Exercise 2.2 (indicator function). For $f_0(x) = x^2$ and $C = [2, 3]$, draw the epigraph of the function given by $f(x) = f_0(x) + \iota_C(x)$. Explain why f is convex and graphically identify its minimizer and minimum value.

Exercise 2.3 (quadratic optimization problems). For the quadratic optimization problem

$$\underset{x \in C}{\text{minimize}}\ 2x_1^2 + 3x_2, \text{ where } C = \{(x_1, x_2) \mid -x_1 + x_2 \le 2,\ x_1 \le 4,\ x_1 \ge 0,\ x_2 \ge 0\},$$

identify Q, c, A, b, D, d in the canonical form

$$\underset{x \in \mathbb{R}^n}{\text{minimize}}\ \tfrac{1}{2}x^\top Q x + c^\top x \quad \text{subject to} \quad Ax = b,\ Dx \le d$$

and verify that Q is positive semidefinite.

Exercise 2.4 (convexity of sum function). Consider the function given by

$$f(x) = f_0(x) + \iota_C(x), \quad \text{where } f_0(x) = h(a^\top x + \alpha).$$

Here, a is an n-dimensional column vector, α is a negative number, the set C is convex, and $h(z) = z^2$. (a) Show that f_0 is convex. (b) Show that f is convex.

Exercise 2.5 (model formulation for Granite Aggregate). Consider the company "Granite Aggregates," which extracts and transports gravel from multiple quarries to construction sites. We seek to formulate a linear optimization model using indices $i = 1, 2, 3, 4, 5$ for quarries, $j = 1, 2, 3, 4$ for construction sites, and $t = 1, 2, 3$ for types of gravel. The decision variables are:

$$x_{ijt} = \text{ tons of gravel of type } t \text{ transported from quarry } i \text{ to construction site } j.$$

All these variables are constrained to be nonnegative. We also have the following data:

$$a_{it} = \text{ tons of gravel of type } t \text{ that's available from quarry } i$$

$$r_{jt} = \text{ tons of gravel of type } t \text{ that's requested at construction site } j.$$

Write linear equalities and/or inequalities for each of the following constraints in terms of these decision variables and data. (a) For every type of gravel, each construction site must receive at least the requested amount. (b) No more than the available amount of gravel of each kind can be transported out of each quarry. (c) How many constraints belong to the system of constraints in (a) and in (b), respectively?

Exercise 2.6 (model formulation for Imports). The company “Imports Inc.” stores imported products in several warehouses before the products are sold to retailers. We seek to formulate a linear optimization model using indices $p = 1, 2, \dots, 10$ for products, $w = 1, 2, \dots, 20$ for warehouses, and $q = 1, 2, 3, 4$ for quarters of a year. The decision variables are:

$$x_{pwq} = \text{number of items of product } p \text{ stored at warehouse } w \text{ at the beginning of } q$$

$$s_{pwq} = \text{number of items of product } p \text{ shipped out of warehouse } w \text{ during quarter } q$$

$$a_{pwq} = \text{number of items of product } p \text{ arriving at warehouse } w \text{ during quarter } q.$$

All these variables are constrained to be nonnegative. Write linear equalities and inequalities for each of the following constraints in terms of these decision variables. (a) For each product and warehouse, the number of items stored in the beginning of a quarter and the number of items arriving during the quarter must balance the numbers departing during the quarter and the numbers carried over to the next quarter. The last quarter of the year carries over to the first quarter of the next year. (b) At every warehouse, the inventory of each product shouldn’t be reduced by more than 40% from the previous quarter. (c) At warehouse 1, only up to 2 items of type 3 can be stored per 5 items of type 2 in any quarter.

Exercise 2.7 (model formulation for transportation company). Consider a transportation company that moves items from customers to destinations through some transshipment center. Let $c = 1, 2, \dots, 347$, $d = 1, 2, \dots, 23$, and $i = 1, 2, \dots, 8$ represent customers, destinations, and transshipment centers, respectively. Let s_c be the number of items to be shipped from customer c and t_d be the number of items to be received at destination d. Each item must go through exactly one transshipment node before it reaches its destination. We have the decision variables

$$x_{ci} = \text{number of items shipped from customer } c \text{ to transshipment center } i$$

$$y_{id} = \text{number of items shipped from transshipment center } i \text{ to destination } d.$$

(a) Write three sets of constraints that ensure flow balance at each customer, each transshipment center, and each destination. (b) The maximum number of items that can be shipped through transshipment center i is K_i, $i = 1, 2, \dots, 8$. Write a set of constraints that enforces this capacity restriction. (c) Suppose that the cost per item shipped from customer c to transshipment center i is α_{ci} and the cost per item shipped from transshipment center i to destination d is β_{id}. Write a minimum cost objective function.

Exercise 2.8 (lasso regression). With data from Exercise 1.5, minimize $\|Dc - y\|_2^2 + \theta\|c\|_1$ to determine polynomials of degree $n = 5, 10, 20$, and 30. Use an equivalent formulation in the form of a quadratic optimization problem. In each case, show the effect of varying θ from zero and up to a large number. Graph the resulting polynomials.

Guide. Introduce $1 + n$ additional decision variables $z_0, z_1, \ldots, z_n$ and note that the term $\theta\|c\|_1$ can be replaced by $\theta \sum_{j=0}^{n} z_j$ and the constraints

$$c_j \leq z_j, \quad -c_j \leq z_j, \qquad j = 0, 1, \ldots, n.$$

A large θ penalizes heavily any nonzero coefficient, which results in sparse polynomials with many zero coefficients.

Exercise 2.9 (subdifferentiability). For the convex functions given by (a) $f(x) = |3x|$ and (b) $f(x) = \max\{x + 1, 2\}$, determine their subgradients using the definition. Illustrate the subgradients.

Exercise 2.10 (extended Fermat rule). Consider the convex function given by $f(x) = \max\{0, 3x\}$. (a) Draw the epigraph of f. (b) Compute the subgradients of f and, assisted by Theorem 2.9, use them to determine all minimizers of f.

Exercise 2.11 (normal cones and optimality). Consider the convex function given by $f(x) = x_1^2 + 3x_2^2$, with $x = (x_1, x_2)$, and the convex set

$$C = \{x \mid x_1 \geq 0, x_2 \geq 1, x_1 + x_2 \leq 2\}.$$

(a) Draw the set C and its normal cone $N_C(x)$ at the point $x = (0, 1)$. (b) Use Theorem 2.13 to confirm that $(0, 1) \in \operatorname{argmin}_C f$.

Exercise 2.12 (linearization). For the model in Example 1.12, construct an approximating problem of the form

$$\underset{x \in C}{\text{minimize}} \max_{i=1,\ldots,m} c_i^{\top} x + b_i$$

using Definition 2.7. Write explicit expressions for the n-dimensional vectors c_i and the scalars b_i. Show that the resulting minimum value is no greater than that in Example 1.12 and is nondecreasing as m grows. Reformulate the problem as an equivalent linear optimization problem.

Exercise 2.13 (least-squares problems). Let A be an $m \times n$-matrix and b be an m-dimensional vector. (a) A regularized least-squares problem, which encourages solutions that have small Euclidean norms, is of the form

$$\underset{x \in \mathbb{R}^n}{\text{minimize}} \|Ax - b\|_2^2 + \theta\|x\|_2^2,$$

for some parameter $\theta \in (0, \infty)$. Show that the problem can be reformulated as a least-squares problem; see Example 2.21. (b) A weighted least-squares problem is of the form

$$\underset{x \in \mathbb{R}^n}{\text{minimize}} \|W(Ax - b)\|_2^2,$$

where W is a diagonal $m \times m$-matrix that potentially weighs the components of the vector $Ax - b$ differently. (An $n \times n$ matrix is *diagonal* if all its off-diagonal elements are zero.) Show that the problem can be reformulated as a least-squares problem; see Example 2.21.

(c) On the data from Exercise 1.5, numerically demonstrate the effect of changing θ and W in the regularized and weighted least-squares problems.

Exercise 2.14 (optimization over a box). Let $f(x) = 2(x_1 - 1)^2 + 3(x_2 + 1)^2 + 4x_3^2$ and $C = [1, 2] \times [0, \infty) \times [-1, 1]$, which is a box in three dimensions. Determine $\operatorname{argmin}_C f$ using the optimality condition in Example 2.18.

Exercise 2.15 (KKT condition). For $C = \{x \mid x_1 + x_2 = 1, x_1 \geq 0, x_2 \geq 0\}$, write the constraints in the canonical form $Ax = b$ and $Dx \leq d$. Identify A, b, D, and d. Determine and visualize the normal cone $N_C(x)$ at every point $x \in C$ using the formula in Proposition 2.22. Use the KKT condition of Theorem 2.23 to confirm that $(1, 0) \in \operatorname{argmin}_C f$ when $f(x) = (x_1 - 1)^2 + x_2$.

Exercise 2.16 (more KKT condition). Consider the problem

$$\underset{x \in \mathbb{R}}{\text{minimize}}\ (x - 1)^2 \ \text{ subject to } \ x \leq 0.$$

Use convexity and the KKT condition from Theorem 2.23 to confirm that $x^\star = 0$ is the only minimizer.

Exercise 2.17 (more KKT condition). Consider the problem

$$\underset{x \in \mathbb{R}^2}{\text{minimize}}\ (x_1 - 1)^2 + x_2^2 \ \text{ subject to } \ x_1 \leq 0.$$

Use convexity and the KKT condition from Theorem 2.23 to confirm that $x^\star = (0, 0)$ is the only minimizer.

Exercise 2.18 (feasible directions). Consider the linear optimization problem

$$\underset{x \in \mathbb{R}^2}{\text{minimize}}\ x_1 + 2x_2 \ \text{ subject to } \ x_1 + x_2 = 1,\ x_1 \geq 0,\ x_2 \geq 0.$$

At $x = (0, 1)$, confirm that $w = (1, -1)$ is a feasible direction.

Exercise 2.19 (basic feasible points and feasible directions). Suppose that a linear optimization problem in standard form has three variables and the constraints

$$2x_1 + 2x_2 - x_3 = 4,\ \ x_1 \geq 0,\ \ x_2 \geq 0,\ \ x_3 \geq 0.$$

(a) Identify the matrix A and the vector b so that the feasible set takes the form $C = \{x \mid Ax = b, x \geq 0\}$. (b) Determine the basic feasible points of C. (c) Determine conditions that a feasible direction w at $x = (2, 0, 0)$ need to satisfy.

Exercise 2.20 (more basic feasible points and feasible directions). Consider the linear optimization problem

$$\underset{x \in \mathbb{R}^2}{\text{minimize}}\ 3x_1 + 2x_2 \ \text{ subject to } \ 2x_1 + x_2 = 1,\ x_1 \geq 0,\ x_2 \geq 0.$$

(a) At $x = (0, 1)$, confirm that $w = (1, -2)$ is a feasible direction. (b) Write the problem in the form

$$\underset{x \in \mathbb{R}^n}{\text{minimize}} \ c^\top x \ \text{ subject to } \ Ax = b, \ x \geq 0,$$

where A is an $m \times n$ matrix. (c) Split the variables into basic and nonbasic, and compute their values.

Exercise 2.21 (revised simplex method). Consider a linear optimization problem in standard form, with five variables, that has been solved to optimality using the revised simplex method. At the obtained minimizer, x_1, x_2, x_3 are basic and x_4, x_5 are nonbasic variables. The following information is available:

$$c_1 = 1, \ c_2 = 2, \ c_3 = 1, \ c_4 = 1, \ B^{-1} = \begin{bmatrix} 0 & 1 & 0 \\ 1 & 0 & 1 \\ 1 & 2 & 0 \end{bmatrix}, \ b = \begin{bmatrix} 1 \\ 1 \\ 1 \end{bmatrix}.$$

(a) How much can c_4 increase before the current minimizer isn't optimal anymore? (b) For what values of b_3 will the current basis matrix produce feasible points?

Exercise 2.22 (step size in revised simplex method). For a linear optimization problem in standard form with six variables, the revised simplex method produces the direction $w = (1, 4, 0, 0, -1, 2)$ from the current point $x = (0, 2, 0, 0, 2, 3)$. Determine the step size in this iteration.

Exercise 2.23 (step size in revised simplex method). For a linear optimization problem in standard form with six variables, the revised simplex method produces the direction $w = (3, 1, 4, 0, 1, 2)$ from the current point $x = (1, 0, 1, 0, 2, 2)$. Determine the step size in this iteration.

Exercise 2.24 (linear constraints). Consider the constraints

$$x_1 + x_2 \leq 1, \ \ x_2 \leq 1, \ \ x_1 \geq 0, \ \ x_2 \geq 0.$$

(a) Reformulate the constraints to fit the form $Ax = b$ and $x \geq 0$ using two additional variables x_3 and x_4. (b) In the reformulation, suppose that x_1 and x_2 are nonbasic variables and x_3 and x_4 are basic. Determine whether the corresponding point is basic feasible. (c) Determine a basic representation (i.e.,, basic and nonbasic variables) for a feasible point with $x_1 = 0$ and $x_2 = 1$. Also, compute the values of the basic variables. How many such basic representations are there?

Exercise 2.25 (extreme points). Consider the set

$$X = \{x \mid x_1 + x_2 \geq 1, \ \tfrac{1}{2}x_1 + x_2 \leq 1, \ x_1 \geq 0, \ x_2 \geq 0\}.$$

(a) Illustrate X with a drawing. (b) Determine all the extreme points of X. (c) Write the constraints defining X in the standard form $Ax = b, \ x \geq 0$ by augmenting the vector x with two additional variables x_3 and x_4. (d) If x_1, x_2 are nonbasic and x_3, x_4 are basic, then determine the values of the latter and check if the resulting point is feasible. (e) Determine the total number of basic representations of all the extreme points identified in (b). (f) Suppose that x_2 and x_4 are nonbasic variables. Determine the value of the basic variables and determine whether the resulting point is basic feasible.

Exercise 2.26 (objective values in revised simplex method). For a linear minimization problem with two nonbasic variables (x_3 and x_4) and six extreme points of the feasible set, the revised simplex method produces the reduced costs $\bar{c}_3 = 1$ and $\bar{c}_4 = 0$ during an iteration. The objective function value at the current point is 4. Give two reasons why the objective function values at the six extreme points can't be 4, 5, 3, 6, 5, and 7.

Exercise 2.27 (alternative support vector machine). For given $(a^\star, \alpha^\star)$, the distance between the two sets $\{x \mid {a^\star}^\top x + \alpha^\star \geq 1\}$ and $\{x \mid {a^\star}^\top x + \alpha^\star \leq -1\}$ is $2/\|a^\star\|_2$. (In Figure 2.19, this is the distance between the dashed lines.) Since the first set corresponds to the points where one hopes to have all the positively labeled observations and similarly with the second set for the negatively labeled observations, we may seek to separate these sets as much as possible to reduce the chance of misclassification of future observations. This leads to a formulation that penalizes large $\|a\|_2$:

$$\underset{(a,\alpha)\in\mathbb{R}^{n+1}}{\text{minimize}} \ \frac{1}{m}\sum\nolimits_{i=1}^{m} \max\left\{0, 1 - y_i(a^\top x^i + \alpha)\right\} + \theta\|a\|_2^2.$$

Derive an equivalent quadratic optimization problem.

Exercise 2.28 (regularized support vector machines). On given training and test data and for a range of θ-values, determine classifiers by solving the regularized hinge-loss problem (2.5). Report the values of the hinge-loss on the training data, hinge-loss on the test data, and the percent correctly classified on the test data. What features in the data set appear more important? Use both a linear-optimization reformulation and the subgradient method. Compare computing times and discuss scalability. For the subgradient method, consider the effect of changing the step size λ.

Exercise 2.29 (subgradient method). For the function $f(x) = \max\{-x - 1, x - 1, 2x - 2\}$, carry out three iteration of the subgradient method from the point $x = 1$. Use step size $\lambda = 1/2$. Stop if you reach a minimizer.

Chapter 3
Overcoming Uncertainty

We often encounter optimization problems with parameters that are unknown at the time when a solution needs to be obtained. A store doesn't know the future demand when it has to decide how to restock its inventory. A classifier is fitted to data that may not represent the context in which it will be applied. Under such circumstances, we would like to formulate a model that accounts for uncertainty and produces good decisions even though future conditions are unknown. This leads to problems of *optimization under uncertainty*. We'll now examine several formulations of such problems and construct algorithms that address the often large-scale nature of the resulting models.

3.A Product Mix Optimization

A firm manufactures and sells four types of wooden surfboards. Each type requires t_{1j} hours for carpentry and t_{2j} hours for finishing, $j = 1, \ldots, 4$. In a time period, there are $d_1 = 6000$ hours available for carpentry and $d_2 = 4000$ for finishing. There's a profit per sale of \$12, \$25, \$21, and \$40 for the four types, respectively. Under the assumption that all units produced can be sold, the goal is to determine how much to manufacture of each type of board during a time period such that the total profit is maximized and the labor constraints are satisfied. This an example of the challenges in the area of *production planning*.

With the hours available collected in the vector $d = (d_1, d_2)$ and the hours needed per surfboard in the matrix

$$T = \begin{bmatrix} t_{11} & t_{12} & t_{13} & t_{14} \\ t_{21} & t_{22} & t_{23} & t_{24} \end{bmatrix} = \begin{bmatrix} 4 & 9 & 7 & 10 \\ 1 & 1 & 3 & 40 \end{bmatrix},$$

we formulate the linear optimization problem

$$\operatorname*{minimize}_{x \in \mathbb{R}^4} \ c^\top x \quad \text{subject to} \quad Tx \le d, \quad x \ge 0, \tag{3.1}$$

J. O. Royset, Optimization Concepts, *Springer Series in Operations Research and Financial Engineering*,
https://doi.org/10.1007/978-3-032-36819-5_3

where $c = (-12, -25, -21, -40)$ due to the switch from maximization to minimization. The first row in the vector inequality $Tx \leq d$ ensures that we don't use more carpentry hours than available and, likewise, the second row constrains the finishing hours. The revised simplex method (Section 2.E) or an interior-point method (Section 2.D) produces a minimizer

$$x^\star = (1333.3,\ 0,\ 0,\ 66.7), \quad \text{with minimum value } -\$18667.$$

This means that the maximum profit is \$18667. The prescribed production quantities 1333.3, 0, 0, and 66.7 for surfboard types 1, 2, 3, and 4, respectively, aren't all integers and this might raise questions about how to implement such a plan in practice. If the model supports long-term planning, where insight is more important than exact production numbers, then the issue might be minor. It's reasonable to avoid restrictions to integers and thus ensure a convex problem. Anyhow, the relatively large values in $x^\star$ make rounding to the nearest integer a satisfactory remedy. We return to the topic of integer restrictions in Chapter 4.

The main take-away from the minimizer $x^\star$ is that surfboards of type 2 and 3 shouldn't be produced at all. Maybe the firm should get rid of those two production lines? Before we jump to such a drastic conclusion, let's get a bit more realistic and account for the fact that the number of hours needed to produce each board may not be known with certainty. Each entry in the matrix T now becomes a random variable and we denote by $\boldsymbol{T}$ the resulting random matrix. For simplicity's sake, let's assume that each entry of $\boldsymbol{T}$ takes on four possible values with equal probability (1/4) and that these entries are statistically independent of one another; see Table 3.1.

entry	possible labor hours			
$\boldsymbol{t}_{11}$	3.60	3.90	4.10	4.40
$\boldsymbol{t}_{12}$	8.25	8.75	9.25	9.75
$\boldsymbol{t}_{13}$	6.85	6.95	7.05	7.15
$\boldsymbol{t}_{14}$	9.25	9.75	10.25	10.75
$\boldsymbol{t}_{21}$	0.85	0.95	1.05	1.15
$\boldsymbol{t}_{22}$	0.85	0.95	1.05	1.15
$\boldsymbol{t}_{23}$	2.60	2.90	3.10	3.40
$\boldsymbol{t}_{24}$	37.0	39.0	41.0	43.0

Table 3.1: Possible values for each entry in the random matrix $\boldsymbol{T}$.

We've eight random variables, each taking four possible values. This yields a total of $4^8 = 65536$ possible T matrices (outcomes or scenarios) and each one of these has equal probability of occurring. Let's denote by $p_i = 1/65536$ the probability of a particular outcome T^i, $i = 1, \ldots, 65536$.

Since the firm must decide on the production plan before the number of hours required for carpentry and finishing are known with certainty, there's the possibility that the plan actually exceeds the number of hours available. Therefore, we must factor in the possibility of having to pay for overtime. This recourse cost is incurred at $a_1 = 5$ dollars per extra carpentry hour and $a_2 = 10$ dollars per extra finishing hour and enters into play after the production plan x has been selected and the times required, T^i, have been observed. At least potentially, the firm makes a different decision about overtime in each one of the 65536 possible outcomes for $\boldsymbol{T}$. Let y_1^i and y_2^i denote the number of overtime hours hired for carpentry and finishing, respectively, when the matrix $\boldsymbol{T}$ turns out to be T^i, and let

$$y = \left(y_1^1, y_2^1, y_1^2, y_2^2, \ldots, y_1^{65536}, y_2^{65536}\right)$$

be the vector of all these overtime decisions. Then, we obtain the model

$$\begin{aligned}
\underset{x\in\mathbb{R}^4,\, y\in\mathbb{R}^{131072}}{\text{minimize}} \; & \sum_{j=1}^{4} c_j x_j + \sum_{i=1}^{65536} p_i (a_1 y_1^i + a_2 y_2^i) \\
\text{subject to } & \sum_{j=1}^{4} t_{1j}^i x_j - y_1^i \le d_1, \quad i = 1, \ldots, 65536 \\
& \sum_{j=1}^{4} t_{2j}^i x_j - y_2^i \le d_2, \quad i = 1, \ldots, 65536 \\
& x_j, y_1^i, y_2^i \ge 0, \quad j = 1, \ldots, 4;\; i = 1, \ldots, 65536,
\end{aligned}$$

which also is a linear optimization problem but of a larger size than the original one due to the introduction of y.

The objective function now being minimized is the sum of the immediate cost (actually, the negative profits) and the *expected recourse cost* because one must consider 65536 possible overtime costs. The constraints involving random quantities are written out explicitly for all the outcomes:

$$\text{carpentry hours } \sum_{j=1}^{4} t_{1j}^i x_j \text{ shouldn't exceed regular plus overtime hours } d_1 + y_1^i,$$

and finishing hours are restricted similarly. Since there's the possibility of making a *recourse decision* $y^i = (y_1^i, y_2^i)$ that depends on the outcomes of the random variables, the model falls in the class of *stochastic optimization problem with recourse.*

Oftentimes, there's more than one source of uncertainty in a problem. For example, due to employee absence, the available hours for carpentry and finishing may also have to be modeled as random variables, say, $\boldsymbol{d}_1$ and $\boldsymbol{d}_2$. Let's adopt the values listed in Table 3.2 as possibilities for these quantities.

entry	possible values			
$\boldsymbol{d}_1$	5873	5967	6033	6127
$\boldsymbol{d}_2$	3936	3984	4016	4064

Table 3.2: Possible values for $\boldsymbol{d}_1$, each occurring with probability 1/4, and for $\boldsymbol{d}_2$, each also occurring with probability 1/4.

We now need to replace d by $d^i = (d_1^i, d_2^i)$ and account for $4^2 = 16$ possible d^i vectors. This gives a total of $\nu = 1048576$ possible outcomes of $(\boldsymbol{T}, \boldsymbol{d})$. With $p_i = 1/\nu$ and $a = (a_1, a_2)$, the model becomes:

$$\begin{array}{llllll}
\text{minimize} & c^\top x & +p_1 a^\top y^1 & +p_2 a^\top y^2 & + \cdots + p_\nu a^\top y^\nu & \\
\text{subject to} & T^1 x & -y^1 & & & \le d^1 \\
 & T^2 x & & -y^2 & & \le d^2 \\
 & \vdots & & & \ddots & \vdots \\
 & T^\nu x & & & -y^\nu & \le d^\nu \\
 & x \ge 0, & y^1 \ge 0, & y^2 \ge 0, & \cdots \quad y^\nu \ge 0. &
\end{array}$$

Although still linear, the model is even larger than the previous one. The sparse, block-angular structure, however, is beneficial and can be leveraged in specialized algorithms such as the L-Shaped method (see, e.g., [30, Section 5.H]). We obtain the minimizer

$$x^{\star\star} = (257,\ 0,\ 665.2,\ 33.8), \quad \text{with minimum value } -\$18051,$$

using the revised simplex method (Section 2.E). This minimizer is *robust* because it's constructed by a process that examines all the 1048576 possibilities and accounts for the resulting recourse costs and the associated probabilities of having to pay those costs. In contrast, our initial solution $x^\star$ doesn't fare well when assessed in the last model:

$$c^\top x^\star + \sum\nolimits_{i=1}^{\nu} p_i a^\top y^i = -\$16942,$$

where $y^i = (y_1^i, y_2^i)$ is the overtime hours required under outcome i, i.e.,

$$y_1^i = \max\Big\{0, \sum\nolimits_{j=1}^{4} t_{1j}^i x_j^\star - d_1^i\Big\} \quad \text{and} \quad y_2^i = \max\Big\{0, \sum\nolimits_{j=1}^{4} t_{2j}^i x_j^\star - d_2^i\Big\}.$$

The expected profit of \$16942 is significantly worse than that from the robust solution $x^{\star\star}$ (\$18051). We note that the profit initially stipulated for $x^\star$ (\$18667) is misleading because it doesn't account for potential overtime cost, which in fact turns out to be \$1725, on average, across the possible outcomes. Thus, the expected profit drops to \$16942.

Moreover, the initial solution $x^\star$ provides the wrong strategic insight by suggesting that the production line for type 3 can be closed down. In reality, boards of type 3 play a central role by filling in the production schedule for some outcomes and therefore should be produced in large numbers—in fact 665, more than any other type! If the production line for such boards had been closed down, then the firm would have missed this profit-boosting option.

3.B Risk Modeling

Expectations arise in formulations where the goal is to optimize the *average* performance across different outcomes or scenarios, but we may also be concerned about "worst case" performance. For example, minimizing expected wait time at a call center wouldn't suffice when customers with above-average wait times simply hang-up, write angry reviews, and take their business elsewhere. An aeronautical engineer shouldn't minimize the average displacement at the tip of an aircraft wing, but instead account for nearly "all possible" displacement levels. We'll develop an approach to optimization under uncertainty that includes expectation-minimization as a special case but also adapts to concerns about risk, safety, reliability, and worst-case performance.

In an application, a random variable $\boldsymbol{\xi}$ might model an unknown quantity such as a future cost or loss, which implies a particular concern about high values of $\boldsymbol{\xi}$; a low value of $\boldsymbol{\xi}$ would be "good news." With this perspective, an assessment of $\boldsymbol{\xi}$ could be based on its 10% or 1% highest outcomes. These worst-case outcomes would then supplement the expected value in our optimization of decisions involving such future costs and losses. Quantiles and superquantiles formalize assessments of this kind.

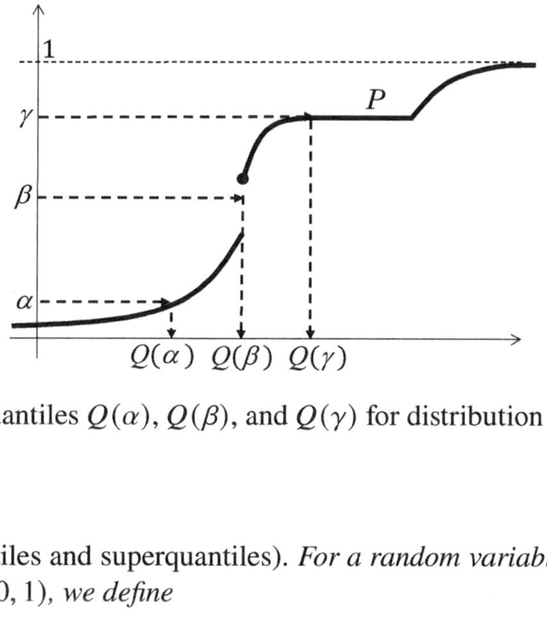

Fig. 3.1: Quantiles $Q(\alpha)$, $Q(\beta)$, and $Q(\gamma)$ for distribution function P.

Definition 3.1 (quantiles and superquantiles). *For a random variable $\boldsymbol{\xi}$ with distribution function P and $\alpha \in (0, 1)$, we define*

$$\alpha\text{-quantile:} \quad Q(\alpha) = \min\left\{\xi \;\middle|\; P(\xi) \geq \alpha\right\}$$

$$\alpha\text{-superquantile:} \quad \bar{Q}(\alpha) = Q(\alpha) + \frac{1}{1-\alpha}\mathbb{E}\big[\max\{0, \boldsymbol{\xi} - Q(\alpha)\}\big].$$

Moreover, we define $\bar{Q}(0) = \mathbb{E}[\boldsymbol{\xi}]$ and $\bar{Q}(1) = \lim_{\alpha \nearrow 1} \bar{Q}(\alpha)$. □

For α in Figure 3.1, the equation $P(\xi) = \alpha$ has a unique solution, which is the quantile $Q(\alpha)$. For β in the figure, $P(\xi) = \beta$ has no solution due to a jump in the distribution function, and one takes the lowest ξ that satisfies the relation $P(\xi) \geq \beta$ as the β-quantile. For γ, there's a whole range of ξ that solves $P(\xi) = \gamma$ and the lowest one is defined as the γ-quantile. In any case, an α-quantile is the dividing line between the best (lowest) $\alpha 100\%$ outcomes and the worst (highest) $(1 - \alpha)100\%$ outcomes of a random variable.

An α-superquantile amounts to the corresponding α-quantile *plus* a nonnegative term expressed by the expected value of the excess of $\boldsymbol{\xi}$ over the quantile. In effect,

$$\alpha\text{-superquantile of } \boldsymbol{\xi} \;=\; \text{average of the worst } (1-\alpha)100\% \text{ outcomes of } \boldsymbol{\xi}.$$

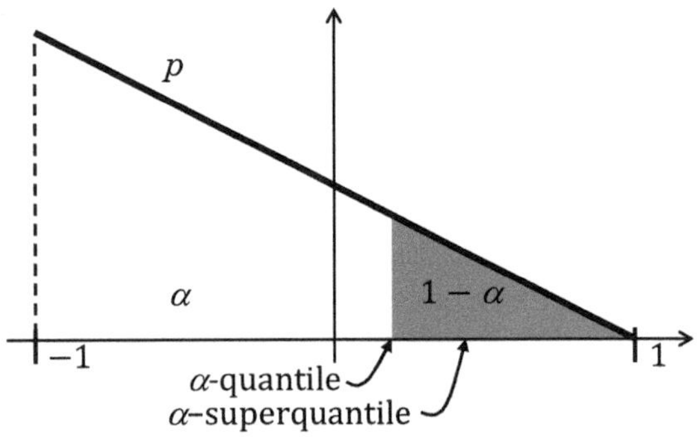

Fig. 3.2: Quantile and superquantile of density function p for a triangular distribution.

Example 3.2 (triangular distribution). The α-quantiles and α-superquantiles of a random variable $\boldsymbol{\xi}$ with the triangular density function p in Figure 3.2 are

$$Q(\alpha) = 1 - 2\sqrt{1-\alpha} \quad \text{and} \quad \bar{Q}(\alpha) = 1 - \tfrac{4}{3}\sqrt{1-\alpha}$$

for $\alpha \in (0,1)$, respectively.

Detail. Let P be the distribution function of $\boldsymbol{\xi}$. Since $p(\xi) = -\xi/2 + 1/2$ for $\xi \in [-1,1]$, the solution of the equation

$$P(\xi) = \int_{-1}^{\xi} p(\xi_0)\, d\xi_0 = \alpha$$

is $Q(\alpha) = 1 - 2\sqrt{1-\alpha}$. Thus,

$$\begin{aligned}
\bar{Q}(\alpha) &= Q(\alpha) + \frac{1}{1-\alpha}\int_{-1}^{1} \max\big\{0, \xi - Q(\alpha)\big\} p(\xi)\, d\xi \\
&= Q(\alpha) + \frac{1}{1-\alpha}\int_{Q(\alpha)}^{1} \xi p(\xi)\, d\xi - \frac{Q(\alpha)}{1-\alpha}\int_{Q(\alpha)}^{1} p(\xi)\, d\xi \\
&= \frac{1}{1-\alpha}\int_{Q(\alpha)}^{1} \xi p(\xi)\, d\xi = 1 - \tfrac{4}{3}\sqrt{1-\alpha},
\end{aligned}$$

after invoking the fact that $\int_{Q(\alpha)}^{1} p(\xi)\, d\xi = 1 - \alpha$ and the formula for $Q(\alpha)$. □

Example 3.3 (normal distribution). The α-quantiles and α-superquantiles of a random variable $\boldsymbol{\xi}$ that's normally distributed with mean μ and variance σ^2 are

$$Q(\alpha) = \mu + \sigma\Phi^{-1}(\alpha) \quad \text{and} \quad \bar{Q}(\alpha) = \mu + \frac{\sigma\varphi(\Phi^{-1}(\alpha))}{1-\alpha}$$

for $\alpha \in (0,1)$, respectively, where φ is the standard normal density function given by

$$\varphi(\xi) = \frac{1}{\sqrt{2\pi}} \exp\left(-\tfrac{1}{2}\xi^2\right)$$

and $\Phi^{-1}(\alpha)$ is the corresponding α-quantile given by the standard normal distribution function Φ. The formulas follow by the same arguments as in Example 3.2. □

The boundary cases $\alpha = 0$ and $\alpha = 1$ represent extreme positions that we may take in assessing a random variable $\boldsymbol{\xi}$: $\bar{Q}(0) = \mathbb{E}[\boldsymbol{\xi}]$ conveys information only about the average performance while $\bar{Q}(1)$ truly represents the worst outcome; $\bar{Q}(1) = 1$ in Example 3.2 and $\bar{Q}(1) = \infty$ in Example 3.3.

Superquantiles appear at first to be more cumbersome than quantiles as they're defined in terms of the latter. However, they can be expressed independently of quantiles by a minimization formula that turns out to be highly compatible with optimization models as we see next; [30, Proposition 3.9] provides a proof.

Proposition 3.4 (quantiles and superquantiles via minimization). *For $\alpha \in (0,1)$ and an integrable random variable $\boldsymbol{\xi}$, with α-quantile and α-superquantile denoted by $Q(\alpha)$ and $\bar{Q}(\alpha)$, respectively, one has*

$$Q(\alpha) \in \operatorname{argmin}_{\gamma\in\mathbb{R}} \left\{\gamma + \frac{1}{1-\alpha}\mathbb{E}\big[\max\{0, \boldsymbol{\xi} - \gamma\}\big]\right\}$$

$$\bar{Q}(\alpha) = \min_{\gamma\in\mathbb{R}} \left\{\gamma + \frac{1}{1-\alpha}\mathbb{E}\big[\max\{0, \boldsymbol{\xi} - \gamma\}\big]\right\}.$$

□

The focus on integrable random variables ensures that $\bar{Q}(\alpha)$ is finite for $\alpha \in (0,1)$, while still addressing most applications. The multitude of minimizers alluded to in the proposition takes place when $P(Q(\alpha)) = \alpha$ and the distribution function P of $\boldsymbol{\xi}$ has a "flat stretch" to the right of $Q(\alpha)$; see the γ-case in Figure 3.1. Regardless, Proposition 3.4 states that both the quantile $Q(\alpha)$ and the superquantile $\bar{Q}(\alpha)$ can be obtained by solving a one-dimensional optimization problem; the former is a minimizer and the latter is the minimum value.

In risk analysis, the α-superquantile of $\boldsymbol{\xi}$ is referred to as the *superquantile risk*. *Conditional value-at-risk* (cvar), *average value-at-risk* (avar), and *expected shortfall* are also used in the literature.

Let's consider optimization models involving superquantiles. Suppose that we're faced with a cost that depends on the value of an m-dimensional random vector $\boldsymbol{\xi}$ and an n-dimensional decision vector x, and is expressed by a function f. Now, $f(\boldsymbol{\xi}, x)$ represents the uncertain cost given decision x. For each decision x, $f(\boldsymbol{\xi}, x)$ is a random variable, assumed to be integrable. We're concerned about its "worst-case" values as quantified by an α-superquantile. In particular, we seek a decision x in a feasible set C for which the α-superquantile of $f(\boldsymbol{\xi}, x)$ is as low as possible. This leads us to the model

$$\underset{x \in C}{\text{minimize}} \ \alpha\text{-superquantile of } f(\boldsymbol{\xi}, x). \tag{3.2}$$

The goal is to determine a feasible decision with the lowest possible cost on average across the worst $(1-\alpha)100\%$ outcomes. The resulting decision would typically be rather different than one obtained by minimizing the expected cost; decisions with the possibility of high values tend to be avoided.

The minimization formula for superquantiles in Proposition 3.4 is useful computationally as it frees us from having to compute quantiles within an algorithm for superquantile risk minimization. After inserting the formula, we realize that (3.2), with $\alpha \in (0, 1)$, can be reformulated as

$$\underset{x \in C, \gamma \in \mathbb{R}}{\text{minimize}} \ \mathbb{E}\left[\gamma + \frac{1}{1-\alpha} \max\left\{0, f(\boldsymbol{\xi}, x) - \gamma\right\}\right],$$

which brings us back to minimizing an expectation function albeit with one additional variable. If $\boldsymbol{\xi}$ is finitely distributed with outcomes represented by the set Ξ and corresponding probabilities $\{p_\xi > 0, \xi \in \Xi\}$, then the problem simplifies further to

$$\underset{x \in C, \gamma \in \mathbb{R}}{\text{minimize}} \ \gamma + \frac{1}{1-\alpha} \sum\nolimits_{\xi \in \Xi} p_\xi \max\left\{0, f(\xi, x) - \gamma\right\}. \tag{3.3}$$

Using Proposition 1.8, we realize that the objective function is convex provided that f is convex as a function of x for all $\xi \in \Xi$. Thus, dealing with uncertainty in this comprehensive manner doesn't muddle up any convexity that might be present in f.

The max-term in (3.3) produces a nonsmooth objective function. While the subgradient method and its extension might apply, it can also be useful to consider the following reformulation:

$$\begin{aligned} &\underset{x \in C, \gamma \in \mathbb{R}, z \in \mathbb{R}^\nu}{\text{minimize}} \ \gamma + \frac{1}{1-\alpha} \sum\nolimits_{\xi \in \Xi} p_\xi z_\xi \\ &\text{subject to } \ f(\xi, x) - \gamma \leq z_\xi \ \text{ and } \ 0 \leq z_\xi \quad \forall \xi \in \Xi, \end{aligned} \tag{3.4}$$

where $z = (z_\xi, \xi \in \Xi)$ are additional variables and ν is the cardinality of Ξ. The last problem is also convex as long as C is convex and f is convex in x for all $\xi \in \Xi$ because the feasible set

$$\left\{(x, \gamma, z) \in C \times \mathbb{R} \times \mathbb{R}^\nu \;\middle|\; f(\xi, x) - \gamma \leq z_\xi, \ 0 \leq z_\xi \ \ \forall \xi \in \Xi\right\}$$

is convex by Proposition 1.7 and we can invoke Proposition 2.4. If f is affine in x for all $\xi \in \Xi$ and C is polyhedral, then the problem can be stated as a linear optimization problem. In any case, we can either solve (3.3), which is nonsmooth due to the max-term, or (3.4) that might be better structured for available algorithms but involves ν additional decision variables and 2ν additional constraints.

Example 3.5 (quantile vs superquantile minimization). Suppose that we face a difficult decision regarding two financial instruments, both associated with uncertainty. The first instrument requires, with probability 0.1, a payment of 2 dollars per dollar committed and, with probability 0.9, yields one dollar per dollar committed. Let the random variable ξ_1 model these losses:

$$\xi_1 = \begin{cases} 2 & \text{with probability } 0.1 \\ -1 & \text{with probability } 0.9. \end{cases}$$

The second instrument is modeled by the random variable ξ_2, which has the same probability distribution as that of ξ_1 but the two random variables are statistically independent. The instruments aren't lucrative; most likely we recover our commitment, but with a small probability we face a large loss. Suppose that we need to allocate 1 million dollars between these two instruments. Since the instruments appear equally unappealing, we might be led to believe that any allocation is fine. This would be a big mistake, but one that remains hidden for an analyst examining quantiles. Superquantiles reveal that the best strategy would be to allocate half a million to each instrument.

Detail. Let $x \in [0, 1]$ be the fraction of our million dollars allocated to the first instrument, the remainder is allocated to the second instrument. Then, the random variable describing our loss is

$$f(\xi, x) = x\xi_1 + (1 - x)\xi_2.$$

Since there are only four possible outcomes of $\xi = (\xi_1, \xi_2)$, we find that

$$f(\xi, x) = \begin{cases} -1 & \text{with probability } 0.81 \\ 2 - 3x & \text{with probability } 0.09 \\ 3x - 1 & \text{with probability } 0.09 \\ 2 & \text{with probability } 0.01. \end{cases}$$

Let's now examine the α-quantile and α-superquantile of $f(\xi, x)$ for $\alpha = 0.9$. The above probabilities lead to

$$Q(0.9) = \begin{cases} 3x - 1 & \text{if } x \in [0, 1/2] \\ 2 - 3x & \text{otherwise,} \end{cases}$$

which is visualized in Figure 3.3. With the quantiles known, Definition 3.1 leads to

$$\bar{Q}(0.9) = Q(0.9) + \frac{1}{1 - 0.9}\mathbb{E}\Big[\max\big\{0, f(\xi, x) - Q(0.9)\big\}\Big] = \begin{cases} -2.7x + 2 & \text{if } x \in [0, 1/2] \\ 2.7x - 0.7 & \text{otherwise,} \end{cases}$$

which is also shown in Figure 3.3. The superquantiles define a convex function in x, while the quantiles don't. The minimization of superquantiles has $x^\star = 0.5$ as minimizer, but the minimization of quantiles results in $x = 0$ and $x = 1$. The decision $x^\star$ involves *hedging*, a well-known strategy in finance to reduce risk. The decision lowers the probability of a loss of 2 million dollars from 0.1 to 0.01 compared to the choice $x = 0$ or $x = 1$. This comes at the expense of reducing the probability of a loss of -1 million dollars from 0.9 to 0.81. The decision $x^\star$ also results in the possibility of a loss of 0.5 million dollars (with probability 0.18), but this might be much more palatable than a 2-million-dollar loss. The minimum value of the 0.9-superquantiles is 0.65 million dollars, which is a more reasonable and conservative assessment of the uncertain future loss than the wildly optimistic -1 million provided by the minimum value of the 0.9-quantiles.

There could be different opinions about the various outcomes from these investments, but it's clear that superquantile minimization has two advantages over quantile minimization: the former preserves convexity in the problem formulation and tends to bring forward nontrivial solutions such as those relying on hedging, while the latter may not do either. □

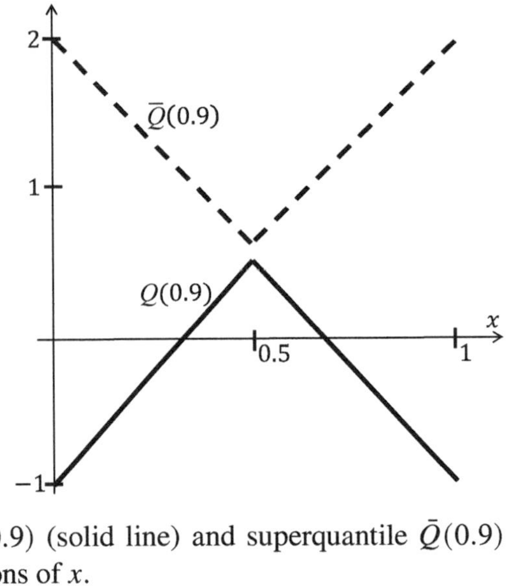

Fig. 3.3: Quantile $Q(0.9)$ (solid line) and superquantile $\bar{Q}(0.9)$ (dashed) of $f(\xi,x)$ in Example 3.5 as functions of x.

Example 3.6 (resilience and superquantiles). Superquantiles highlight the level of resilience in a system and identify how "bad" it can get. This is especially important in assessment of earthquake damage where losses are usually moderate, but they can become catastrophic. From a study of the cumulative loss due to earthquake damage during the next 50 years in the greater Vancouver region in Canada [19], we obtain 100000 loss values that one can view as outcomes for a random variable modeling the cumulative loss; see the histogram of Figure 3.4. The cumulative loss is most like less than one billion Canadian dollars. In fact, 90% of the 100000 outcomes are below 4.39

billion, i.e., the 0.9-quantile of the cumulative loss is 4.39 billion. Although sizable, it seems that the region is quite resilient to earthquakes. It turns out that this conclusion is flawed and a study of superquantiles paints a gloomier picture.

Detail. The largest outcome is actually 373 billion; Figure 3.4 should have been extended much to the right. Superquantiles quantify this long right-tail. Specifically, the 0.9-superquantile of the cumulative loss is 28.92 billion. This means that when the cumulative loss exceeds 4.39 billion, it does so substantially resulting in a loss of 28.92 billion on average. The region might not be resilient to earthquake losses after all. The picture remains the same if we consider the 0.95-quantile and the 0.95-superquantile, which are 14.00 and 49.88 billion, respectively. □

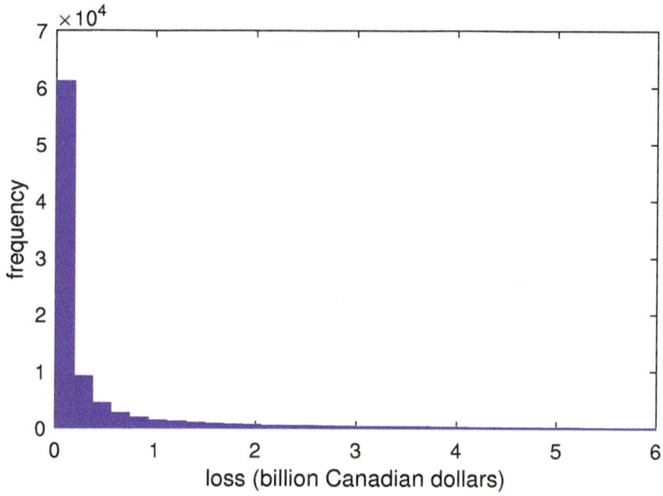

Fig. 3.4: Histogram of 100000 outcomes of the cumulative loss due to earthquake damage in the Vancouver region of Canada.

A difference between quantiles and superquantiles emerges: A quantile is determined by the number of outcomes to its right, but the distances to the right can be arbitrary without changing the quantile. In contrast, a superquantile equals the corresponding quantile plus a term that averages the outcomes to the right of that quantile and thus is sensitive to values far to the right; [26] and [30, Chapter 8] include further discussion of risk modeling.

3.C Models of Uncertainty

A central component of a stochastic optimization problem is the probability distribution of the random vector $\boldsymbol{\xi}$ defining expectation functions and superquantiles. Since it determines the possible values of $\boldsymbol{\xi}$ and the associated probabilities, the choice of probability distribution typically influences the resulting decision and becomes an important part of the modeling process. The choice of probability distribution is usually informed by data about the random phenomena that ξ represents. However, when the data is biased, corrupted, or simply unavailable in sufficient numbers, we need to introduce assumptions corresponding to our understanding of the situation. In this section, we'll construct models of uncertainty by estimating probability density functions.

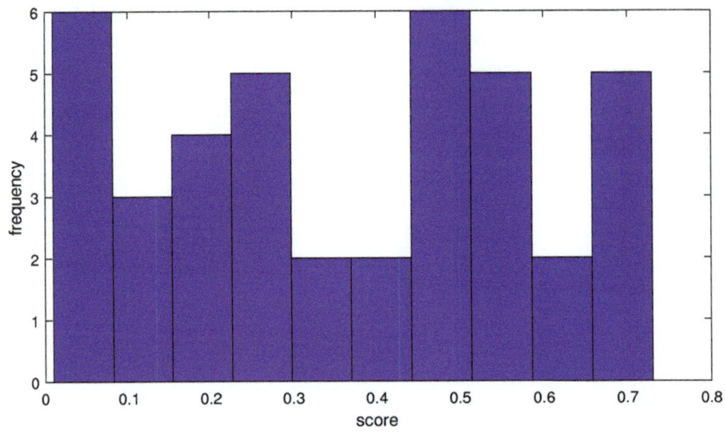

Fig. 3.5: Histogram of 40 scores about the performance of a technology.

Example 3.7 (uncertainty modeling in the US Army). Periodically, the US Army decides in which promising new technologies it should invest. The technologies could be hardware such as improved night-vision goggles, means to extract water from air, body sensors, and other futuristic gadgets. The decision needs to be made before the actual performance and usefulness of the technologies are fully known. The US Army, as many large organizations, relies on subject-matter experts to predict parameter values for which there's little or no data. Suppose that 40 experts are given a presentation about a technology and then asked to individually assign a score between 0.00 (poor) and 1.00 (excellent) that reflects the technology's future performance; see Figure 3.5 for a histogram summarizing the resulting 40 data points. Clearly, the opinions about the technology vary greatly, with the lowest score being 0.01 and the highest 0.73, but the data still provides a starting point from which we can develop a probability distribution for a random variable ξ that represents our (uncertain) assessment of the technology.

Detail. Given the data $\xi^1, \dots, \xi^{40}$ in Figure 3.5, we could have modeled ξ as having a finite distribution with $\Xi = \{\xi^1, \dots, \xi^{40}\}$ and equal probabilities $p_i = 1/40$. This model of uncertainty assumes that one of the subject-matter experts is correct; a value not proposed by any expert can't occur. Generally, a model that *only* relies on historical data fails to account for the possibility that the future might not be an instance of the past! A possible approach to address this concern and to ensure that we plan for outcomes (scenarios) that appear reasonable, but may never have occurred, is to select a density function for ξ *informed* by the data.

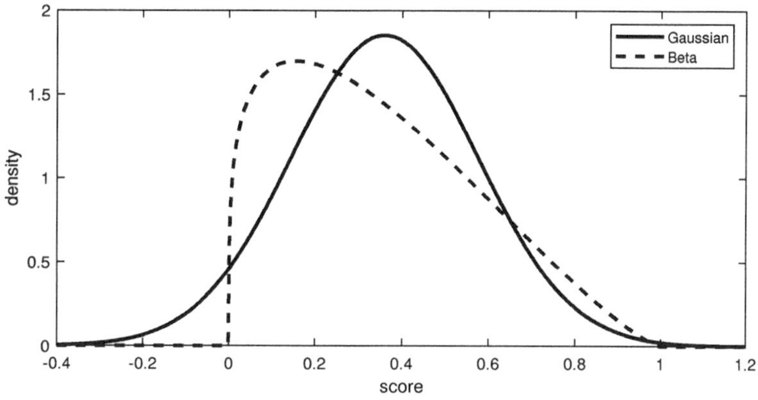

Fig. 3.6: Density functions modeling the uncertain performance of a technology.

Using the maximum likelihood criterion as laid out in Section 1.E, we can fit a density function to the 40 data points. The best-fitting Gaussian and beta density functions are shown by solid and dashed lines, respectively, in Figure 3.6. Although one can debate which of the two density functions is more appropriate, they both assign positive probabilities to outcomes that the finite-distribution model deems impossible to occur. A stochastic optimization model incorporating one of these density functions would account for a wider array of possible performance values for the technology, which might produce a better decision in practice; see [36] for a detailed discussion.

Still, the density functions in Figure 3.6 exhibit some questionable characteristics. The Gaussian density function assigns positive probabilities to values below zero and above one, which contradicts the fact that scores are certainly between 0.00 and 1.00. Neither the Gaussian nor the beta density function reflects the presence of several peaks of the histogram in Figure 3.5. □

As a model of uncertainty about a parameter, we would like to choose a density function that's closely tailored to the available information and assumptions. The maximum likelihood approach from Section 1.E provides a general framework for estimating density functions and permits a variety of side conditions. Given univariate data $\xi^1, \dots, \xi^\nu$, we recall from Section 1.E that a maximum likelihood estimate is a minimizer of the model

$$\operatorname*{minimize}_{p \in \mathcal{P}} -\frac{1}{\nu} \sum\nolimits_{i=1}^{\nu} \ln p(\xi^i), \tag{3.5}$$

where $\mathcal{P}$ is a class of candidate density functions. The usual classes of density functions such as Gaussian, beta, exponential, and so forth may not satisfy our assumptions about the random phenomenon at hand. In Example 3.7, neither the Gaussian nor the beta density function meets our expectations. A more flexible class of functions are the first-order *epi-splines*, which are piecewise affine functions of the form

$$p(\xi) = a_j \xi + b_j \quad \text{when } \xi \in (\tau_j, \tau_{j+1}), \ \ j = 1, \dots, q,$$

where $-\infty < \tau_1 < \tau_2 < \cdots < \tau_{q+1} < \infty$ are mesh points and $\{a_j, b_j,\ j = 1, \ldots, q\}$ are coefficients to be determined; see Figure 3.7.

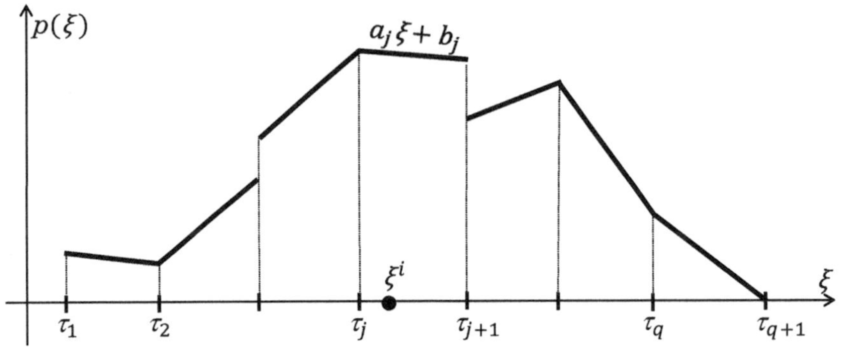

Fig. 3.7: Density function expressed as an epi-spline.

The value of p exactly at a mesh point τ_j is usually pinned down by insisting on the density function being continuous or some other rule. The problem of minimizing over $p \in \mathcal{P}$ then becomes one of minimizing over the vectors of coefficients $a = (a_1, \ldots, a_q)$ and $b = (b_1, \ldots, b_q)$. Let j_i be the index j corresponding to the interval in which ξ^i is located (cf. Figure 3.7). Then, $p(\xi^i) = a_{j_i}\xi^i + b_{j_i}$ and (3.5) specializes to

$$\begin{aligned}
&\underset{a,b\in\mathbb{R}^q}{\text{minimize}} -\frac{1}{\nu}\sum\nolimits_{i=1}^{\nu} \ln(a_{j_i}\xi^i + b_{j_i})\\
&\text{subject to} \qquad a_j\tau_j + b_j \geq 0,\quad j = 1, \ldots, q\\
&\qquad\qquad\qquad a_j\tau_{j+1} + b_j \geq 0,\quad j = 1, \ldots, q\\
&\sum\nolimits_{j=1}^{q} \tfrac{1}{2}a_j(\tau_{j+1}^2 - \tau_j^2) + b_j(\tau_{j+1} - \tau_j) = 1.
\end{aligned}$$

The inequalities ensure that the resulting density function is nonnegative. The equality constraint implements the requirement that a density function integrates to one. The exact formula follows from evaluating the integral

$$\int_{\tau_1}^{\tau_{q+1}} p(\xi)\, d\xi = \sum\nolimits_{j=1}^{q} \int_{\tau_j}^{\tau_{j+1}} (a_j\xi + b_j)\, d\xi.$$

In addition to these basic conditions, we can include a number of additional constraints to ensure that the resulting density function complies with our knowledge and assumptions. For example, the constraints

$$a_{j-1}\tau_j + b_{j-1} = a_j\tau_j + b_j,\quad j = 2, \ldots, q$$

ensure a continuous density function. The slope in one interval changes with at most $\delta \geq 0$ relative to the next one when

$$a_j - \delta \leq a_{j-1} \leq a_j + \delta, \;\; j = 2, \ldots, q$$

and this introduces a "smooth" look when q is large and δ is small. If already continuous, the density function becomes nonincreasing if $a_j \leq 0$ for all $j = 1, \ldots, q$. Since all these constraints are linear equations and inequalities, they define a polyhedral feasible set. The objective function is convex because the negative logarithm function is convex (under the convention that $-\ln \alpha = \infty$ when $\alpha \leq 0$) and we can invoke Proposition 1.8. For further theoretical and computational considerations regarding epi-splines, we refer to [28, 29].

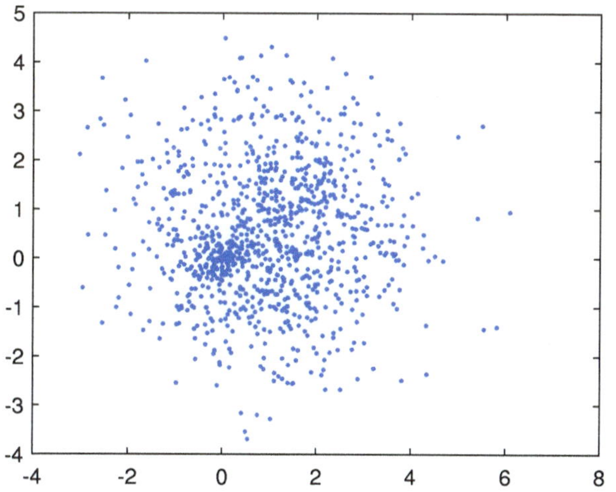

Fig. 3.8: Positions of 1000 robotic bugs. Axes are labeled in miles.

Example 3.8 (robotic swarm). Consider a group (swarm) of 1000 aerial autonomous systems. These may be tiny "bugs" that are pollinating fruit trees or sensing air quality in an urban area. We would like to determine a probability distribution of the Euclidean distance from every bug to its nearest neighboring bug. Figure 3.8 shows the positions of the bugs at one time. With complete knowledge about everybody's position, it's easy to compute the nearest-neighbor distance for every bug; Figure 3.9 summarizes the distances in a histogram. Nearly all bugs are closer than 0.2 miles of some other bug.

The challenge in practice is that the bugs are tiny and carry small batteries and no GPS device. They're programmed to only report nearest-neighbor distances occasionally. Our goal is to determine a density function that resembles Figure 3.9 using just a small number of such reports.

Detail. We follow the epi-spline approach laid out before the example. Distances are nonnegative so we set $\tau_1 = 0$. Our prior experience indicates that distances are typically less than one mile and we set $\tau_{q+1} = 1$. Moreover, let $\tau_j = (j-1)/q$, with $q = 1000$. Using $\nu = 100$ reports, the maximum likelihood model, with continuity and change-in-slope (smoothing) constraints ($\delta = 0.5$), results in the density function in Figure 3.10, which is quite close to the "truth" in Figure 3.9.

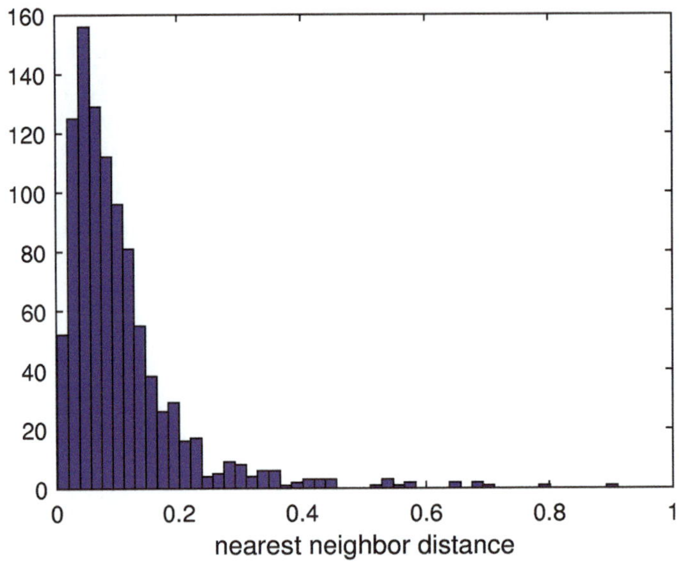

Fig. 3.9: Histogram of nearest-neighbor distance for robotic bugs. The horizontal axis has unit of miles and the vertical axis gives the number of bugs.

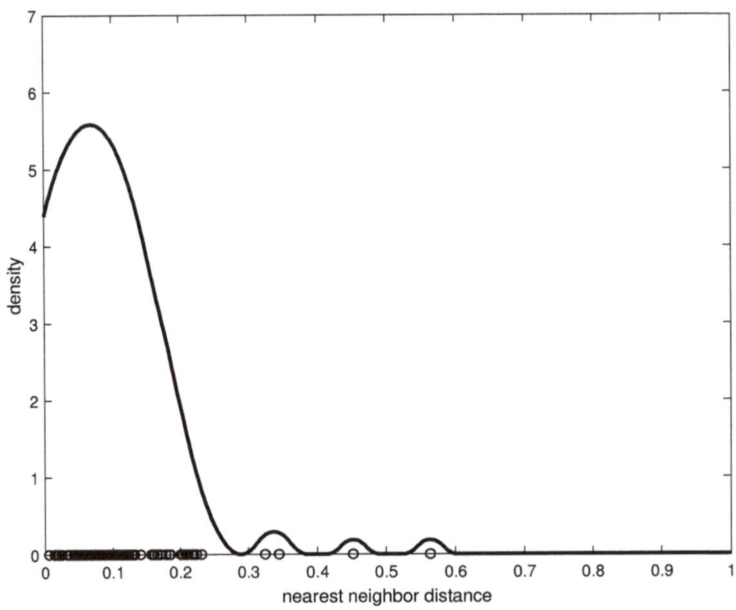

Fig. 3.10: Estimate of nearest-neighbor distance using 100 reports indicated by circles.

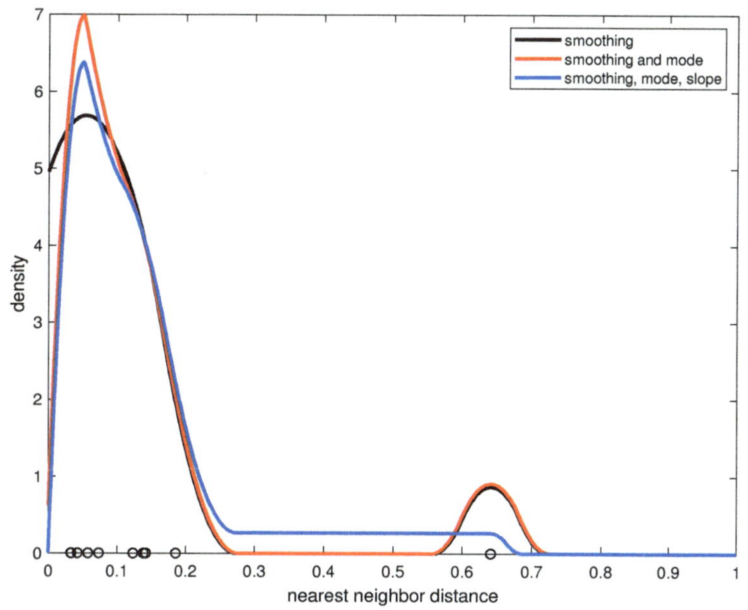

Fig. 3.11: Estimate of nearest-neighbor distance using 10 reports indicated by circles.

With only 10 reports, however, the model results in the density function illustrated with a black line in Figure 3.11. There's a large bump around 0.65 miles, which we can safely say is "incorrect" based on our (perfect) knowledge from Figure 3.9. We seem to have *over-fitted* the data. An approach to combat over-fitting is to introduce additional constraints. Let's assume that a mode (peak) of the density function is located at $\xi^\star = 0.05$ miles. This requirement is enforced by the constraints

$$a_j\tau_j + b_j \leq a_{j^\star}\xi^\star + b_{j^\star} \quad \text{and} \quad a_j\tau_{j+1} + b_j \leq a_{j^\star}\xi^\star + b_{j^\star}, \quad j = 1, \ldots, q,$$

where $j^\star$ specifies the interval containing $\xi^\star$. Moreover, since there seems to be more rapid changes in slope to the left of the mode than to the right, we modify the change-in-slope (smoothing) constraints to have $\delta = 5$ for intervals to the left of $\xi^\star$ and $\delta = 0.5$ to the right of $\xi^\star$. The resulting estimate is the red line in Figure 3.11. Although the bump at 0.65 miles persists, the density function drops to zero at zero miles. This seems reasonable as the bugs are programmed to avoid collision. With this insight, we impose $p(0) = 0$, i.e., $b_1 = 0$. We also enforce unimodality, i.e., the slope needs to be nonnegative to the left of $\xi^\star$ and nonpositive to the right:

$$a_j \geq 0 \text{ when } \tau_j < \xi^\star \quad \text{and} \quad a_j \leq 0 \text{ when } \tau_j > \xi^\star.$$

The resulting estimate is given by the blue line in Figure 3.11. Indeed, the density function is unimodal, with a mode at $\xi^\star$, and we now have a reasonable tail, with no misleading bump. □

The example illustrates that it's possible to obtain quality estimates of a density function even with little data as long as additional constraints are brought in. Experienced analysts may identify reasonable assumptions that would guide the choice of constraints. Or, we can adopt a process of trial and error on different data sets to determine the more suitable set of constraints—a process called cross-validation in statistics.

3.D Stochastic Gradient Descent and Machine Learning

Stochastic optimization problems quickly become large scale due to the need for considering many possible values of random vectors, for example as produced by Monte Carlo sampling or a sizeable data set. Although the algorithms of Chapters 1 and 2 might apply, it's convenient and sometimes essential to have fast specialized algorithms that utilize the structure of these problems. Machine learning is one area where problem instances tend to become large due to sizable data sets (see Section 1.B, Section 1.E, and Section 2.F) but also due to many trainable parameters.

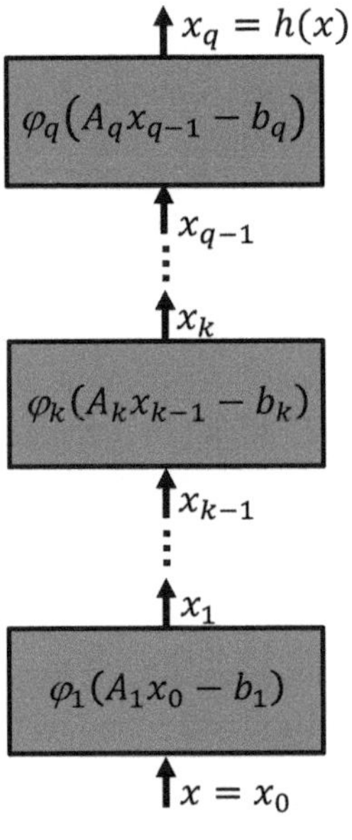

Fig. 3.12: Structure of simple neural network with trainable parameters $A_1, \ldots, A_q$ and $b_1, \ldots b_q$.

Example 3.9 (optimization of neural networks). Image recognition and natural language processing are challenging learning problems due to the vast variety of situations that an automated system may encounter. Even in a limited context such as understanding handwritten text, a system needs to tackle the diversity of personal "styles" our elementary school teachers failed to eradicate! Machine learning systems based on *neural networks* solve large-scale, highly-structured optimization problems.

Detail. The situation resembles the data fitting in Section 1.B and classification in Example 1.13, but now the data $\{x^i, y^i\}_{i=1}^m$ may represent a collection of black-and-white images so that y^i is a vector predicting the label of the ith image (e.g., dog, cat, person, etc.) and x^i is a vector specifying the gray-scale for each pixel. Again, the goal is to fit a function to the data, but now using a sophisticated class of functions, much beyond polynomials. These functions are defined by a neural network that successively transform an input vector x to an output vector $h(x)$. For example, the functions may be of the form

$$h(x) = x_q, \quad \text{with } x_k = \varphi_k(A_k x_{k-1} - b_k), \quad k = 1, \dots, q, \quad \text{and } x_0 = x,$$

where A_k is a $d_k \times d_{k-1}$-matrix and b_k is a d_k-dimensional vector, both varying in size across the layers $k = 1, \dots, q$; see Figure 3.12. These matrices and vectors contain the trainable parameters to be optimized in the same manner as the polynomial coefficients in Section 1.B. The mappings φ_k is a given *activation function* and is typically relatively simple. They may transform each component α of their input according to the formulas $\max\{0, \alpha\}$ or $1/(1 + e^{-\alpha})$.

If the fitting criteria is "average loss" expressed by some function f, then the resulting optimization model becomes

$$\underset{A_1,\dots,A_q,b_1,\dots,b_q}{\text{minimize}} \ \frac{1}{m} \sum_{i=1}^{m} f\big(h(x^i), y^i\big), \tag{3.6}$$

where the trainable parameters $A_1, \dots, A_q, b_1, \dots, b_q$ are the decision variables; they fully determine h. The number of data points ν tends to 10^5 or larger. The number of decision variables could be in the millions, billions, and even trillions, especially for natural language processing. In that context, the data is also nearly limitless consisting of all text available on the internet. Further complications arise from the fact that (3.6) isn't convex except in rather specific instances such as when $q = 1$, φ_1 is affine, and f is convex in its first argument, i.e., the situation in Section 1.B. Still, problems of this general form are solved, at least *approximately*, on a routine basis across a range of application domains. A main reason for this is that the problem possesses structure that allows the use of specialized algorithms. □

Example 3.10 (optimization of transformers). The neural network in Example 3.9 (see Figure 3.12) is a fully connected feedforward network, but it won't suffice in most modern setting such as for natural language processing. *Transformers* are components of more advanced neural networks, especially tailored to settings where the input is a sequence of data points such as a paragraph of text. They're key to large language models. However, from the optimization perspective this is merely a change to a more complicated objective function involving very many trainable parameters as decision variables.

Detail. The input to a transformer layer in a neural network is an $n \times d$-matrix X, where n is the length of the input sequence (e.g., number of tokens in the text input) and d is the dimension of the embedding representing each token. To produce the output of the layer, one passes through the following steps:

$$H^0 = X + P, \quad Q = H^0 W^1, \quad K = H^0 W^2, \quad V = H^0 W^3,$$

where P is a given matrix representing positional encoding, W^1, W^2, W^3 are $d \times d_k$-matrices to be optimized, and d_k is some number. Next, one computes $S = QK^\top/\sqrt{d_k}$ and the $n \times n$-matrix A with elements

$$A_{ij} = \frac{\exp(S_{ij})}{\sum_{j'=1}^{n} \exp(S_{ij'})},$$

where S_{ij} are the elements of S. We continue by computing

$$Y = AVW^4,$$

where W^4 is a $d_k \times d$-matrix to be optimized.

For any d-dimensional row vector $v = [v_1 \ \ldots \ v_d]$, we define the mapping

$$\Psi(v) = \big(\Psi_1(v), \ldots, \Psi_d(v)\big),$$

where

$$\Psi_i(v) = \gamma_i \frac{v_i - \mu(v)}{\sqrt{\sigma^2(v) + 10^{-5}}} + \beta_i,$$

with trainable parameters γ_i and β_i, and

$$\mu(v) = \frac{1}{d} \sum_{i=1}^{d} v_i, \qquad \sigma^2(v) = \frac{1}{d} \sum_{i=1}^{d} \big(v_i - \mu(v)\big)^2.$$

We use this mapping to compute the matrix H^1 with rows

$$H_i^1 = \Psi(H_i^0 + Y_i),$$

where H_i^0 and Y_i are the ith rows of H^0 and Y, respectively. Next, we compute

$$H^2 = \varphi(H^1 W^5 + B^1) W^6 + B^2,$$

where φ is an activation function of the kind mentioned in Example 3.9 but now applied to each element of the matrix $H^1 W^5 + B^1$. The matrices W^5 and W^6 are of dimensions $d \times d_h$ and $d_h \times d$, respectively, and will be optimized. The same holds for the $n \times d_h$-matrix B^1 and the $n \times d$-matrix B^2. Finally, we compute the output $n \times d$-matrix H^3 from the transformer layer, which has rows

$$H_i^3 = \Psi(H_i^1 + H_i^2),$$

where H_i^2 is the ith row of H^2. In total, the layer involves the trainable parameters:

$$d \times d_k\text{-matrices } W^1, W^2, W^3, \quad d_k \times d\text{-matrix } W^4$$
$$\text{scalars } \gamma_1, \ldots, \gamma_d, \ \beta_1, \ldots, \beta_d$$
$$d \times d_h\text{-matrix } W^5, \quad d_h \times d\text{-matrix } W^6$$
$$n \times d_h\text{-matrix } B^1, \quad n \times d\text{-matrix } B^2.$$

A particular neural network may have many such layers, even up to 100 layers, each with its own trainable parameters. All these parameters become the decision variables in an optimization problem similar to (3.6) for training the neural network on the available data. This gives rise to a complicated objective function, but it's still just another objective function! Finding a minimizer of this function is challenging because of its nonconvexity and the extraordinarily many variables. To get a sense, a typical GPT-style transformer for a large language model may have $d = 4096$ or 12288, $d_h = 4d$, and $d_k = 64$ or 128, producing an optimization problem with maybe a trillion variables. □

In view of (3.6) and Sections 1.B, 1.E, and 2.F, we recognize that learning problems take the form of minimizing a *sum* over the data. The situation is similar in optimization under uncertainty (Section 3.A), and even in risk modeling of Section 3.B. While some of these situations also involve constraints, we omit that complication and consider the problem

$$\underset{x \in \mathbb{R}^n}{\text{minimize}}\ \varphi(x) = \frac{1}{m} \sum\nolimits_{i=1}^{m} f_i(x), \tag{3.7}$$

where the function f_i could be the error (or loss) of data point i under a neural network specified by x. (Inputs to neural networks are often denoted by x in the literature as in Example 3.9, but we here return to our usual notation with x representing decision variables.) If each f_i is smooth, then the gradient descent method of Section 1.F applies. If f_i is nonsmooth but convex, then the subgradient method applies (Section 2.G). These methods require the computation of (sub)gradients of the objective function, which for (3.7) amounts to computing (sub)gradients of m functions in each iteration. We'll develop extensions of these methods that implement a remarkable coarse approximation of the (sub)gradients of the objective function.

We recall that the gradient descent method applied to (3.7) computes

$$x^{\nu+1} = x^\nu + \lambda^\nu w^\nu, \quad \text{where } w^\nu = -\frac{1}{m} \sum\nolimits_{i=1}^{m} \nabla f_i(x^\nu) \text{ and } \lambda^\nu > 0.$$

As an approximation, we may select an index $i_\nu \in \{1, \ldots, m\}$ and set

$$w^\nu = -\nabla f_{i_\nu}(x^\nu).$$

Thus, we approximate the average of $\{\nabla f_i(x^\nu), i = 1, \ldots, m\}$ by a single term. At first, this seems hopelessly inaccurate, but in fact the resulting method can work. It gives rise to the *stochastic gradient descent method*.

Example 3.11 (expectation by stochastic gradient descent). Suppose that $\boldsymbol{\xi}$ is an n-dimensional random vector with outcomes $\xi_1, \dots, \xi_m$, each occurring with probability $1/m$. Let the functions defining φ in (3.7) be

$$f_i(x) = \tfrac{1}{2}\|\xi_i - x\|_2^2, \quad i = 1, \dots, m.$$

A minimizer of φ is characterized by the Fermat rule $\nabla\varphi(x) = 0$ because f_i is convex. This vector equation simplifies to

$$\sum\nolimits_{i=1}^{m} (\xi_i - x) = 0$$

and thus

$$x = \frac{1}{m}\sum\nolimits_{i=1}^{m} \xi_i.$$

We recover the expected value of $\boldsymbol{\xi}$ by solving (3.7).

It turns out that we obtain the minimizer by also following the approximation scheme indicated above if the step size is selected appropriately.

Detail. As suggested, we adopt the iterative scheme with

$$x^{\nu+1} = x^\nu + \lambda^\nu w^\nu, \quad \text{where } w^\nu = -\nabla f_{i_\nu}(x^\nu), \quad \nu = 0, 1, 2, \dots,$$

and $\{i_0, i_1, \dots\}$ is a sample generated independently and uniformly over $\{1, \dots, m\}$. In this case, we obtain the gradient

$$\nabla f_i(x) = x - \xi_i$$

so that

$$x^{\nu+1} = x^\nu - \lambda^\nu(x^\nu - \xi_{i_\nu}) = \lambda^\nu \xi_{i_\nu} + (1 - \lambda^\nu)x^\nu.$$

Applying this expression recursively, we obtain that

$$\begin{aligned}
x^{\nu+1} &= \lambda^\nu \xi_{i_\nu} \\
&\quad + (1 - \lambda^\nu)\lambda^{\nu-1}\xi_{i_{\nu-1}} \\
&\quad + (1 - \lambda^\nu)(1 - \lambda^{\nu-1})\lambda^{\nu-2}\xi_{i_{\nu-2}} \\
&\quad \vdots \\
&\quad + (1 - \lambda^\nu)(1 - \lambda^{\nu-1}) \cdot \ldots \cdot (1 - \lambda^1)\lambda^0 \xi_{i_0} \\
&\quad + (1 - \lambda^\nu)(1 - \lambda^{\nu-1}) \cdot \ldots \cdot (1 - \lambda^1)(1 - \lambda^0)x^0.
\end{aligned}$$

Suppose that $\lambda^\nu = 1/(\nu + 1)$ and $x^0 = 0$. Then, for $i = 1, 2, \dots, \nu$,

$$(1-\lambda^\nu)(1-\lambda^{\nu-1})\cdot\ldots\cdot(1-\lambda^{\nu-i+1})\lambda^{\nu-i} = \frac{\nu}{\nu+1}\frac{\nu-1}{\nu}\cdot\ldots\cdot\frac{\nu-i+1}{\nu-i+2}\frac{1}{\nu-i+1}$$
$$= \frac{1}{\nu+1},$$

which together with the expression for $x^{\nu+1}$ give

$$x^{\nu+1} = \frac{1}{\nu+1}\sum\nolimits_{k=0}^{\nu} \xi_{i_k}. \tag{3.8}$$

Thus, the iterative scheme computes the average of the sample points, which is known to converge to the expected value of $\boldsymbol{\xi}$ as $\nu \to \infty$ by the law of large numbers. We recover the minimizer of φ despite the seemingly coarse approximations in the iterative scheme. Interestingly, the method estimates $\mathbb{E}[\boldsymbol{\xi}]$ iteratively without the need for storing the sample. As soon as $x^{\nu+1}$ has been computed, ξ_{i_ν} (and x^ν) can be discarded. When n and ν are large, this can be a significant advantage. □

These ideas extend much beyond the setting of the example. In particular, (3.7) can very well involve a nonsmooth function despite the fact that the deep-rooted name stochastic *gradient* descent method indicates otherwise. The presence of nonsmooth *and* nonconvex functions f_i, however, go beyond the present scope and we assume that f_i is convex in the next algorithm; see [30, Chapter 4] and [12] for extensions.

Stochastic Gradient Descent Method.

Data. $x^0 \in \mathbb{R}^n$ and $\lambda^\nu \in (0,\infty)$, $\nu = 0,1,2,\ldots$.
Step 0. Set $\nu = 0$.
Step 1. Generate i_ν by sampling uniformly from $\{1,\ldots,m\}$.
Step 2. With $w^\nu \in \partial f_{i_\nu}(x^\nu)$ and some step size λ^ν, set

$$x^{\nu+1} = x^\nu - \lambda^\nu w^\nu.$$

Step 3. Replace ν by $\nu+1$ and go to Step 1.

An implementation of the method would require a rule for selecting step sizes—the simple constant rule is often used—and might also involve slightly different rules in Step 1 such as not resampling an index before all indices have been sampled. It's also common to sample more than one index in each iteration (mini-batching) so that $w^\nu \in \partial f_{i_\nu}(x^\nu)$ in Step 2 becomes

$$w^\nu = \frac{1}{s}\sum\nolimits_{i\in S_\nu} w_i^\nu, \quad \text{with } w_i^\nu \in \partial f_i(x^\nu),$$

where S_ν is a subset of $\{1,\ldots,m\}$ with cardinality s, typically 64 or 128.

There are several possible convergence results that can be derived for this algorithm; see, e.g., [12]. If f_i is real-valued and convex with bounded subgradients for all i and (3.7) has a minimizer, then the average point

$$\bar{x} = \frac{1}{\nu} \sum\nolimits_{k=0}^{\nu-1} x^k$$

after ν iterations has an error

$$\varphi(\bar{x}) - \min \varphi \leq \frac{\beta}{\sqrt{\nu}} \quad \text{on average} \tag{3.9}$$

under a particular (constant) step size. Here, β is a positive constant that depends on the size of the subgradients and the distance from the initial point x^0 to a minimizer of φ. While similar to a convergence result for the subgradient method (Section 2.G), this error bound involves two "averages." First, the statement pertains to $\bar{x}$, the average point of all the iterates up to iteration ν. The inherent randomness in the stochastic gradient descent method makes it more difficult to say something meaningful about the last iteration, and this prompts such averaging. Second, the error (3.9) holds on average across many runs of the method. Since the method is random, each run tends to be different. On average, however, the error is of the same order as for the subgradient method; see [30, Section 3.G] and [18, Chapter 4] for a more detailed analysis and extensions.

The stochastic gradient descent method is a main algorithm for solving large-scale learning problems due to its low computational cost per iteration and other desirable properties when applied to functions f_i defined in terms of neural networks. For example, the various data points can be stored on different servers without difficulty as only one data point (or a mini-batch) is accessed at a time. Still, its slow rate of convergence (only $1/\sqrt{\nu}$) implies the need for many iterations. It's common to do millions of iterations even for moderately sized learning problems.

Example 3.12 (support vector machines). Classification can be carried out by means of the hinge-loss problem

$$\underset{(a,\alpha)\in\mathbb{R}^{n+1}}{\text{minimize}} \; \frac{1}{m} \sum\nolimits_{i=1}^{m} f_i(a,\alpha), \quad \text{where } f_i(a,\alpha) = \max\left\{0, 1 - y_i(a^\top x^i + \alpha)\right\};$$

recall Section 2.F. It involves a sum of real-valued convex functions as in (3.7). Application of the stochastic gradient descent method to this problem requires a subgradient $w \in \partial f_i(a,\alpha)$ for some randomly selected index $i \in \{1, 2, \dots, m\}$. We deduce from Example 2.32 that

$$\partial f_i(a,\alpha) = \begin{cases} \{0\} & \text{if } 1 - y_i(a^\top x^i + \alpha) < 0 \\ \left\{(-y_i x^i, -y_i)\right\} & \text{if } 1 - y_i(a^\top x^i + \alpha) > 0 \\ \left\{\beta(-y_i x^i, -y_i) \;\middle|\; \beta \in [0,1]\right\} & \text{otherwise.} \end{cases}$$

□

Example 3.13 (text analytics). Classification problems appear in the analysis of large collections of documents. For instance, a social network may need to monitor posts for offensive language, hate speech, and other expressions that violate their member policy. The text in a post can be associated with an n-dimensional vector x, where each component x_j represents the frequency of a particular word or phrase. If a data set with labels (violation, not-violation) is available, then we can construct a support vector machine (cf. Section 2.F and Example 3.12) that categorizes any body of text.

Detail. As an illustration, consider the blog post:

> I like optimization theory more than I thought. It's the most useful class I am taking this semester. I certainly like it more than statistics.

Let $j = 1, 2, 3, 4$ correspond to "like," "more," "I," and "class," respectively. These words appear 2, 2, 4, and 1 times. Considering a "bag-of-words" model of the text, we let x_j be the relative frequency of the jth word so that $x = (2/9, 2/9, 4/9, 1/9)$. (Modern language models represent text in a much more sophisticated manner; see Example 3.10.) In this case, a human "judge" will probably label the body of text as not-violation. If a large collection of such documents has been labeled, then a support vector machine can use the resulting data set to construct a function given by $h(x) = a^\top x + \alpha$ that predicts the label of any document, of course, possibly with some errors. Easily, the size of x is tens of thousands and the data set may contain millions of labeled documents. Thus, an algorithm that can handle large-scale problems becomes essential. The stochastic gradient descent method is attractive because its rate of convergence doesn't depend on the number of terms summed up in an expectation function and it accesses only one document (or a mini-batch) in each iteration. □

Although the analysis of the stochastic gradient descent method guides the choice of step size (see, e.g., [30, Section 3.G]), it still needs to be tuned in practice. A common approach is to hold the step size fixed for a number of iterations and then reduce it by a factor, and this process is repeated 3-10 times. We can also attempt to automatically adapt the algorithm to the inherent variability and scale of a problem instance. An example of such a strategy follows below, which also introduces the idea of averaging gradients across iterations; see [17].

Accelerated Stochastic Gradient Descent with ADAM.

Data. $x^0 \in \mathbb{R}^n$, $\lambda = 10^{-3}$, $\alpha = 0.9$, $\beta = 0.999$, $\varepsilon = 10^{-8}$.
Step 0. Set $\nu = 0$ and $\mu_j^0 = \sigma_j^0 = 0$, $j = 1, \dots, n$.
Step 1. Generate a sample $i_\nu \in \{1, \dots, m\}$.
Step 2. Set $v = (v_1, \dots, v_n) \in \partial f_{i_\nu}(x^\nu)$.
Step 3. For $j = 1, \dots, n$, set

$$\mu_j^{\nu+1} = \frac{\alpha\mu_j^\nu + (1-\alpha)v_j}{1-\alpha^{\nu+1}} \quad \text{and} \quad \sigma_j^{\nu+1} = \frac{\beta\sigma_j^\nu + (1-\beta)(v_j)^2}{1-\beta^{\nu+1}}.$$

Step 4. Set $w = (w_1, \ldots, w_n)$, with

$$w_j = \frac{\mu_j^{\nu+1}}{\sqrt{\sigma_j^{\nu+1}} + \varepsilon}, \quad j = 1, \ldots, n.$$

Step 5. Set $x^{\nu+1} = x^{\nu} - \lambda w$, replace ν by $\nu + 1$, and go to Step 1.

Here, $\alpha^{\nu+1}$ is indeed α raised to power $\nu + 1$ and $\beta^{\nu+1}$ is similarly understood; $(v_j)^2$ means v_j squared. Steps 3 and 4 compute a weighted and scaled average of previous subgradients in an effort to modulate their often volatile values. The scaling can be viewed as an attempt to compute expectation and variance of each component of the subgradient. This motivates the acronym ADAM, which stands for adaptive moment estimation.

3.E Simple Recourse Problems

Optimization problems frequently include nontrivial constraints. While the rather general stochastic gradient descent method can be extended to handle some constraints (see, e.g., [30, Section 3.G]), there's also merit in algorithms that leverage the special structure of a problem directly. We'll now examine the structure that arises in some stochastic optimization problems.

In applications, we often end up with problems whose decision processes fit the following pattern:

$$\text{decision: } x \;\rightsquigarrow\; \text{observation: } \xi \;\rightsquigarrow\; \text{recourse cost evaluation.}$$

The recourse cost evaluation may or may not be "simple" and it's useful to make this distinction in the selection of solution procedures. In some instances, the recourse cost evaluation requires solving another optimization problem, which we return to in Section 3.G. If it doesn't and the evaluation can be expressed in terms of an explicit formula, then the problem is said to be one with *simple recourse*. More precisely, a *stochastic optimization problem with simple recourse* takes the form

$$\boxed{\underset{x \in C}{\text{minimize}}\; f_0(x) + \mathbb{E}\big[f(\boldsymbol{\xi}, x)\big]}$$

where C is a subset of $\mathbb{R}^n$ and the function f is given by an explicit formula that's easy to evaluate. The function takes two arguments: an m-dimensional vector ξ representing uncertain parameters and an n-dimensional vector x of decision variables.

The newsvendor problem from (1.9) is of this type with

$$f_0(x) = (o - s)x, \quad C = \mathbb{R}, \quad f(\xi, x) = \max\{0, s(x - \xi)\}.$$

The product mix problem in Section 3.A has $f_0(x) = c^\top x$, $C = [0, \infty)^4$, and

$$f(\xi, x) = a_1 \max\left\{0, \sum_{j=1}^{4} t_{1j}x_j - d_1\right\} + a_2 \max\left\{0, \sum_{j=1}^{4} t_{2j}x_j - d_2\right\},$$

with ξ now being the vector with elements defined by the uncertain parameters $t_{11}, \dots, t_{14}, t_{21}, \dots, t_{24}, d_1, d_2$. These parameters have $\nu = 1048576$ possible outcomes

$$\{t_{11}^i, \dots, t_{14}^i, t_{21}^i, \dots, t_{24}^i, d_1^i, d_2^i, \; i = 1, \dots, \nu\}$$

as summarized in Tables 3.1 and 3.2. Thus, the expectation

$$\mathbb{E}\big[f(\boldsymbol{\xi}, x)\big] = \sum_{i=1}^{\nu} p_i \left(a_1 \max\left\{0, \sum_{j=1}^{4} t_{1j}^i x_j - d_1^i\right\} + a_2 \max\left\{0, \sum_{j=1}^{4} t_{2j}^i x_j - d_2^i\right\}\right),$$

with $p_i = 1/\nu$, matches the terms $p_1 a^\top y^1 + p_2 a^\top y^2 + \cdots + p_\nu a^\top y^\nu$ from Section 3.A. The recourse variables $y^1, \dots, y^\nu$ represent exactly the amount of overtime needed. Consequently, the product mix problem is a stochastic optimization problem with simple recourse.

We focus on a computationally attractive subclass of problems that stems from the linear optimization problem

$$\underset{x \in \mathbb{R}^n}{\text{minimize}} \; c^\top x \quad \text{subject to} \quad Ax = b, \; Tx = \xi, \; x \geq 0,$$

where A and T are $m_1 \times n$ and $m_2 \times n$-matrices, respectively, and ξ is an m_2-dimensional vector of parameters. The difficulty is, however, that the value of ξ is unknown. To model this situation, we assign an additional cost to a decision x when Tx fails to match the actual value of ξ. The cost is described by a convex function h, the *recourse cost function*. We assume that the vector of parameters is modeled by the random vector $\boldsymbol{\xi}$ with a finite distribution over the set of outcomes (scenarios) Ξ and the associated probabilities are p_ξ. This leads to a stochastic optimization problem with simple recourse:

$$\underset{x \in C}{\text{minimize}} \; c^\top x + \sum_{\xi \in \Xi} p_\xi f(\xi, x), \tag{3.10}$$
$$\text{with } C = \{x \mid Ax = b, x \geq 0\} \quad \text{and} \quad f(\xi, x) = h(\xi - Tx),$$

which is our focus in this section. The problem seeks to find a balance between the two terms in the objective function: an "immediate cost" $c^\top x$ and an expected "future" cost due to Tx not matching ξ.

We can reformulate (3.10) in a way that might be analytical and computational beneficial. If $\boldsymbol{\xi}$ represents demand for the output of certain economic activities, then $z = Tx$ is the vector being *tendered* by the decision maker to match the uncertain demand $\boldsymbol{\xi}$. We refer to z as a *tender*. Introducing the tender as a vector of decision variables in the formulation, we obtain the equivalent formulation

$$\underset{x\in\mathbb{R}^n, z\in\mathbb{R}^{m_2}}{\text{minimize}}\ c^\top x + \sum\nolimits_{\xi\in\Xi} p_\xi h(\xi - z) \ \text{ subject to } \ Ax = b,\ Tx = z,\ x \geq 0. \tag{3.11}$$

Since (3.10) and (3.11) are equivalent, there seems to be no point in this reformulation. In some applications, however, it turns out that the latter formulation can make use of *separable* properties in h. Suppose that

$$h(y) = \sum\nolimits_{i=1}^{m_2} h_i(y_i), \tag{3.12}$$

which means that the recourse cost function separates into the sum of one-dimensional functions. Separability is quite common in practice because each component $\boldsymbol{\xi}_i$ of $\boldsymbol{\xi}$ might represent the demand for a specific product and the associated cost of not meeting this demand could be independent of those for other products. The newsvendor problem of Section 1.C falls, trivially, in this category because it involves only one product. The product mix problem of Section 3.A also fits this mold when the only uncertainty is about the number of hours of carpentry and finishing. Indeed, with $\xi = (d_1, d_2)$ and $z = Tx$, the product mix problem has

$$f(\xi, x) = h(\xi - Tx) = h(\xi - z) = a_1 \max\{0, z_1 - \xi_1\} + a_2 \max\{0, z_2 - \xi_2\}$$

and we recognize $h_i(y_i)$ from the separable form in (3.12) as $a_i \max\{0, y_i\}$; the recourse cost function separates into two one-dimensional functions. In contrast, $f(\xi, x)$ fails to separate into a sum of one-dimensional functions because the matrix T links the different components of x. With deterministic rows $T_1 = [4\ \ 9\ \ 7\ \ 10]$ and $T_2 = [1\ \ 1\ \ 3\ \ 40]$ in the matrix T, the product mix problem then takes the form:

$$\underset{x\in\mathbb{R}^4, z\in\mathbb{R}^2}{\text{minimize}}\ c^\top x + \mathbb{E}\big[a_1 \max\{0, z_1 - \boldsymbol{\xi}_1\}\big] + \mathbb{E}\big[a_2 \max\{0, z_2 - \boldsymbol{\xi}_2\}\big]$$
$$\text{subject to } T_1 x = z_1,\ T_2 x = z_2,\ x \geq 0.$$

Here, the assumption about a deterministic matrix T is critical because otherwise it can't be "pulled out" from the expectation using a tender.

Separability renders all operations much easier because one deals with a juxtaposition of one-dimensional cases. Expectations require only the (marginal) probability distribution of each random variable $\boldsymbol{\xi}_i$ rather than the joint distribution of the random vector $\boldsymbol{\xi}$. One-dimensional integrals might even be available explicitly.

Example 3.14 (expected recourse cost; continuous distribution). Suppose that $\boldsymbol{\xi}$ is an integrable random variable with density function p and the recourse cost function has $h(y) = \max\{\alpha y, \beta y\}$ with $\alpha \leq \beta$. Then,

$$\varphi(z) = \mathbb{E}\big[h(\boldsymbol{\xi} - z)\big] = \alpha \int_{-\infty}^{z} (\xi - z)p(\xi)\, d\xi + \beta \int_{z}^{\infty} (\xi - z)p(\xi)\, d\xi$$
$$= \beta\big(\mathbb{E}[\boldsymbol{\xi}] - z\big) + (\beta - \alpha)\Big(zP(z) - \int_{-\infty}^{z} \xi p(\xi)\, d\xi\Big),$$

where P is the distribution function of $\boldsymbol{\xi}$. The expectation function φ has a rather explicit form, which at least can be evaluated using numerical integration. Moreover, the gradient $\nabla\varphi(z) = (\beta - \alpha)P(z) - \beta$ is directly given in terms of P. Interestingly, the function φ is smooth despite h being nonsmooth. This is the case because integration actually has a "smoothing" effect on its integrand. We refer to [30, Example 3.33] for further details. □

Example 3.15 (aircraft allocation). Facing uncertain passenger demand, an airline seeks to allocate aircraft of various types among its routes in such a way as to minimize operating cost plus expected revenue loss from flying with less than full planes. Let

$$i = 1, \dots, m_1 \quad \text{represent aircraft types}$$
$$j = 1, \dots, m_2 \quad \text{represent airline routes.}$$

The net operating cost of a full aircraft of type i on route j is

$$c_{ij} = o_{ij} - t_{ij} r_j,$$

where o_{ij} is the fixed expense of flying an aircraft of type i on route j, t_{ij} is the corresponding passenger capacity, and r_j is the revenue per passenger on route j. Let b_i be the number of available aircraft of type i. The passenger demand on the m_2 different routes is uncertain and let's model it by a random vector $\boldsymbol{\xi} = (\boldsymbol{\xi}_1, \dots, \boldsymbol{\xi}_{m_2})$ assumed to be finitely distributed with outcomes Ξ and associated probabilities $p_\xi > 0$, $\xi \in \Xi$. This example summarizes the first published study of a nontrivial stochastic optimization problem with recourse [11]; the authors computed a solution for the data in Exercise 3.8 by hand!

Following the approach in Section 3.A, we distinguish between decisions made before we know the passenger demand and recourse decisions evaluating additional costs tailored to the demand. This leads to the decision variables

x_{ij} representing the number of aircraft of type i on route j

$y_j(\xi)$ representing the number of empty seats on route j under demand ξ

and the *resource allocation problem*:

$$\begin{aligned}
\underset{x,y}{\text{minimize}} \quad & \sum\nolimits_{i=1}^{m_1} \sum\nolimits_{j=1}^{m_2} c_{ij} x_{ij} + \sum\nolimits_{\xi\in\Xi} p_\xi \sum\nolimits_{j=1}^{m_2} r_j y_j(\xi) \\
\text{subject to} \quad & \sum\nolimits_{j=1}^{m_2} x_{ij} \le b_i, \;\; i = 1, \dots, m_1 \\
& \sum\nolimits_{i=1}^{m_1} t_{ij} x_{ij} - y_j(\xi) \le \xi_j, \;\; j = 1, \dots, m_2, \; \xi \in \Xi \\
& x_{ij} \ge 0, \;\; i = 1, \dots, m_1, \; j = 1, \dots, m_2 \\
& y_j(\xi) \ge 0, \;\; j = 1, \dots, m_2, \; \xi \in \Xi.
\end{aligned}$$

The objective function has two terms. The first one gives the net operating cost for full flights, which hopefully is a negative number indicating a *potential* profit. The second term adds the expected lost revenue for empty seats. The first set of constraints limits the number of aircraft that can be allocated. The second set of constraints simply calculates the number of empty seats on each route in each scenario.

Let ν be the number of possible outcomes in Ξ. Then, the model has $m_1 + m_2\nu$ linear inequalities and $m_1m_2 + m_2\nu$ variables subject to nonnegativity constraints. Since ν is the number of possible outcomes for an m_2-dimensional random vector, it can easily be astronomical in which case a direct solution using the revised simplex method of Section 2.E becomes impossible. Fortunately, we can obtain an alternative, separable formulation with number of variables and constraints independent of ν.

Detail. Let z_j be the total number of seats made available on route j. Then, the problem is equivalently stated as

$$\begin{aligned} \underset{x,z}{\text{minimize}} \ & \sum_{i=1}^{m_1}\sum_{j=1}^{m_2} c_{ij}x_{ij} + \sum_{j=1}^{m_2}\sum_{\xi\in\Xi} p_\xi \max\{-r_j(\xi_j - z_j),\ 0\} \qquad (3.13)\\ \text{subject to} \ & \sum_{j=1}^{m_2} x_{ij} \le b_i, \quad i = 1,\dots,m_1\\ & \sum_{i=1}^{m_1} t_{ij}x_{ij} - z_j = 0, \quad j = 1,\dots,m_2\\ & x_{ij} \ge 0, \quad i = 1,\dots,m_1, \ j = 1,\dots,m_2. \end{aligned}$$

There's no revenue loss if the demand ξ_j on route j exceeds z_j, but for each empty seat there's a cost r_j representing lost revenue and this is accounted for by the max-term in (3.13). The model has m_1 linear inequalities, m_2 linear equalities, and $m_1m_2 + m_2$ variables, some of which are nonnegative. The dependence on the number of scenarios ν on the number of variables and constraints has been removed. The objective function is of the separable form discussed above, which has the added benefit that there's no need for the joint probability distribution of $\boldsymbol{\xi}$; only the marginal ones for the demands on the individual routes enter. Thus, we obtain the simplification

$$\sum_{\xi\in\Xi} p_\xi \max\{-r_j(\xi_j - z_j),\ 0\} = \sum_{k=1}^{\nu_j} p_j^k \max\{-r_j(\xi_j^k - z_j),\ 0\},$$

where $\{\xi_j^k, k = 1,\dots,\nu_j\}$ are the possible demands for route j and $\{p_j^k > 0, k = 1,\dots,\nu_j\}$ are the corresponding probabilities. When introduced in (3.13), this leads to many fewer terms in the objective function.

A disadvantage of the model is its nonlinearity caused by the max-term in (3.13), but this is compensated by the massive reduction in model size when ν is large. We might consider applying an extension of the stochastic gradient descent method that projects infeasible solutions back onto the feasible set [30, Section 3.G]. However, it requires time-consuming projections on a polyhedral set, which could make it less competitive.

Another possibility reformulates (3.13) using linear expressions. Specifically, we obtain

$$
\begin{aligned}
\underset{x,w}{\text{minimize}} \; & \sum_{i=1}^{m_1} \sum_{j=1}^{m_2} c_{ij} x_{ij} + \sum_{j=1}^{m_2} w_j \\
\text{subject to} \quad & \sum_{j=1}^{m_2} x_{ij} \leq b_i, \; i = 1, \dots, m_1 \\
-r_j \bar{\xi}_j^k + r_j \bar{p}_j^k & \sum_{i=1}^{m_1} t_{ij} x_{ij} \leq w_j, \; j = 1, \dots, m_2, \; k = 1, \dots, \nu_j - 1 \\
-r_j \bar{\xi}_j + r_j & \sum_{i=1}^{m_1} t_{ij} x_{ij} \leq w_j, \; j = 1, \dots, m_2 \\
& x_{ij} \geq 0, \; w_j \geq 0, \; i = 1, \dots, m_1, \; j = 1, \dots, m_2,
\end{aligned}
$$

where we use the shortcut notation

$$
\bar{p}_j^k = \sum_{l=1}^{k} p_j^l, \quad \bar{\xi}_j^k = \sum_{l=1}^{k} p_j^l \xi_j^l, \quad \bar{\xi}_j = \sum_{l=1}^{\nu_j} p_j^l \xi_j^l.
$$

We refer to [30, Example 3.37] for further details. The linear optimization problem has $m_1 + \sum_{j=1}^{m_2} \nu_j$ inequalities, $m_1 m_2 + m_2$ variables, and some bounds. The situation is dramatically improved compared to the starting point. Although the new model is larger than (3.13), the nonlinearity has been removed and we can employ the revised simplex method (Section 2.E). The choice between a small nonlinear model or a larger linear one is usually guided by the specifics of the problem instance and the type of hardware and software available. Regardless, the example illustrates the important fact that a real-world problem can be modeled in different ways, each coming with its own pros and cons pertaining to size and structure. □

3.F Control of Water Pollution

Eutrophication management deals with pollution control of lakes and rivers. A lake under study has four sections: section 1 in the west is surrounded by recreational areas, sections 2 and 3 are subject to run-off from fertilized farmland, and section 4 in the east borders an industrial zone; see Figure 3.13. To reduce pollution, we can install sewage treatment plants along the rivers feeding the lake and build reed basins, which absorb cl-phosphates.

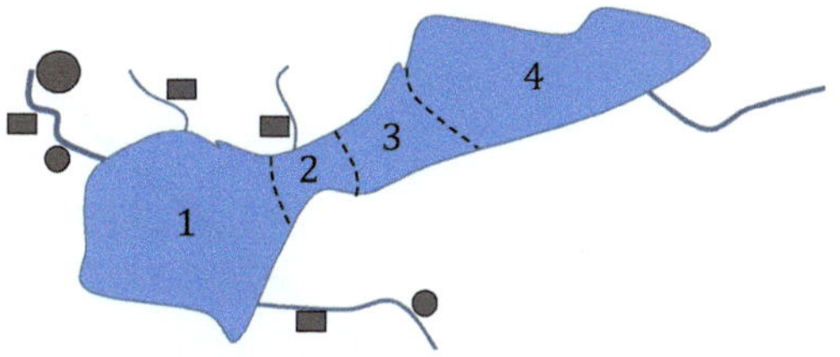

Fig. 3.13: Lake with treatment plants (squares) and reed basins (circles).

The amount of pollution depends on the atmospheric conditions about which there's enough data to reliably estimate probability distributions. In the lake, transport of the pollution occurs mostly from west to east as the major inlets are in the western part of the lake and the only outlet is on the eastern end.

For $j = 1, \dots, n$, let x_j be the size of the treatment plant or reed basin at location j; s_j is an upper bound on x_j. Let d_i be the amount of exogenous pollution entering section i of the lake and $T_i x$ be the net amount of pollution removed from section i by hydro-dynamical transport, with T_i being an n-dimensional row vector. Thus, $d_i - T_i x$ is the amount of pollution in section i, which we assume has been "normalized" so that negative values are considered excellent water quality levels. Since $d = (d_1, \dots, d_4)$ and $T_1, \dots, T_4$ depend on the weather, they're highly uncertain and we model them using the random vector

$$\boldsymbol{\xi} = (\boldsymbol{T}_1, \dots, \boldsymbol{T}_4, \boldsymbol{d}^\top),$$

which we assume has a finite number of outcomes denoted by $\xi = (T_1, \dots, T_4, d^\top)$. A constraint $a^\top x \leq \alpha$ models a budgetary limitation on the choice of x.

In this engineering design setting and many other real-world problems there's a desired state (excellent water quality), but decision makers realize that it might neither be possible nor economical to insist on achieving that state under all circumstances (weather pattern). We would like to explore the trade-off between cost and system performance in the presence of uncertainty. One approach could be to formulate a model in which the 0.95-superquantile of $\boldsymbol{d}_i - \boldsymbol{T}_i x$ is constrained to be below a thresholds. This would result in decisions that guarantee certain levels of water quality in the worst 5% weather conditions; see Section 3.B. If strict regulatory requirements specify such thresholds, then this approach would be reasonable. In the following we develop a "softer" approach where poor water quality is permitted but penalized. This is meaningful if the responsible authority for the lake is simply fined, and not prosecuted, for poor water quality.

For each section $i = 1, \dots, 4$ and positive parameters β_i and τ_i, we adopt the recourse cost function

$$h_i(y) = \begin{cases} 0 & \text{if } y < 0 & \text{(excellent water quality)} \\ (2\beta_i)^{-1} y^2 & \text{if } y \in [0, \beta_i \tau_i] & \text{(acceptable water quality)} \\ \tau_i y - \frac{1}{2}\beta_i \tau_i^2 & \text{if } y > \beta_i \tau_i & \text{(unacceptable water quality).} \end{cases}$$

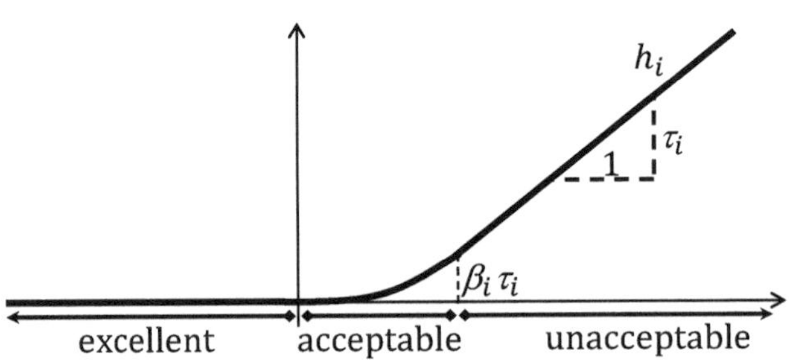

Fig. 3.14: Recourse cost function modeling penalties for poor water quality.

These recourse cost functions are easily interpreted (see Figure 3.14) and can be tied to fines levied in case of poor water quality. We see that

$$h_i(d_i - T_i x)$$

is zero when the water quality is excellent in section i under weather condition ξ but quantifies an increasingly large penalty (cost) as the water quality level drops. The total expected penalty (cost) across all sections is

$$\varphi(x) = \sum_{i=1}^{4} \mathbb{E}\big[h_i(\boldsymbol{d}_i - \boldsymbol{T}_i x)\big].$$

The derivative of h_i becomes

$$h_i'(y) = \begin{cases} 0 & \text{if } y < 0 \\ \beta_i^{-1} y & \text{if } y \in [0, \beta_i \tau_i] \\ \tau_i & \text{otherwise,} \end{cases}$$

which is continuous so h_i is actually smooth and also convex. Thus, the function φ has gradient

$$\nabla\varphi(x) = -\sum_{i=1}^{4} \mathbb{E}\big[\boldsymbol{T}_i^\top h_i'(\boldsymbol{d}_i - \boldsymbol{T}_i x)\big].$$

In addition to the expected penalty, there are direct costs associated with building the treatment plants and reed basins given by the function f_0 in n dimensions. This leads to the formulation

$$\underset{x\in\mathbb{R}^n}{\text{minimize}}\ f_0(x) + \sum_{i=1}^{4} \mathbb{E}\big[h_i(\boldsymbol{d}_i - \boldsymbol{T}_i x)\big] \ \text{ subject to } \ a^\top x \le \alpha,\ 0 \le x \le s,$$

which is a stochastic optimization problem with simple recourse. The objective function is convex (smooth) whenever f_0 is convex (smooth). Since the vectors $\boldsymbol{T}_1, \dots, \boldsymbol{T}_4$ are random, the resulting *technology matrix* $\boldsymbol{T}$, with these as rows, is random and this prevents reformulations in terms of tenders. The evaluation of the expected recourse cost now requires multi-dimensional integration; it's no longer possible to simplify with a series of one-dimensional integrals. Generally, the nature of the technology matrix, so named because it quantifies how an input x is converted into an output $\boldsymbol{T}x$, largely governs the difficulty of simple recourse problems.

3.G Linear Recourse Problems

Stochastic optimization problems with simple recourse may involve a random vector $\boldsymbol{d}$ representing uncertain demand and a random technology matrix $\boldsymbol{T}$ transforming the decision vector x into a random vector $\boldsymbol{T}x$ to be compared with $\boldsymbol{d}$; see Section 3.E. This leads to a decision process in two stages:

$$\text{decision: } x \;\rightsquigarrow\; \text{observation: } (T,d) \;\rightsquigarrow\; \text{evaluation: } y = d - Tx.$$

The first-stage decision x is taken before the parameters (T, d) are known and the second-stage decision y simply evaluates the degree by which we comply with the requirement $Tx = d$. Let's now examine situations with more complex second-stage decisions and consider the decision process:

$$\text{decision: } x \;\rightsquigarrow\; \text{observation: } (W,T,d) \;\rightsquigarrow\; \text{decision: } y \geq 0,\ Wy = d - Tx,$$

where the second-stage decision y isn't simply about discrepancy between Tx and d, but also depends on an uncertain *recourse matrix* W. In fact, there might be many such satisfactory y among which we would like to take the one with the lowest cost, a quantity that also might be uncertain. This results in the *stochastic optimization problem with linear recourse*:

$$\boxed{\operatorname*{minimize}_{x \in C_1}\ f_0(x) + \mathbb{E}\big[f(\boldsymbol{\xi}, x)\big]} \tag{3.14}$$

where $C_1 \subset \mathbb{R}^{n_1}$ defines first-stage constraints, f_0 is a function in n_1 dimensions representing the first-stage cost, and f is a *recourse cost function* taking as input a q-dimensional vector of uncertain parameters and an n_1-dimensional vector of first-stage decision variables. Its value $f(\xi, x)$ is defined as the minimum value of the linear optimization problem

$$\operatorname*{minimize}_{y}\ a^\top y \text{ subject to } Wy = d - Tx,\ y \geq 0. \tag{3.15}$$

The second-stage decision is now associated with its own cost vector a, an n_2-dimensional column vector. The recourse matrix W is of dimension $m_2 \times n_2$ and the technology matrix T is of dimension $m_2 \times n_1$. The demand is represented by the m_2-dimensional column vector d. All these quantities might be uncertain and let's collect them in the vector

$$\xi = (a^\top, W_1, \ldots, W_{m_2}, T_1, \ldots, T_{m_2}, d^\top),$$

where W_i and T_i are the ith rows of W and T, respectively. The random vector $\boldsymbol{\xi}$ is then of dimension $q = n_2 + (1 + n_2)m_2 + n_1 m_2$.

The stochastic optimization problem with linear recourse (3.14) incorporates both anticipation *and* adaptation: The first-stage decision x is selected while anticipating future conditions and the second-stage decision y adapts to the revealed value of $\boldsymbol{\xi}$. For example, there might be a trade-off between capital investments and the cost of day-to-day operations, between the design and layout of an overall system and the adjustments necessary to respond to particular scenarios.

A stochastic optimization problem with simple recourse is the special case when the second-stage linear optimization problem (3.15) has a minimizer that can be obtained explicitly for every (ξ, x) without having to invoke the revised simplex method or similar algorithms; say $f(\xi, x) = h(d - Tx)$ for some simple function h.

A main concern in stochastic optimization problems with linear recourse is that certain first-stage decisions (capital investments) could cause difficulties at the second stage; day-to-day operations may become infeasible, i.e., there are no feasible solutions in (3.15). Alternatively, there could be a shortage of constraints, permitting the objective function value in (3.15) to be arbitrarily low. In these cases, the minimum value in (3.15) isn't a well-defined real value. To capture such special cases, we introduce the following generalization of a minimum value.

Infimum. The *infimum* of a nonempty set $Y \subset [-\infty, \infty]$, denoted by $\inf Y$, is the *highest* scalar α with $\alpha \le \beta$ for all $\beta \in Y$. Moreover, $\inf \emptyset$ is defined as ∞. When Y contains $\inf Y$, one refers to $\inf Y$ as the *minimum* of Y and denotes it by $\min Y$. The distinction is necessary because neither $\mathbb{R}$ nor the interval $(0, 1)$ has a minimum, but $\inf \mathbb{R} = -\infty$ and $\inf (0, 1) = 0$ are well defined.

For a function f and a set C, we write

$$\inf_C f = \inf_{x \in C} f(x) = \inf \{ f(x) \,|\, x \in C \}$$

to denote the infimum of f on C. If the set $C = \mathbb{R}^n$ and thus causing no restrictions to x, then we simply write

$$\inf f = \inf_x f(x) = \inf_{\mathbb{R}^n} f$$

for the infimum of f. For example, $f(x) = e^x$ has $\inf f = 0$, while $f(x) = x$ has $\inf f = -\infty$. The set $C = \emptyset$ produces $\inf_C f = \infty$ regardless of f.

If $f(\bar{x}) < \infty$ and $f(\bar{x}) = \inf_C f$, then $\inf_C f$ is the *minimum* (or *minimum value*) of f on C, $\bar{x}$ is a *minimizer* of f on C, and we may write "min" in place of "inf."

Returning to (3.15), we can use the infimum-terminology to define more precisely:

$$f(\xi, x) = \inf_y \{ a^\top y \,|\, Wy = d - Tx, \; y \ge 0 \}. \tag{3.16}$$

Thus, if the linear optimization problem is infeasible for given ξ and x, then $f(\xi, x) = \infty$. If it's unbounded, then $f(\xi, x) = -\infty$. The recourse cost function is now well defined regardless of the situation.

Let's focus on identifying when the *expected* recourse cost $\mathbb{E}[f(\boldsymbol{\xi}, x)]$ is finite and thereby avoid unpleasant surprises in the second stage. We're especially concerned about when $\mathbb{E}[f(\boldsymbol{\xi}, x)] = \infty$ because then x is infeasible even if it satisfies the conditions in C_1.

To understand the situation, let's consider a fixed decision $\bar{x}$. If $f(\boldsymbol{\xi}, \bar{x})$ is integrable, then $\mathbb{E}[f(\boldsymbol{\xi}, \bar{x})]$ is finite by definition. Thus, the difficulty arises when $f(\boldsymbol{\xi}, \bar{x})$ isn't integrable. This would certainly be the case if $f(\xi, \bar{x}) = \infty$ for some outcome ξ with positive probability p_ξ. Then, $\mathbb{E}[f(\boldsymbol{\xi}, \bar{x})] = \infty$ because the formula for the expectation includes the term $p_\xi f(\xi, \bar{x})$. The recourse cost function $f(\xi, \bar{x}) = \infty$ when there's no $y \ge 0$ satisfying $Wy = d - T\bar{x}$, i.e., (3.15) is infeasible.

Generally, for a linear recourse problem, we say that the recourse cost function imposes *induced constraints* when

$$C_2 = \{x \mid \mathbb{E}[f(\xi, x)] < \infty\}$$

is a strict subset of $\mathbb{R}^{n_1}$. Together with C_1, the induced constraints furnish the feasible set $C_1 \cap C_2$ of the problem. In contrast to C_1, which we assume is explicitly specified in terms of equality and inequality constraints, C_2 isn't always easily characterized. It depends on an intricate interplay between the various components of (3.15) as well as the probability distribution of ξ.

A problem has *complete recourse* if $C_2 = \mathbb{R}^{n_1}$, i.e., there are no induced constraints. It has *relatively complete recourse* if $C_2 \supset C_1$, in which case there might be induced constraints, but they're superfluous due to the explicit constraints.

Example 3.16 (induced constraints). Consider a recourse cost function given by

$$f(\xi, x) = \inf_y \{y \mid y = x - \xi, \ y \geq 0\},$$

with the random variable ξ being uniformly distributed on $[1, 2]$. The set of induced constraints is then $C_2 = [2, \infty)$.

Detail. The recourse cost $f(\xi, x) = x - \xi$ if $x \geq \xi$ and $f(\xi, x) = \infty$ otherwise. Thus,

$$\mathbb{E}\big[f(\xi, x)\big] = \begin{cases} \infty & \text{if } x < 2 \\ x - \frac{3}{2} & \text{if } x \geq 2. \end{cases}$$

The second formula for $x \geq 2$ follows from the calculations

$$\mathbb{E}\big[f(\xi, x)\big] = \int_1^2 f(\xi, x)\, d\xi = \int_1^2 (x - \xi)\, d\xi = x - \left(\tfrac{1}{2}2^2 - \tfrac{1}{2}1^2\right) = x - \tfrac{3}{2}.$$

The problem of minimizing $\mathbb{E}[f(\xi, x)]$ subject to $x \in C_1 = [0, \infty)$ has neither complete nor relatively complete recourse because $C_2 = [2, \infty)$. However, minimize $\mathbb{E}[f(\xi, x)]$ subject to $x \in [2, \infty)$ has the same minimizer and minimum value as well as relatively complete recourse. In fact, every stochastic optimization problem with linear recourse can in principle be reformulated as one with relatively complete recourse by incorporating the induced constraints in C_1. However, there might be no viable procedure for carrying out this reformulation. □

There are several specialized algorithms that apply to the stochastic optimization problem with linear recourse (3.14); see, e.g., [30, Section 5.H], which presents an algorithm that iteratively discovers induced constraints and adds them to the explicit constraints. However, when ξ has a finite distribution with outcomes Ξ and corresponding probabilities $p_\xi, \xi \in \Xi$, the path followed in Section 3.A is always available: replicate the second-stage decision variables. This results in a reformulation of (3.14):

$$\underset{x \in C_1, y_\xi, \xi \in \Xi}{\text{minimize}} \ f_0(x) + \sum_{\xi \in \Xi} p_\xi a_\xi^\top y_\xi \tag{3.17}$$
$$\text{subject to} \quad W_\xi y_\xi = d_\xi - T_\xi x, \ y_\xi \geq 0 \ \ \forall \xi \in \Xi.$$

Here, the vector y_ξ represents the second-stage decisions in case of scenario ξ and $a_\xi, W_\xi, T_\xi, d_\xi$ are the corresponding vectors and matrices in that scenario. Thus, the optimization model specifies both a first-stage decision x *and* contingency plans for the second-stage decisions in every scenario. Any x for which a scenario ξ produces $W_\xi y \neq d_\xi - T_\xi x$ regardless of $y \geq 0$ is now ruled out explicitly. There's no concern about induced constraints because of the shift to an expanded formulation. If f_0 and C_1 are convex, then (3.17) is convex. However, computational tractability quickly becomes an issue as the number of scenarios grows.

Example 3.17 (network capacity expansion). In a power transmission network, there's net demand ξ_i at vertex $i \in V = \{1, \dots, m\}$, which could be negative reflecting excess supply of power. Overall in the network, supply exceeds demand but the capacities of the transmission lines, represented by edges in the network, need to be expanded from the current levels. Let a_e be the current capacity on edge e and x_e be the corresponding planned capacity expansion, which is limited by v_e and associated with the cost $c_e x_e$. Our goal is to optimize the capacity expansion while accounting for a budget of β. The problem is historically motivated. In the 2001 California energy crisis, some of the blackouts were blamed on the lack of capacity of the transmission lines between Southern and Northern California.

The power flow in a transmission network is governed by laws of physics that are reasonably well approximated by a linear relation between the net demand vector $\xi = (\xi_1, \dots, \xi_m)$ and a vector of voltage angles $y = (y_1, \dots, y_m)$ as expressed via the given m-dimensional vectors $b_i, i \in V$. Specifically, we would need the voltage angles to satisfy

$$b_i^\top y \geq \xi_i \quad \forall i \in V. \tag{3.18}$$

A difference in voltage angle between two vertices connected by an edge requires a certain capacity that's captured with the constraints

$$|y_i - y_j| \leq a_e + x_e \quad \forall e = (i, j) \in E,$$

where E is the set of n edges in the network. The net demand is uncertain due to variability in consumption and renewable power generation and we model it as a random vector $\boldsymbol{\xi}$ with a finite distribution; outcomes are given by the set Ξ and $\{p_\xi > 0, \xi \in \Xi\}$ are the corresponding probabilities. Since it wouldn't be economical to attempt to prevent violation of the constraints (3.18) for every outcome $\xi \in \Xi$, we adopt a nonnegative penalty of δ_i per unit violation of the constraint at vertex i. This leads to the formulation

$$\underset{x \in \mathbb{R}^n}{\text{minimize}} \sum_{\xi \in \Xi} p_\xi f(\xi, x) \ \text{ subject to } \sum_{e \in E} c_e x_e \leq \beta, \ 0 \leq x_e \leq v_e \ \forall e \in E,$$

where the recourse cost function

$$f(\xi, x) = \inf_{y \in \mathbb{R}^m} \left\{ \sum\nolimits_{i \in V} \delta_i \max\{0, \xi_i - b_i^\top y\} \,\middle|\, |y_i - y_j| \le a_e + x_e \ \ \forall e = (i,j) \in E \right\}.$$

We seek to determine a first-stage decision x (transmission capacity expansion) that minimizes expected penalties across all outcomes of $\boldsymbol{\xi}$. The second-stage decision about voltage angles is made after we observe the demand. Although not exactly in the form of the recourse cost function (3.16) of a problem with linear recourse, we can accomplish this with a reformulation.

Detail. Let's introduce the additional variables z_i^+ and z_i^- to represent the positive and negative parts of $\xi_i - b_i^\top y$ and these can be used to construct a linear objective function. We can express the variable y_i in terms of its positive part y_i^+ and negative part y_i^-. Let y^+ and y^- be the vectors of these positive and negative parts so that $y = y^+ - y^-$. After using w_e^+ and w_e^- to convert the constraints on voltage angles into equalities, we obtain

$$\begin{aligned} f(\xi, x) = \inf \Big\{ \sum\nolimits_{i \in V} \delta_i z_i^+ \,\Big|\, & \xi_i - b_i^\top (y^+ - y^-) = z_i^+ - z_i^- \ \ \forall i \in V \\ & y_i^+ - y_i^- - y_j^+ + y_j^- + w_e^+ = a_e + x_e \ \ \forall e = (i,j) \in E \\ & y_i^+ - y_i^- - y_j^+ + y_j^- - w_e^- = -a_e - x_e \ \ \forall e = (i,j) \in E \\ & y_i^+, y_i^-, z_i^+, z_i^-, w_e^+, w_e^- \ge 0 \ \ \forall i \in V,\ e \in E \Big\}. \end{aligned}$$

The minimization is now over nonnegative variables subject to equality constraints and we recognize that it's of the form (3.16); the recourse cost vector a in the notation of that equation consists of δ_i and zeros, the recourse matrix W collects all the coefficients in front of the variables and is deterministic, the demand vector d consists of ξ_i and a_e, and the technology matrix T is deterministic with one and zero elements.

For any ξ and $x \ge 0$, $f(\xi, x) < \infty$ because one can set

$$z_i^+ - z_i^- = \xi_i, \ \ y_i^+ = y_i^- = 0 \ \ \forall i \in V; \qquad w_e^+ = w_e^- = a_e + x_e \ \ \forall e \in E$$

to produce a feasible point. Since $\boldsymbol{\xi}$ is finitely distributed, this means that $\mathbb{E}[f(\boldsymbol{\xi}, x)] < \infty$ as well. (In fact, the objective function is always nonnegative, which means that $\mathbb{E}[f(\boldsymbol{\xi}, x)]$ is actually finite.) Thus, the problem has relatively complete recourse. The problem doesn't have complete recourse, however, because for $x_e < -a_e$ there's no feasible second-stage decision, which can already be seen by the constraint $|y_i - y_j| \le a_e + x_e$.

We can reformulate the problem as a linear optimization problem following the principle behind (3.17), i.e., replicate each of the second-stage variables to make them scenario dependent. This results in the formulation

$$
\begin{aligned}
\text{minimize } \sum_{\xi \in \Xi} p_\xi \sum_{i \in V} \delta_i z_i^+(\xi) \quad \text{subject to} \quad & \sum_{e \in E} c_e x_e \le \beta \\
& 0 \le x_e \le v_e \quad \forall e \in E \\
\xi_i - b_i^\top \big(y^+(\xi) - y^-(\xi)\big) = z_i^+(\xi) - z_i^-(\xi) & \quad \forall i \in V,\ \xi \in \Xi \\
y_i^+(\xi) - y_i^-(\xi) - y_j^+(\xi) + y_j^-(\xi) + w_e^+(\xi) = a_e + x_e & \quad \forall e = (i,j) \in E,\ \xi \in \Xi \\
y_i^+(\xi) - y_i^-(\xi) - y_j^+(\xi) + y_j^-(\xi) - w_e^-(\xi) = -a_e - x_e & \quad \forall e = (i,j) \in E,\ \xi \in \Xi \\
y_i^+(\xi), y_i^-(\xi), z_i^+(\xi), z_i^-(\xi), w_e^+(\xi), w_e^-(\xi) \ge 0 & \quad \forall i \in V,\ e \in E,\ \xi \in \Xi.
\end{aligned}
$$

Although we've ended up with a linear optimization problem, its size is a concern as power transmission networks are large making the replication of variables and constraints across outcomes even more detrimental. Modern implementations of the revised simplex method (Section 2.E) and other algorithms scan a formulation for redundancies and may reduce the size by eliminating some variables and constraints. This extends their reach somewhat, but it's often necessary to leverage the structure of the problem in more detail as discussed in [30, Sections 5.H and 10.E]. □

3.H Summary and Study Guide

In this chapter, we learned how to deal with uncertainty, which is a main issue in many applications.

- Decision making under uncertainty can often be modeled by a two-stage formulation. A first-stage decision x represents our (strategic) decision "now" when uncertain parameters are unknown, for example about future demand. A second-stage decision y represents our (tactical) decision in the future after the uncertainty has been resolved. It models recourse decisions that we'll undertake in an effort to adjust to the actual situation.
- Simple recourse: the second-stage decision is "simple" and easily computed, for example by comparing the amount provided Tx to the amount demanded d.
- Linear recourse: the second-stage decision is given by a linear optimization problem. Since a linear optimization problem may have no minimizers, it could induce additional constraints. These are constraints on the first-stage decision vector x that aren't known explicitly. They identify x-vectors that are so bad that we can't recover in the second stage by choosing a suitable y.
- When facing an uncertain future cost, it could make sense to consider the expected (average) cost and this is the common approach for simple and linear recourse models. However, when we seek to be more conservative, it might be better to consider quantiles and superquantiles. A 0.9-quantile of a random variable specifies a threshold it's only a 10% chance to exceed. A 0.9-superquantile specifies the average value above that threshold. It therefore has the interpretation of being the average of the worst 10% outcomes of the random variable. Mathematically, superquantiles are better

behaved than quantiles. They preserve properties such as convexity and avoid paradoxes associated with quantiles. Thus, we prefer using superquantiles in optimization models.

- The stochastic gradient descent method minimizes an objective function of the form $(1/m)\sum_{i=1}^{m} f_i(x)$ by taking advantage of the sum structure to reduce computing time per iteration. Instead of computing the gradient of the objective function, the method selects one term (or a small batch of terms) and computes the gradient of the resulting simplified function. This produces a direction that's good on average but may not be a descent direction. The reduced cost per iteration compared to the gradient descent method makes the stochastic gradient descent method the workhorse for solving large-scale problems in machine learning. Despite its name, the stochastic gradient descent method is neither a descent method nor a gradient method (because it handles nonsmoothness by using subgradients).
- There are many versions of the stochastic gradient descent method including ADAM, which is widely used in practice. Most versions require tuning of algorithmic parameters such as the step size to achieve reasonable performance.

3.I Exercises

Exercise 3.1 (stochastic resources). Consider the product mix problem when the only uncertainty is about the number of hours available for carpentry and finishing; c, a, and deterministic T are as in Section 3.A. The random variable $\boldsymbol{d}_1$ takes on the values 4800, 5500, 6050, and 6150, each with probability $1/4$, while the random variable $\boldsymbol{d}_2$ has values 3936, 3984, 4016, and 4064, also with probability $1/4$. The two random variables are statistically independent. Solve both the stochastic optimization problem and the deterministic version (3.1), with $d = (5625, 4000)$, the expected values of $\boldsymbol{d}_1$ and $\boldsymbol{d}_2$. Compare the minimizers.

Exercise 3.2 (superquantiles). Consider a random variable with two possible outcomes 0 and 1, occurring with probability 0.4 and 0.6, respectively. Compute its 0.8-quantile and 0.8-superquantile.

Exercise 3.3 (investment planning). The amount of money you'll need in retirement is uncertain as it depends on your health, longevity, and many other factors. Let's model it by a random variable $\boldsymbol{\xi}_0$ given in millions of current dollars. Presently, you've saved 2 million dollars. (For simplicity, we assume you'll not be able to contribute more to your savings before retirement.) You can place your savings in the bank, with a fixed yield, and/or invest in two possible funds with uncertain yield. We model these yields by the random variables $\boldsymbol{\xi}_1, \boldsymbol{\xi}_2, \boldsymbol{\xi}_3$. There are 12 possible values of the random vector $(\boldsymbol{\xi}_0, \boldsymbol{\xi}_1, \boldsymbol{\xi}_2, \boldsymbol{\xi}_3)$, each equally likely; see Table 3.3. Your goal is to determine an investment plan $x = (x_1, x_2, x_3)$ that minimizes the superquantile risk of the shortfall defined as $\boldsymbol{\xi}_0 - \boldsymbol{\xi}_1 x_1 - \boldsymbol{\xi}_2 x_2 - \boldsymbol{\xi}_3 x_3$, where x_1, x_2, and x_3 are the allocations to bank deposit, the first fund, and the second fund, respectively.

For each of $\boldsymbol{\xi}_0, \boldsymbol{\xi}_1, \boldsymbol{\xi}_2, \boldsymbol{\xi}_3$, compute the mean, standard deviation, and α-quantiles as well as α-superquantiles for $\alpha = 0, 0.5, 0.75, 0.9, 1$. Compute the same quantities for each of the random variables $\xi_0 - 2\xi_1$, $\xi_0 - 2\xi_2$, and $\xi_0 - 2\xi_3$. Discuss the meaning of these numbers. Formulate and solve a model of the form (3.3) or (3.4) using $\alpha = 0, 0.2, 0.4, 0.6, 0.8, 0.9, 0.99$. Report the resulting optimized investment plans. For each plan, compute the outcomes of the shortfall across the scenarios and compare with the outcomes for the plans $(2,0,0)$, $(0,2,0)$, and $(0,0,2)$. For the various plans, also compute the probability of going broke, i.e., having a positive shortfall.

	possible outcomes of random variables											
ξ_0	2.3	2.2	1.9	2.0	2.1	2.2	1.9	2.0	2.0	2.1	1.9	2.3
ξ_1	1.0	1.0	1.0	1.0	1.0	1.0	1.0	1.0	1.0	1.0	1.0	1.0
ξ_2	1.6	1.5	1.4	1.4	1.3	1.3	1.2	1.1	0.9	0.8	0.7	0.7
ξ_3	0.9	0.9	1.0	0.9	1.0	1.1	1.1	1.2	1.1	1.2	1.1	1.2

Table 3.3: The 12 possible outcomes for the random vector in Exercise 3.3.

Exercise 3.4 (distribution of age at failure). Let's carry out *failure analysis* for a type of engineering system, which has a lifetime of 10 years, but occasionally experiences repairable failures during that time. We would like to estimate the chance of failure when the system is at a certain age using the following historical data of the age (in years) of a system at failure $\{0.1, 0.02, 9.3, 8.9, 3.2, 7.8, 9.5, 1.1, 9.7, 0.5\}$. The data as well as engineering insight tell us that failure is more likely for a new system (due to burn-in issues) and also for an aging system (due to wear and tear). For this reason, let's estimate a u-shaped density function as a model of uncertainty about the age of failure. Using epi-splines and the maximum likelihood procedure from Section 3.C, write constraints that ensure a u-shaped density function. Solve the resulting estimation problem.

Guide. The 10-year lifetime of the system indicates a density function defined on $[0, 10]$. A u-shaped function must be convex. This can be achieved by imposing continuity constraints as well as conditions on the slopes in each interval.

Exercise 3.5 (stochastic gradient descent). For the problem of minimizing $f(x) = (1/m)\sum_{i=1}^{m}(x - i)^2$ over $x \in \mathbb{R}$. Provide the details of one iteration of the stochastic gradient descent method as applied to this problem.

Exercise 3.6 (stochastic gradient descent). Consider the function given by

$$f(x) = 3x_1^2 + x_2^2 + 0.1x_1x_2 + (x_1 - 1)^2 + 2(x_2 - 2)^2.$$

Implement the gradient descent and stochastic gradient descent methods and apply them to f starting from the point $(2, 2)$. Experiment with a range of fixed step sizes. Plot the trajectories the methods take from the initial point $(2, 2)$. Discuss the variation you see across the trajectories. Compare the results with the minimizer of f obtained using the Fermat rule and convexity.

Exercise 3.7 (stochastic gradient descent for support vector machines). Implement the stochastic gradient descent method using a fixed step size λ and apply it to the regularized hinge-loss problem (2.5) in Section 2.F for some data set; see also Example 3.12. For a given number of iterations (such as 100000), compute the objective function value

$$\frac{1}{m}\sum\nolimits_{i=1}^{m} f_i(\bar{a},\bar{\alpha}) + \theta\|\bar{a}\|_1, \quad \text{where } (\bar{a},\bar{\alpha}) = \frac{1}{\bar{\nu}+1}\sum\nolimits_{\nu=0}^{\bar{\nu}}(a^\nu,\alpha^\nu)$$

is the average across all the iterations. Repeat the calculations 30 times and compute the average of the 30 objective function values. Replicate the process for different λ and θ. Test the classifiers on a separate but similar data set and compute the number of correct classifications for each one.

aircraft	capacities: t_{ij}				
1	16	15	28	23	81
2	–	10	14	15	57
3	–	5	–	7	29
4	9	11	22	17	55
route:	1	2	3	4	5

aircraft	operating cost: o_{ij}				
1	18	21	18	16	10
2	–	15	16	14	9
3	–	10	–	9	6
4	17	16	17	15	10
route:	1	2	3	4	5

Table 3.4: Capacities and operating costs, with "–" indicating unavailable.

route	demand : probability				
1	200 : 0.2	220 : 0.05	250 : 0.35	270 : 0.2	300 : 0.2
2	50 : 0.3	150 : 0.7			
3	140 : 0.1	160 : 0.2	180 : 0.4	200 : 0.2	220 : 0.1
4	10 : 0.2	50 : 0.2	80 : 0.3	100 : 0.2	340 : 0.1
5	580 : 0.1	600 : 0.8	620 : 0.1		

Table 3.5: Demand ξ_j^k and probability p_j^k, $k = 1,\dots,\upsilon_j$, for all $j = 1,\dots,5$.

Exercise 3.8 (aircraft allocation; computations). In your own words, explain the indices, variables, data, objective function, and constraints of the first optimization model in Example 3.15. How do you justify the model? Discuss assumptions, computational aspects, and potential insights the model can provide.

Using the last formulation in Example 3.15, solve the aircraft allocation problem with four aircraft types ($i = 1,\dots,4$), five routes ($j = 1,\dots,5$), and the data $b = (10, 19, 25, 15)$, $r = (13, 13, 7, 7, 1)$, and as given in Tables 3.4 and 3.5. Formulate and solve the corresponding problem when uncertainty is ignored and the demand is set to the expected demand. Compare the two solutions and their costs.

Exercise 3.9 (from simple to linear recourse). Show that a simple recourse problem with

$$f(\xi, x) = \sum_{i=1}^{m_2} \max\big\{\alpha_i(\xi_i - T_i x),\ \beta_i(\xi_i - T_i x)\big\},$$

where $\xi = (\xi_1, \ldots, \xi_{m_2})$ and $\alpha_i \leq \beta_i$, can be reformulated as a problem with linear recourse.

Guide. It suffices to write f in the form (3.16). □

Exercise 3.10 (sensitivity analysis versus stochastic optimization). Stranda Furniture Company manufactures wooden desks, tables, and chairs. A desk sells for \$60, a table sells for \$40, and a chair sells for \$10. The manufacture requires lumber, carpentry labor, and finishing labor. The price per board feet of lumber is \$2 and each desk, table, and chair require 8 board feet, 6 board feet, and 1 board feet, respectively. The cost per hour of carpentry is \$5.2, with the requirement for each desk, table, and chair being 2, 1.5, and 0.5 hours, respectively. The cost per hour of finishing work is \$4, with the requirement for each desk, table, and chair being 4, 2, and 1.5 hours, respectively. The demand for desks, tables, and chairs is 150, 125, and 300, respectively. (This case study stems from [14].)

(a) Compute the per-item profit, i.e., sales price minus production cost, for each desk, table, and chair. What conclusion may you reach from these numbers?

(b) Identify a linear optimization problem that minimizes the production cost minus the revenue from sales. Solve the resulting model.

(c) Suppose that the demand is uncertain with three possible scenarios. "Low:" 50 desks, 20 tables, 200 chairs. "Likely:" 150 desks, 110 tables, 225 chairs. "High" 250 desks, 250 tables, 500 chairs. Solve the model in (b) for each of the three scenarios and discuss the results.

(d) Using the scenarios from (c), construct a linear optimization problem that minimizes the expected production cost minus the revenue from sales while assuming that all decisions must be made prior to knowing the demand. Solve the resulting model and discuss the results.

(e) Adjust the model in (d) to the case when acquisition of material needs to be made *before* the demand is known but the allocation of carpentry and finishing labor can be done *after* we learn the demand. Solve the model and discuss the results.

Exercise 3.11 (infimum). Determine the infimum, $\inf f$, of the functions given by: (a) $f(x) = -x^2$; (b) $f(x) = 1/x$ if $x > 0$ and $f(x) = \infty$ otherwise; (c) $f(x) = x$ if $x > 0$ and $f(x) = \infty$ otherwise.

Exercise 3.12 (induced constraints). Consider the problem

$$\underset{x \in \mathbb{R}}{\text{minimize}}\ x + \tfrac{1}{2} f_1(x) + \tfrac{1}{2} f_2(x),$$

where $f_i(x) = \inf_y\{y \mid 2i - x \leq y,\ y \leq 0\}$ for $i = 1, 2$. Identify the explicit as well as the induced constraints on x.

Chapter 4
Special Structures: Integers and Networks

Convexity is a key structural property of an optimization problem that facilitates the development of efficient algorithms. Integer restrictions and network structures likewise have important algorithmic consequences. In this chapter, we'll introduce the wide variety of applications that can benefit from integer restrictions as a modeling tool for complex logical relations between decision variables. We'll also present the fundamental algorithms for solving the resulting optimization models and discuss the increase in computational complexity as compared to convex problems. We'll summarize the relief provided by network structure, which allows some problems with integer restrictions to be solved by the revised simplex method or other specialized algorithms. The chapter ends with a discussion of multi-stage decision making and reinforcement learning.

4.A Formulations using Integer Restrictions

There are many practical situations where decision variables represent indivisible quantities: the number of newspapers to have in inventory (cf. Example 1.3) and the number of units to ship (cf. Example 2.5). We ignored this fact in the earlier examples, with the hope that the resulting solutions would still be meaningful for decision makers. There was a good reason for this "negligence." Changing the constraints from $x \in [0, \infty)$ to the *integer constraint* $x \in \{0, 1, 2, \dots\}$ would introduce nonconvexity, making much of the theory and algorithms from the earlier chapters inapplicable. Practically, a variable that's optimized to the value 125.4 can probably be rounded to 125 with few consequences. (We shouldn't forget that our optimization models are just *models* of the real world and their precision shouldn't be overestimated.) The situation is different for yes-no decisions such as building a new warehouse or not. A decision of this kind is best represented by a variable x that's restricted to the set $\{0, 1\}$.

In this section, we discuss modeling of yes-no decisions as well as more intricate logical conditions using integer constraints.

J. O. Royset, Optimization Concepts, *Springer Series in Operations Research and Financial Engineering*,
https://doi.org/10.1007/978-3-032-36819-5_4

Example 4.1 (transportation problem with minimum quantities). We recall the situation from Example 2.5: a minimizer recommends a low shipping volume of 50 units from Washington State to San Francisco. This might be undesirable for a variety of practical reasons not accounted for in the model, and management may require that any shipment involves at least 100 units. We can incorporate this change using the additional decision variables

$$y_{ij}, \quad i = 1, \ldots, m, \; j = 1, \ldots, n$$

and the additional constraints

$$w_i y_{ij} \geq x_{ij} \geq 100 y_{ij} \quad \text{and} \quad y_{ij} \in \{0, 1\}, \quad i = 1, \ldots, m, \; j = 1, \ldots, n. \tag{4.1}$$

The variable $y_{ij} = 1$ if it's decided to move units between warehouse i and demand location j, and is zero otherwise. This enforces exactly the right conditions on x_{ij}:

$$w_i y_{ij} \geq x_{ij} \geq 100 y_{ij} \quad \text{becomes} \quad \begin{cases} w_i \geq x_{ij} \geq 100 & \text{if } y_{ij} = 1 \\ 0 \geq x_{ij} \geq 0 & \text{if } y_{ij} = 0. \end{cases}$$

The condition $y_{ij} \in \{0, 1\}$ is an example of an integer constraint. Here it takes the form of a *binary constraint.*

Detail. The feasible set defined by the constraints in Example 2.5 and the additional constraints in (4.1) is nonconvex due to the integer constraints. However, the problem remains highly structured and this can be utilized by algorithms as we see in the next sections. Table 4.1 provides a minimizer for the specific problem at hand under the additional constraints. Its cost is higher than that of Table 2.1 (without integer constraints), but now we avoid shipping low volumes. □

	Demand locations			Cost
Warehouses	San Francisco	Chicago	New York	
Washington State	0	305	0	\$1450.50
North Carolina	320	0	270	

Table 4.1: Shipping quantities subject to volume constraint.

Example 4.1 gives rise to a *mixed-integer optimization problem* because it involves variables that can take any real value, sometimes called *continuous variables*, and other variables that are restricted to integers, i.e., *integer variables* or in the case of the example actually *binary variables.* A problem might also only involve integer or binary variables as we see next in two examples.

Example 4.2 (knapsack problem). A hiker seeks to optimally fill a knapsack of weight capacity α with items from a list of n items. Each item has a weight a_j, but also a value

$-c_j$ representing the utility for the hiker. No item can be packed partially so we adopt a binary variable x_j, with $x_j = 1$ representing item j being packed and $x_j = 0$ specifying the item being left behind. We seek to maximize the utility of the selected items. After shifting to our usual minimization framework, we obtain the model

$$\boxed{\underset{x \in \{0,1\}^n}{\text{minimize}}\ c^\top x \ \ \text{subject to}\ \ a^\top x \leq \alpha}$$

It specifies the best packing plan within the weight limit. Due to its canonical form, the model is referred to as a *knapsack problem*.

Since the model only involves binary variables, we can in principle enumerate all the possibilities, check their feasibility and retain one with the best objective function value. There are 2^n possibilities, which grows quickly as n increases. For example, with $n = 30$, there are $2^{30} \approx 10^9$ possibilities. If it takes one millisecond to check each one of these possibilities, then the overall computations would take 12 days! Thus, the simplicity of the model is deceiving. Integer constraints introduce a fundamental challenge due to their nonconvexity.

Detail. A solution approach to the knapsack problem could be to replace the binary restriction $x \in \{0, 1\}^n$ by the convex constraint $x \in [0, 1]^n$, and then round the solution. This replacement has the advantage that we obtain a linear optimization problem and can apply the simplex or interior point methods. Often however, this approach falls short. For $n = 3$, $\alpha = 10$, and

$$c = (-3, -2, -2); \qquad a = (5.01,\ 5,\ 5),$$

the unique minimizer of the knapsack problem is

$$x^\star = (0, 1, 1), \quad \text{with minimum value } -4.$$

The convex constraint $x \in [0, 1]^n$ results in

$$\bar{x} = (1,\ 5/5.01,\ 0), \quad \text{with minimum value } -3 - 2 \cdot 5/5.01 \approx -4.996.$$

After rounding $\bar{x}$, we obtain $(1, 1, 0)$, which is qualitatively different than $x^\star$. In fact, it's infeasible because

$$a^\top \begin{bmatrix} 1 \\ 1 \\ 0 \end{bmatrix} = 10.01 > 10.$$

One can also try various *heuristics*. These are algorithms without the usual performance and convergence guarantees of other optimization algorithms; see, e.g., [34] for an overview. An example of a heuristic is the *greedy algorithm*: compute the ratio c_j/a_j for each j. Select j with the lowest ratio, add the corresponding item to the knapsack, replace α by $\alpha - a_j$, and remove j from the list of items. Repeat this process while α remains positive. The algorithm simply selects the items that in isolation appears best. The ratio c_j/a_j is sometimes refer to as "bang-for-buck." For the numerical instance above, we obtain

$$c_1/a_1 = -0.6, \quad c_2/a_2 = -0.4, \quad c_3/a_3 = -0.4.$$

Thus, the greedy algorithm produces the solution $(1,0,0)$, which is feasible but its value of -3 is suboptimal. □

Example 4.3 (assignment problem). Consider a university staff assigning students to dorm rooms or an engineer assigning machines to jobs on a factory floor. Similarly, a dispatcher has to assign delivery drivers to routes and airlines need to assign crews to flights. The *assignment problem* models these situations using a variable x_{ij}, which is given the value 1 if agent (student/machine/driver/crew) i is assigned task (room/job/route/flight) j. The cost or disutility of assigning agent i to task j is c_{ij}. The problem can be formulated as follows:

$$\begin{array}{lll}
\underset{x}{\text{minimize}} & \sum_{i=1}^{n}\sum_{j=1}^{n} c_{ij}x_{ij} & \\
\text{subject to} & \sum_{j=1}^{n} x_{ij} = 1, & i = 1,\dots,n \\
& \sum_{i=1}^{n} x_{ij} = 1, & j = 1,\dots,n \\
& x_{ij} \in \{0,1\}, & i = 1,\dots,n, \;\; j = 1,\dots,n
\end{array}$$

The first set of constraints assigns each agent to one task, while the second one ensures that each task is covered by one agent. As in the case of Example 4.2, there's a finite number of possibilities but it tends to be even larger since the number of variables grows as n^2. □

We hint to the many conditions that can be modeled using integer variables and *linear* equality and inequality constraints in the next abstract example. The focus on linearity is primarily motivated by the fact that such constraints can be handled with relative ease; see Chapter 2 and the sections below.

Example 4.4 (logical constraints). The condition that a binary variable z should represent an "inclusive or" of the two binary variables y and x can be written using the constraints

$$x \le z, \quad y \le z, \quad x + y \ge z. \tag{4.4}$$

Likewise, z being an "exclusive or" of y and x is ensured by

$$z \le x + y, \quad z \ge x - y, \quad z \ge -x + y, \quad z \le 2 - x - y. \tag{4.5}$$

Detail. An "inclusive or" means that $x = 1$, or $y = 1$, or $x = y = 1$ if and only if $z = 1$. We can verify (4.4) by checking the different cases. If $x = 1$, then the first constraint in (4.4) ensures that $z = 1$; recall that all the variables are binary. If $y = 1$, then the second constraint does the job. If both x and y are 1, then again $z = 1$. For the converse, if $z = 1$, then x or y or both are 1.

An "exclusive or" means that $x = 1$, or $y = 1$, but not $x = y = 1$, if and only if $z = 1$. We see that (4.5) is correct by checking the various cases. □

We end this series of examples by a situation that's closer to real-world applications. As the applications become more complex, it's important to be disciplined when we formulate a problem. The next example introduces a set of conventions that helps us write clear, human-readable formulations.

Example 4.5 (gasoline loading). A gasoline truck has C compartments that can be filled with P gasoline products. To avoid mixing, we can't fill the same compartment with two different products. For each product $p = 1, \dots, P$, there's a particular order quantity "order_p," measured in gallons, that we seek to satisfy. However, we allow a shortfall of up to "maxshort_p." The capacity of compartment c is "cap_c," $c = 1, \dots, C$. We would like to determine the loading plan that minimizes the shortfall, i.e., the amount of ordered product that we fail to load. We refer to [8], from where this example is taken, for further details.

We write down a model for this problem using *standard form of notation*, which attempts to clarify each component of an optimization model. It recommends more descriptive names than i, j or x, y to enhance readability and understanding; see [35] for a comprehensive discussion.

Detail. The standard notation organizes the model formulation into six parts: indices, sets, data, variables, objective function, and constraints:

Indices

c	compartment	(~ 10)
p	product	(~ 5)

Sets

$\mathcal{C}$	set of compartments	$\mathcal{C} = \{1, \dots, C\}$
$\mathcal{P}$	set of products	$\mathcal{P} = \{1, \dots, P\}$

Data

order_p	order quantity of product p	(gallons)
maxUnder_p	max underfilling of product p	(gallons)
cap_c	capacity of compartment c	(gallons)

Variables

QUANT_{pc}	quantity of product p in compartment c	(gallons)
FILL_{pc}	1 if product p is in compartment c; 0 otherwise	
SHORT_p	underfilling of product p	(gallons)

Objective function

$$\text{minimize} \sum_{p \in \mathcal{P}} \text{SHORT}_p$$

Constraints

$$\sum_{p \in \mathcal{P}} \text{FILL}_{pc} \leq 1 \qquad \forall c \in C \tag{4.6}$$

$$\text{SHORT}_p + \sum_{c \in C} \text{QUANT}_{pc} = \text{order}_p \qquad \forall p \in \mathcal{P} \tag{4.7}$$

$$\text{QUANT}_{pc} \leq \text{cap}_c \, \text{FILL}_{pc} \qquad \forall p \in \mathcal{P}, c \in C \tag{4.8}$$

$$\text{SHORT}_p \leq \text{maxUnder}_p \qquad \forall p \in \mathcal{P} \tag{4.9}$$

$$\text{QUANT}_{pc} \geq 0 \qquad \forall p \in \mathcal{P}, c \in C$$

$$\text{SHORT}_p \geq 0 \qquad \forall p \in \mathcal{P}$$

$$\text{FILL}_{pc} \in \{0, 1\} \qquad \forall p \in \mathcal{P}, c \in C.$$

We also recommend including a description of each nontrivial constraint. Constraints (4.6) ensure that at most one product is filled in each compartment. Constraints (4.7) compute the underfilling of each product. Constraints (4.8) prevent overfilling a compartment. Constraints (4.9) ensure that the shortfall of each product is within the allowable range.

The standard notation conveys some helpful details. The list of indices includes typical cardinalities ($\sim 10, \sim 5$). They immediately indicate to the reader the model size. The data and variables are specified with units, which help verify constraints (left-hand side and right-hand side should have the same units). The data is given with descriptive names and the same holds for the variables. To distinguish between data and variables, we use capital letters for the latter. □

4.B Branch-and-Bound Algorithm

To solve a mixed-integer optimization problem, we've always the following conceptual algorithm available to us: remove any integer restrictions but bound the variables if possible. For example, let

$$x \in \{0, 1\} \text{ be replaced by } x \in [0, 1]$$
$$x \in \{0, 1, 2, \dots\} \text{ be replaced by } x \in [0, \infty)$$
$$x \in \{\dots, -1, 0, 1, \dots\} \text{ be replaced by } x \in \mathbb{R}.$$

Removing integer constraints in this manner produces a *relaxation*, with the resulting problem having minimum value no greater than the minimum value of the actual problem. Solving the relaxation can already be informative as it produces an optimistic lower bound on the minimum value of the actual problem. Presumably, the relaxation is simpler to solve because the nonconvex integer constraints are replaced by convex constraints. However, a minimizer of the relaxation will typically fail to satisfy the integer restrictions.

Next, one can start trying different integer values for the variables restricted by integer constraints. When those integer variables have been fixed to some feasible value, we can optimize the remaining continuous variables. This *restriction* has fewer variables than the actual problem, and no integer constraints. Any minimizer for the restriction is feasible for the actual problem and produces an upper bound on the minimum value of the actual problem. This bound combined with the one from the relaxation bracket the minimum value of the actual problem. If the bracket is sufficiently narrow, then we may be satisfied with the feasible decision we've obtained. If we aren't, we can try fixing the integer variables to some other values, which results in another restriction we can solve to obtain another upper bound.

If the integer variables are bounded, then this process will eventually produce the minimizers of the actual problem since there are only a finite number of possible values for the integer variables. However, this isn't a practical approach. If $x \in \{0, 1\}^n$, then there are 2^n possible values for x. It quickly becomes impossible to cycle through all the possibilities.

The *branch-and-bound algorithm* implements the ideas of relaxations and restrictions with a small but important adjustment. It improves the lower bound by fixing some of the integer variables and relaxing the other ones. This produces bounds that can be used to *prune* the number of problems that needs to be solved. The process of constructing different problems through such fixing and relaxation is referred to as *partitioning*. It results in a collection of subproblems to be solved.

Example 4.6 (partitioning). Consider the integer linear optimization problem

$$\underset{x \in C}{\text{minimize}} \ -x_1 + 3x_2 \quad \text{subject to} \quad x_1,\ x_2 \text{ integers}$$
$$\text{where } C = \{x \mid x_1 + 4x_2 \le 9.9,\ 2x_1 + x_2 \le 7.8,\ x_1 \ge 0,\ x_2 \ge 0\}.$$

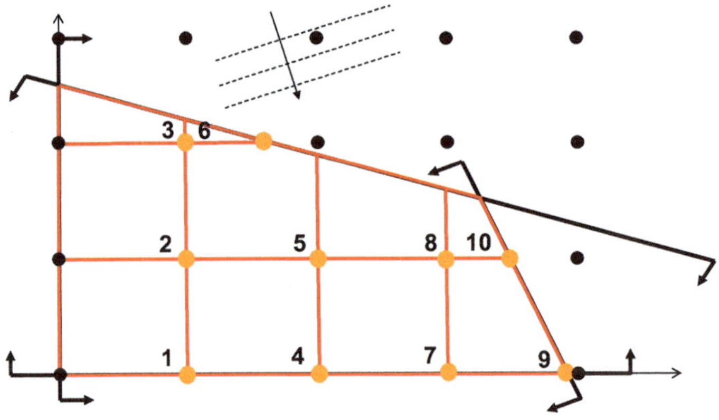

Fig. 4.1: Full partitioning of the polyhedral set C in Example 4.6. Dots represent integer points. Dashed lines indicate contours of the objective function.

Figure 4.1 shows the polyhedral set C partitioned into 10 different subareas. Subarea 1 in the figure restricts x to $[0,1]^2$. Subarea 2 is $[0,1] \times [1,2]$. Not all areas are squares; subareas 3, 8, and 9 are trapezoids, subarea 6 is a triangle, and subarea 10 is a general polyhedron. Let C_i be subarea i, m_i be the minimum value over subarea i and $\hat{x}^i$ be the corresponding minimizer marked with orange dots in Figure 4.1, i.e.,

$$m_i = \min_{x \in C_i} -x_1 + 3x_2 = -\hat{x}_1^i + 3\hat{x}_2^i.$$

The minimum value of the actual problem is no smaller than $\min_{i=1,\dots,10} m_i$. While this seems silly as one obtains the same lower bound on the minimum value of the actual problem by minimizing $-x_1 + 3x_2$ over C in one shot. However, we can do better by ignoring some partitions. For instance, one could ignore partitions 4, 5, and 6 and only minimize over the other ones. We can't "miss" any minimizer of the actual problem by doing so; there are none with $x_1 \in (1,2)$; see Figure 4.2. Likewise, we may ignore subareas 2, 5, 8, and 10 as indicated in Figure 4.3.

We conclude that $\min_{i=1,3,7,9} m_i$ is a valid lower bound on the minimum value of the actual problem. While in the present numerical example this lower bound is identical to $\min_{x \in C} -x_1 + 3x_2$ because the smallest value is in the bottom right corner, other instances might be different. A clever partitioning may strengthen the lower bound on the minimum value of the actual problem. □

The branch-and-bound algorithm systematically partitions $\mathbb{R}^n$, keeping track of the various subproblems that have been solved and that need to be solved. Instead of a formal algorithm, we discuss the main steps in an example.

Example 4.7 (branch-and-bound algorithm). We return to the integer linear optimization problem in Example 4.6 and systematically implement the ideas laid out there.

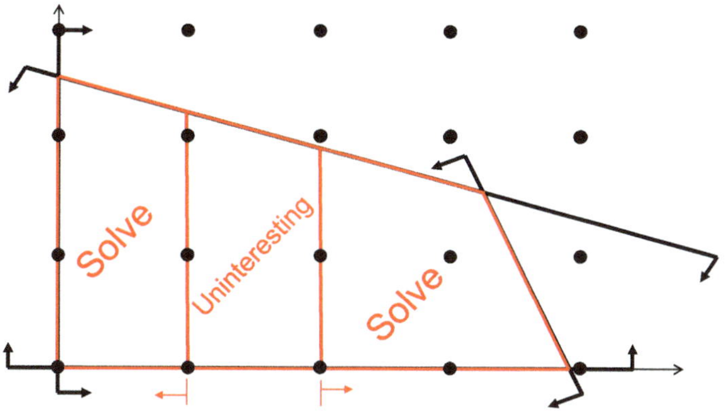

Fig. 4.2: Partitioning of the polyhedral set C in Example 4.6 while identifying an uninteresting portion.

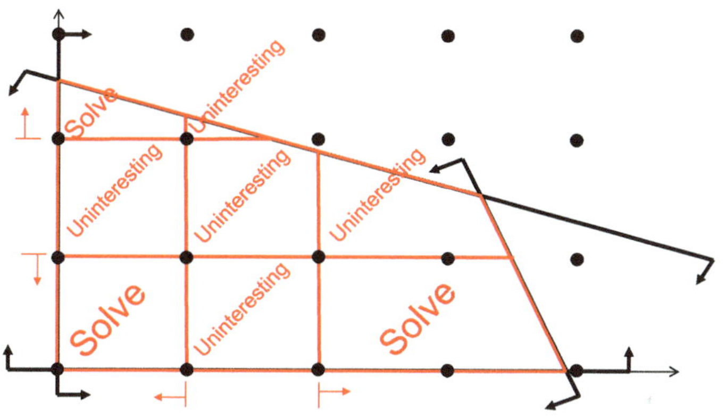

Fig. 4.3: Partitioning of the polyhedral set C in Example 4.6 while identifying uninteresting portions.

Step A. Root node relaxation. We start by minimizing $-x_1 + 3x_2$ subject to $x \in C$, which is referred to as the *root node* relaxation. The minimizer is $(3.9, 0)$ in the lower right corner of C in Figure 4.1, with value $-3.9 + 3 \cdot 0 = -3.9$. This value is a lower bound on the minimum value of the actual problem. Since the minimizer isn't feasible in the actual problem, it doesn't furnish an upper bound on the minimum value of the actual problem.

Step B. Branching. We next partition C into two subareas as indicated in Figure 4.2. This produces the two subproblems

$$\operatorname*{minimize}_{x \in C} \{-x_1 + 3x_2 \mid x_1 \leq 1\} \quad \text{and} \quad \operatorname*{minimize}_{x \in C} \{-x_1 + 3x_2 \mid x_1 \geq 2\}.$$

The first one has minimizer $(1, 0)$ with value $-1 + 3 \cdot 0 = -1$ and the second one has the same minimizer and minimum value as at the root note. Since the first minimizer $(1, 0)$ is feasible in the actual problem, its value furnishes an upper bound on the minimum value of the actual problem. We've now bracketed the minimum value of the actual problem between −3.9 and −1. The optimality gap is $-1-(-3.9) = 2.9$. Equally important, there's no need to partition further the left-most subarea of Figure 4.2 because all feasible points with $x_1 \leq 1$ in the actual problem can't be better.

Step C. Branching on the right. Under the restriction $x_1 \geq 2$ already in place, we may partition "vertically" by considering

$$\underset{x\in C}{\text{minimize}}\,\{-x_1+3x_2 \mid x_1 \geq 2, x_2 \leq 1\} \quad \text{and} \quad \underset{x\in C}{\text{minimize}}\,\{-x_1+3x_2 \mid x_1 \geq 2, x_2 \geq 2\}.$$

The first problem has the same minimizer as the root node, while the second one is infeasible. The discovery of an infeasible subproblem allows us to conclude that there's no need to partition its (empty) feasible set any further. The lowest minimum value across each subarea furnishes a lower bound on the minimum value of the actual problem: $\min\{-1, -3.9, \infty\} = -3.9$, which is the same lower bound as from the root node relaxation. The optimality gap remains unchanged.

Step D. Branching further. We consider further partitioning and the problems

$$\underset{x\in C}{\text{minimize}}\,\{-x_1+3x_2 \mid x_1 \in [2,3], x_2 \leq 1\} \quad \text{and} \quad \underset{x\in C}{\text{minimize}}\,\{-x_1+3x_2 \mid x_1 \geq 4, x_2 \leq 1\}.$$

The first one has minimizer $(3,0)$ and minimum value $-3+3\cdot 0 = -3$, which is feasible in the actual problem. Thus, $(3,0)$ is a new incumbent that furnishes an upper bound of −3. The gap improves to $-3-(-3.9) = 0.9$. The second problem is infeasible. Looking across all subproblems, we see that the lower bound improves to $\min\{-1, -3, \infty, \infty\} = -3$. The optimality gap is zero and the algorithm terminates.

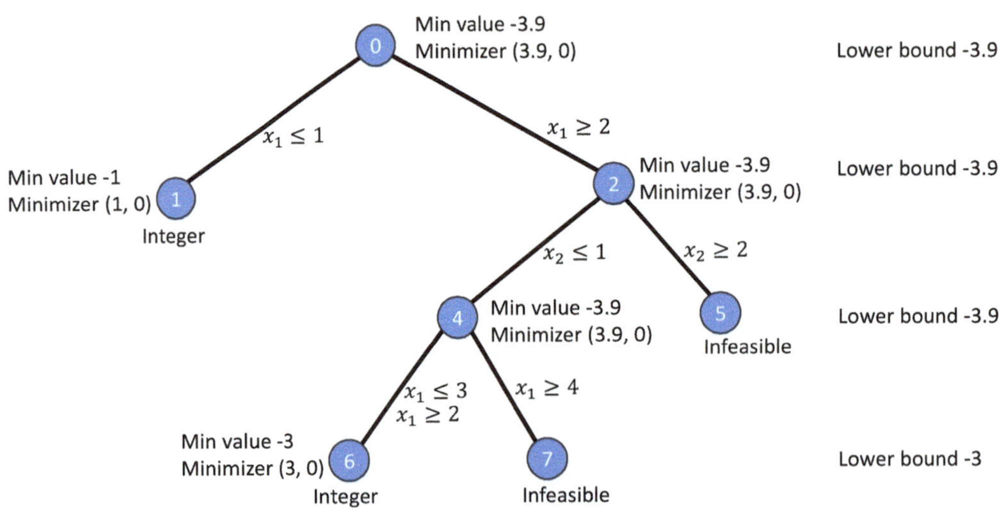

Fig. 4.4: Branch-and-bound tree for Example 4.7.

Figure 4.4 visualizes the computations we've done as a *branch-and-bound tree*. Whenever a node in the tree produces a feasible solution of the actual problem or the subproblem is infeasible, there's no need for branching further below that node. We experience this at nodes 1, 5, 6, and 7. There's one more possibility: if the subproblem at a node produces a minimum value that's greater or equal to the best upper bound seen thus far, then again there's no need for branching further below that node because all minimum values for the corresponding subproblems can't improve on the upper bound. They all involve further restrictions.

We compute lower bounds on the minimum value of the actual problem by looking "across" the branch-and-bound three, which then corresponds to a particular partitioning of the feasible set. The root node (node 0) always produces a lower bound. In this example, the value is -3.9. Nodes 1 and 2 define a partitioning and produces the lower bound $\min\{-1, -3.9\}$. Nodes 1, 4, and 5 define another partitioning, with a lower bound of $\min\{-1, -3.9, \infty\}$. Nodes 1, 6, 7, and 5 define yet another partitioning and give the lower bound of $\min\{-1, -3, \infty, \infty\}$.

There several decisions that need to be made during the algorithm. How should we partition the feasible set of the actual problem? This is referred to as the *branching strategy* and determines the subproblems that we'll encounter. Then there are choices about the order in which to solve the subproblems. One could proceed down in the branch-and-bound tree, which may lead to a good upper bound, or one could attempt to complete a "level" in the tree in an attempt to improve the lower bound. The computing time for the branch-and-bound algorithm tends to vary a lot among seemingly similar problem instances, in part, because there's no universally good answer to these questions. Regardless, the algorithm is usually employed with a positive tolerance so that it stops when the difference between the upper and lower bounds is sufficiently small; see [20] for many details. □

4.C Cutting Planes

Integer constraints introduce nonconvexity, with significant computational implications when solved using the branch-and-bound algorithm. We may need to solve an exponentially large number of subproblems. However, there's another possibility leveraging a fundamental theorem. It relies on the notion of a convex hull.

Definition 4.8 (convex hull). *The* convex hull *of a set $C \subset \mathbb{R}^n$ is another set*

$$\operatorname{con} C = \Big\{ \sum\nolimits_{i=1}^m \lambda_i x^i \;\Big|\; \sum\nolimits_{i=1}^m \lambda_i = 1,\ \lambda_i \geq 0,\ x^i \in C,\ \textit{for all } i = 1, \dots, m \;\textit{and all natural numbers } m \Big\}.$$

□

The convex hull $\operatorname{con} C$ is convex regardless of the convexity of C because it involves "averaging" of points in C. In fact, $\operatorname{con} C = C$ if C is convex. Figure 4.5 illustrates two possibilities. Informally, $\operatorname{con} C$ is the smallest convex set that contains C and thus amounts to "filling in" any holes or inward cusps in C.

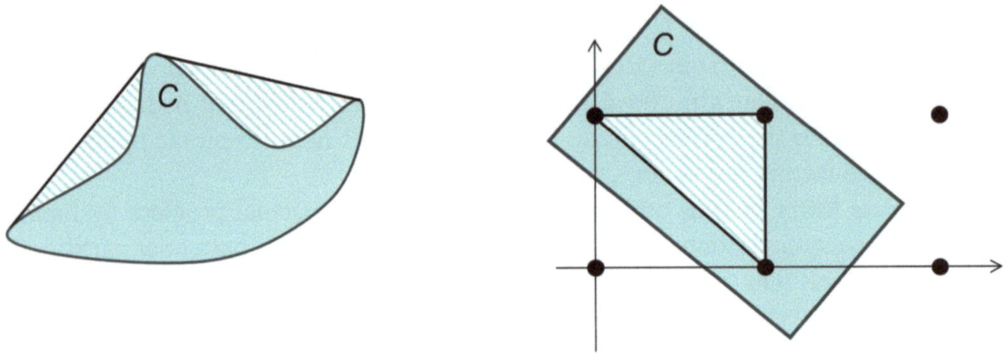

Fig. 4.5: Left: A nonconvex set C (solid color) and its convex hull. Right: A polyhedral set C in the form of a rectangle and the convex hull of the intersection $C \cap \{0, 1\}^2$, which forms a triangle.

Using the notion of a convex hull, we can reformulate any nonconvex optimization problem into a convex one and this enables an alternative solution approach for integer constrained optimization problems.

Theorem 4.9 (reduction to convex optimization). *For any set $C \subset \mathbb{R}^n$ and n-dimensional vector c, one has*

$$\inf\{c^\top x \mid x \in C\} = \inf\{c^\top x \mid x \in \operatorname{con} C\}.$$

Proof. Since $C \subset \operatorname{con} C$, the relation holds with $\geq$ trivially. To establish the relation with $\leq$, we consider two cases. First, suppose that $\operatorname{con} C = \emptyset$. Then, the right-hand side is ∞ and the relation holds with equality. Second, suppose that $\operatorname{con} C \neq \emptyset$. Let $\bar{x} \in \operatorname{con} C$ be arbitrary. *Caratheodory's theorem* states that there exist $\lambda_0, \lambda_1, \ldots, \lambda_n \geq 0$ and $x^0, x^1, \ldots, x^n \in C$ satisfying

$$\sum_{j=0}^{n} \lambda_j = 1 \quad \text{and} \quad \sum_{j=1}^{n} \lambda_j x^j = \bar{x}.$$

Thus, the objective function value

$$c^\top \bar{x} = c^\top \sum_{j=0}^{n} \lambda_j x^j = \sum_{j=0}^{n} \lambda_j c^\top x^j \geq \sum_{j=0}^{n} \lambda_j \inf\{c^\top x \mid x \in C\} = \inf\{c^\top x \mid x \in C\}.$$

Since $\bar{x} \in \operatorname{con} C$ is arbitrary, we also have that

$$\inf\{c^\top x \mid x \in \operatorname{con} C\} \geq \inf\{c^\top x \mid x \in C\}.$$

This completes the proof. □

The theorem gives rise to a promising direction. We can solve the convex problem of minimizing $c^\top x$ subject to $x \in \operatorname{con} C$ and obtain a minimizer $x^\star$ and the minimum value

μ. Then, μ is also the minimum value of the actual problem over just C by Theorem 4.9. If $x^\star \notin C$, then we've reduced the space of possible solutions because we now can limit the search to $x \in C$ with $c^\top x = \mu$.

In the important special case when C is the intersection of a polyhedral set and the integer points $\{0, 1, 2, \dots\}^n$, then minimizing $c^\top x$ subject to $x \in \operatorname{con} C$ is a linear optimization problem because $\operatorname{con} C$ is also polyhedral. This has the additional advantage that one can use the revised simplex method (cf. Section 2.E) and leverage a convenient property:

> The revised simplex method computes a minimizer of $c^\top x$ over $x \in \operatorname{con} C$ that's an extreme point of $\operatorname{con} C$ and every such extreme point must be in C.

Consequently, in this case, optimization over $\operatorname{con} C$ using the revised simplex method immediately produces the minimum value *and* a minimizer of the actual problem over C.

The focus on a linear objective function in Theorem 4.9 isn't restrictive because of a general fact. For any function f in n dimensions and $C \subset \mathbb{R}^n$, one has

$$x^\star \in \operatorname{argmin}_{x \in C} f(x) \quad \text{if and only if} \quad (x^\star, \gamma) \in \operatorname{argmin}_{x,\gamma} \big\{ \gamma \,\big|\, f(x) \leq \gamma,\ x \in C \big\}.$$

Thus, we can always reformulate a problem into one with a linear objective function by introducing an additional variable.

The main practical challenge with Theorem 4.9 is that we typically don't have a convenient description of $\operatorname{con} C$. Nevertheless, the theorem indicates that it's worthwhile to spend computational effort to determine $\operatorname{con} C$, at least partially. Current state-of-the-art algorithms for solving mixed-integer optimization problems spend at least some time on estimating the convex hull of a feasible set and combines it with the branch-and-bound algorithm. While it's rarely computationally viable to construct the convex hull fully, solvers generate collections of inequalities at various stages of the branch-and-bound tree that hopefully improve lower bounds even if they don't fully recover the convex hull; see, e.g., [20].

The goal of these additional inequalities is to *strengthen* a formulation of a mixed-integer optimization problem in the following sense. If

$$C = \big\{ (x, y) \in S \,\big|\, x \in \mathbb{R}^n, y \in \{0, 1, 2, \dots\}^m \big\}$$

is the feasible set of the problem, then we say that

$$C' = \big\{ (x, y) \in S' \,\big|\, x \in \mathbb{R}^n, y \in \{0, 1, 2, \dots\}^m \big\}$$

provides a *stronger formulation* of C if $C = C'$ and

$$\big\{ (x, y) \in S' \,\big|\, x \in \mathbb{R}^n, y \geq 0 \big\} \text{ is a strict subset of } \big\{ (x, y) \in S \,\big|\, x \in \mathbb{R}^n, y \geq 0 \big\}.$$

This means that while C is identical to C', the relaxation of C' is better than that of C.

Figure 4.6 illustrates the situation with a stronger formulation in the right portion of the figure obtained by adding the inequalities in red. We observe that the longer red line can't be moved further down without excluding the feasible point $(1, 2)$.

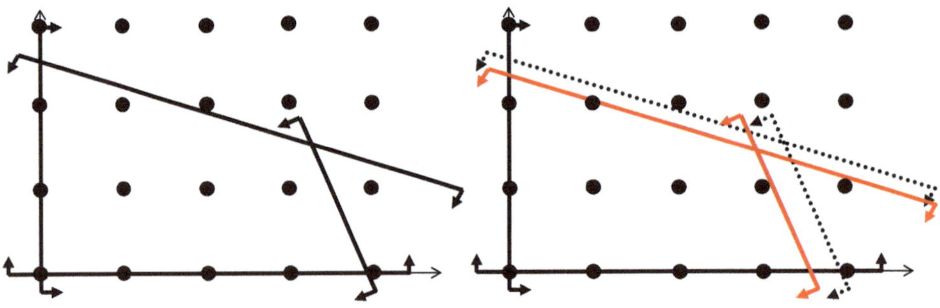

Fig. 4.6: Left: a feasible set is an intersection of a polyhedral set and 10 integer points. Right: a stronger formulation with a smaller polyhedral set obtained by adding two inequalities (red lines).

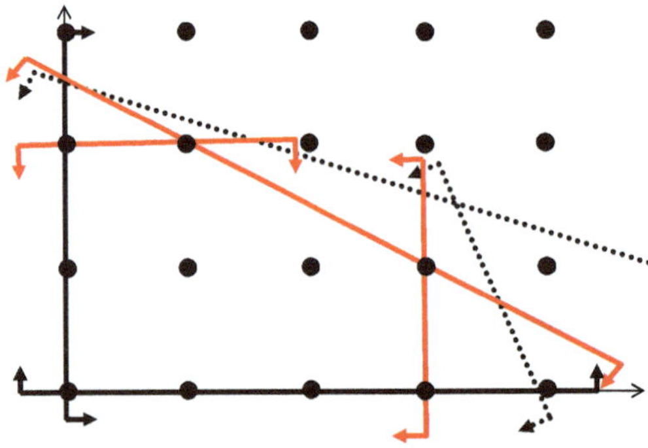

Fig. 4.7: Strongest possible formulation of the feasible set in Figure 4.6(left).

The convex hull of a feasible set furnishes the strongest possible formulation, but it's rarely readily available. In the context of Figure 4.6, the addition of three carefully selected inequalities defines the convex hull of the feasible set; see Figure 4.7. While these inequalities are easily identified here, the computational cost to do the same in higher dimensions is significant.

Example 4.10 (big-M formulation). Consider a situation with a continuous variable $x \in [0, \alpha]$ and a binary variable $y \in \{0, 1\}$. If $y = 0$, then we would like to have $x = 0$. If $y = 1$, then x can be any number in $[0, \alpha]$. These requirements are fully captured by

$$x \leq My$$

regardless of the choice of $M \geq \alpha$. Thus, one might be led to believe that any such number will do the job equally well.

Figure 4.8 shows that $M = \alpha$ produces a stronger formulation than $M > \alpha$. The reason is that the triangle defined by $x \geq 0$, $y \in [0, 1]$, and $x \leq \alpha y$ is the convex hull of the feasible set in this case. (The feasible set is the union of origin and the line segment $[0, \alpha] \times \{1\}$, and is marked with red in Figure 4.8.) With $M > \alpha$, the conditions $x \geq 0$, $y \in [0, 1]$, and $x \leq My$ define a larger trapezoid. When formulating mixed-integer optimization problems, one would like to avoid "big-M" of this kind, especially in complex models where it's hard to determine suitably low values of M. □

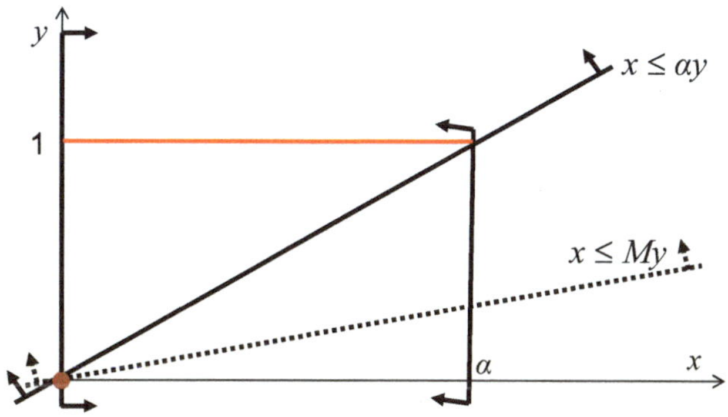

Fig. 4.8: Big-M formulation using M much larger than α.

4.D Network Problems

Many practical problems involve networks of vertices and edges with various data. Finding a shortest path between two locations in a road network falls in this category; see Example 2.1. The transportation problem (Example 2.5) and assignment problem (Example 4.3) can also be viewed as involving networks, but of a simpler kind.

Example 2.1 provides insight about the inherent benefits derived from a network structure. The vertex-edge incidence matrix A and the vector b consist only of the numbers -1, 0, and 1. They define the constraint $Ax = b$, which enforces flow-balance at each vertex. Since the example seeks a shortest path, it may seem necessary to also enforce the constraint $x \in \{0, 1\}^n$ so that each decision variable x_e equals 1 or 0, i.e., edge e is used or not in the path. However, the model in Example 2.1 merely imposes the constraint $x \in [0, 1]^n$. The reason we can

replace the nonconvex set $\{0, 1\}^n$ by the convex set $[0, 1]^n$

and obtain a linear optimization problem stems from the special structure of $Ax = b$ and the linear objective function $c^\top x$, with nonnegative coefficient vector c. In this case, there's no incentive to have $x_e > 0$ for an edge e not on a shortest path. Even if $c_e = 0$, a positive flow x_e will have to be traced back to the start vertex s and forward to the terminal vertex t due to the flow-balance constraints $Ax = b$. This additional flow adds a positive quantity to the objective function. Consequently, it can't be part of a minimizer. (Here, we assume the problem is nontrivial with a positive-length shortest path.) This doesn't mean, however, that every minimizer of the model in Example 2.1 is a binary vector.

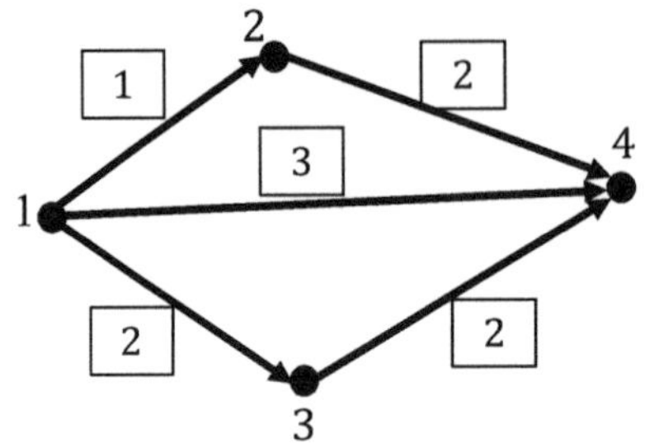

Fig. 4.9: A network with two optimal paths (1-2-4, 1-4) and one suboptimal path (1-3-4) between vertex 1 and vertex 4. Edge lengths are given inside boxes.

Figure 4.9 shows a network with three paths. Paths 1-2-4 and 1-4 are optimal, with a shortest path length of 3. Path 1-3-4 is suboptimal with length 4. Thus,

$$x = (x_{12}, x_{13}, x_{14}, x_{24}, x_{34}) = (1, 0, 0, 1, 0)$$

is a minimizer in the model from Example 2.1. Likewise, $(0, 0, 1, 0, 0)$ is a minimizer. But the model has other minimizers too. The flow-balance constraints $Ax = b$ amount to

$$\begin{aligned} x_{12} + x_{13} + x_{14} &= 1 \\ -x_{12} + x_{24} &= 0 \\ -x_{13} + x_{34} &= 0 \\ -x_{14} - x_{24} - x_{34} &= -1. \end{aligned}$$

These constraints are satisfied for any

$$\bar{x} = (\alpha,\ 0,\ 1 - \alpha,\ \alpha,\ 0), \quad \text{with } \alpha \in [0, 1].$$

The objective function value of $\bar{x}$ is

$$c^\top \bar{x} = 1\alpha + 2 \cdot 0 + 3(1 - \alpha) + 2\alpha + 2 \cdot 0 = 3,$$

which means that $\bar{x}$ is also a minimizer.

These observations hold in general. If $x^\star$ and $x^{\star\star}$ are minimizers in the linear optimization problem from Example 2.1 and they're binary vectors, then the new point

$$x^\lambda = (1 - \lambda)x^\star + \lambda x^{\star\star}, \quad \text{with } \lambda \in (0, 1),$$

is also a minimizer because the feasibility checks

$$Ax^\lambda = (1-\lambda)Ax^\star + \lambda Ax^{\star\star} = (1-\lambda)b + \lambda b = b$$
$$x_j^\lambda \geq (1-\lambda)0 + \lambda 0 = 0, \quad j = 1,\dots,n$$
$$x_j^\lambda \leq (1-\lambda)1 + \lambda 1 = 1, \quad j = 1,\dots,n$$

are satisfied and the objective function value

$$c^\top x^\lambda = (1-\lambda)c^\top x^\star + \lambda c^\top x^{\star\star} = (1-\lambda)c^\top x^{\star\star} + \lambda c^\top x^{\star\star} = c^\top x^{\star\star}$$

matches the one for the minimizer $x^{\star\star}$. The model in Example 2.1 therefore allows for splitting of flow among shortest paths, but this doesn't prevent us from identifying one or more shortest paths by examining a minimizer. While $x^\star$ and $x^{\star\star}$ are extreme points of $C = \{x \in [0,1]^n \mid Ax = b\}$, the new point x^λ isn't one by Definition 2.28. This means that the revised simplex method can't return x^λ. In fact, it returns a binary vector $x \in \{0,1\}^n$ even though C doesn't include binary restrictions. The special constraints caused by the network structure of the problem—the matrix A is so-called *totally unimodular*—result in binary solutions automatically.

We extend the discussion of network problems by considering two additional examples.

Example 4.11 (maximum flow). In a directed graph (V, E) with vertex set V and edge set E, there are two special vertices $s \neq t$: s is the source vertex and t is the sink vertex. A network operator would like to maximize the flow of a commodity from s to t across the edges, which have finite capacities, and this produces the *maximum-flow problem*. Each edge $e \in E$ has a capacity $u_e > 0$. We also define the outgoing and incoming edges for vertex i as

$$E_i^+ = \{(i,j) \in E \mid j \in V\} \quad \text{and} \quad E_i^- = \{(j,i) \in E \mid j \in V\}, \qquad (4.10)$$

respectively. The variable y_e denotes the flow on edge $e \in E$ and y_a the flow on an artificial "return edge" $a = (t, s) \notin E$; see Figure 4.10.

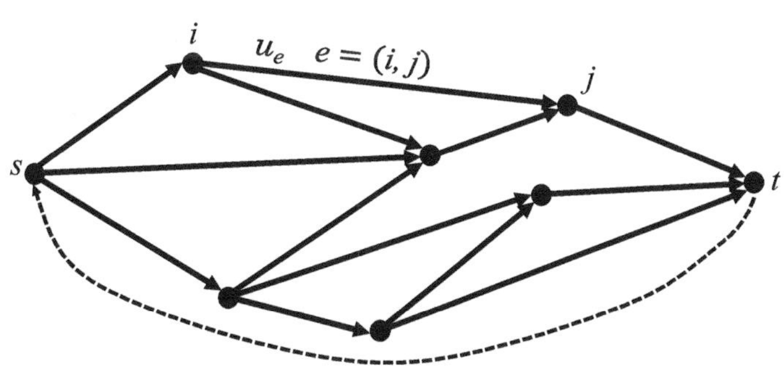

Fig. 4.10: Network with edge capacities for the maximum-flow problem, including the return arc $a = (t, s)$.

Let n be the number of edges in E and $y = (y_e, e \in E)$ be the vector of edge flows. Then, the maximum flow through the network is given by the formulation

$$\begin{aligned} &\underset{y_a, y}{\text{maximize}}\ y_a \\ &\text{subject to } \sum\nolimits_{e \in E_i^+} y_e - \sum\nolimits_{e \in E_i^-} y_e = \delta_i y_a \ \ \forall i \in V \\ &\qquad 0 \le y_e \le u_e \ \ \forall e \in E \end{aligned}$$

where $\delta_s = 1$, $\delta_t = -1$, and $\delta_i = 0$ for all $i \in V \setminus \{s, t\}$; see, for example, [1, Chapters 6-7]. The equality constraints ensure flow-balance at each vertex. The inequality $y_e \le u_e$ limits the flow on edge e to its capacity. To recover our usual setting of minimization, we can simply replace the objective function y_a by $-y_a$. □

Example 4.12 (minimum cost flow). In a directed graph (V, E) with vertex set V and edge set E, flow appears at vertex i in the amount of b_i, where positive b_i indicates a source node and negative b_i indicates a sink note. If the flow represents transportation of goods, then source nodes might correspond to warehouses or factories and sink nodes correspond to stores or consumers. Each edge $e \in E$ has a positive capacity u_e. We seek a flow across the network from source nodes to sink nodes at minimum cost specified by the per-unit cost of c_e on edge e. This results in the *minimum-cost flow network problem*, which can be formulated as the linear optimization problem

$$\begin{aligned} &\underset{x}{\text{minimize}} \sum\nolimits_{e \in E} c_e x_e \\ &\text{subject to } \sum\nolimits_{e \in E_i^+} x_e - \sum\nolimits_{e \in E_i^-} x_e = b_i \ \ \forall i \in V \\ &\qquad 0 \le x_e \le u_e \ \ \forall e \in E \end{aligned}$$

Example 4.11 defines the sets E_i^+ and E_i^- in (4.10). This problem arises in production and distribution planning and generalizes the shortest-path problem from Example 2.1 and the maximum-flow problem from Example 4.11. □

The network problems in Examples 2.1, 4.11, and 4.12 are well-structured linear optimization problems. The simplex method of Section 2.E can be specialized for these cases to produce the *network simplex method* [7, Section 7.3], with a potential for computational savings.

4.E Dijkstra's Algorithm

Optimization problems involving networks can often be solved much more efficiently by specialized algorithms than using linear optimization formulations and the revised simplex method, interior points methods, or even the network simplex method. *Dijkstra's algorithm* is one such specialized algorithm for the shortest-path problem. Let's recall the setting of Example 2.1: A directed graph $G = (V, E)$ consists of a vertex set V and an edge set E; $s \in V$ is the start vertex. Each edge $(i, j) \in E$ possesses length $c_{ij} \in [0, \infty)$. Let m be the number of vertices and n the number of edges.

With E_i^+ again denoting the edges going out from vertex i, i.e.,

$$E_i^+ = \{(i, j) \in E \mid j \in V\},$$

we obtain the following algorithm, which doesn't rely on the linear optimization formulation in Example 2.1.

Dijkstra's Algorithm.

Step 0. Set $S = \emptyset, \bar{S} = V$; $d(i) = \infty$ for all $i \in V$, $d(s) = 0$; $P(s) = 0$; $\nu = 1$.
Step 1. If $\nu = m$, then Stop.
Step 2. Select vertex

$$i \in \operatorname{argmin}_{j \in \bar{S}} d(j).$$

Step 3. Replace S by $S \cup \{i\}$ and $\bar{S}$ by $\bar{S} \setminus \{i\}$.
Step 4. For each $(i, j) \in E_i^+$ with

$$d(j) > d(i) + c_{ij},$$

set $d(j) = d(i) + c_{ij}$ and $P(j) = i$.
Step 5. Replace ν by $\nu + 1$ and go to Step 1.

Since the number of iterations is m and the intermediate calculations also scale linearly with m, the runtime complexity of the algorithm is proportional to m^2. For further details and explanation of why the algorithm indeed terminates with a shortest path from s to every other vertex, we refer to Sections 4.6-8 in [1]. We limit the discussion to an example.

Example 4.13 (shortest paths using Dijkstra's). We seek to find the shortest path from vertex 1 to vertex in 8 in the network of Figure 4.11. The edge lengths are given by the boxed numbers in the figure. From Example 2.1, we know the shortest path uses vertices 1-3-6-8 and its length is 7 because the network coincides with that from Figure 2.1. Now, let's use Dijkstra's algorithm to obtain this shortest path in a scalable manner.

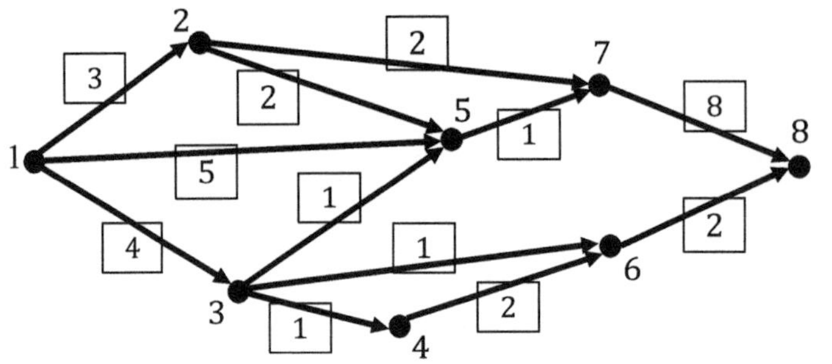

Fig. 4.11: The network from Example 2.1 for which we seek the shortest path from vertex 1 to vertex 8. Boxed numbers give edge lengths.

Detail. We initialize the algorithm in Step 0 by setting $S = \emptyset$, $\bar{S} = \{1, \ldots, 8\}$, $d(i) = \infty$, $i = 2, \ldots, 8$, $d(1) = 0$, $P(1) = 0$, and $\nu = 1$. We can't stop in Step 1 because $\nu < 8$. In Step 2, we select vertex $1 \in \mathrm{argmin}_{j \in \bar{S}}\, d(j)$. Step 3 updates $S = \{1\}$ and $\bar{S} = \{2, \ldots, 8\}$. Step 4 considers the three edges $E_1^+ = \{(1,2), (1,5), (1,3)\}$ and checks

$$
\begin{array}{lllll}
j = 2: & d(2) = \infty; & d(1) + c_{12} = 3 & \text{set:} & P(2) = 1,\ d(2) = 3 \\
j = 5: & d(5) = \infty; & d(1) + c_{15} = 5 & \text{set:} & P(5) = 1,\ d(5) = 5 \\
j = 3: & d(3) = \infty; & d(1) + c_{13} = 4 & \text{set:} & P(3) = 1,\ d(3) = 4.
\end{array}
$$

We're finished with the first iteration and set $\nu = 2$.

In Step 2, we select vertex $2 \in \mathrm{argmin}_{j \in \bar{S}}\, d(j)$. Step 3 updates $S = \{1, 2\}$ and $\bar{S} = \{3, \ldots, 8\}$. Step 4 considers the two edges $E_2^+ = \{(2,7), (2,5)\}$ and checks

$$
\begin{array}{lllll}
j = 7: & d(7) = \infty; & d(2) + c_{27} = 5 & \text{set:} & P(7) = 2,\ d(7) = 5 \\
j = 5: & d(5) = 5; & d(2) + c_{25} = 5 & \text{no update.} &
\end{array}
$$

We're finished with the second iteration and set $\nu = 3$.

In Step 2, we select vertex $3 \in \mathrm{argmin}_{j \in \bar{S}}\, d(j)$. Step 3 updates $S = \{1, 2, 3\}$ and $\bar{S} = \{4, \ldots, 8\}$. Step 4 considers the three edges $E_3^+ = \{(3,5), (3,6), (3,4)\}$ and checks

$$
\begin{array}{lllll}
j = 5: & d(5) = 5; & d(3) + c_{35} = 5 & \text{no update} & \\
j = 6: & d(6) = \infty; & d(3) + c_{36} = 5 & \text{set:} & P(6) = 3,\ d(6) = 5 \\
j = 4: & d(4) = \infty; & d(3) + c_{34} = 5 & \text{set:} & P(4) = 3,\ d(4) = 5.
\end{array}
$$

We're finished with the third iteration and set $\nu = 4$.

In Step 2, there are several choices: $d(4) = d(5) = d(6) = d(7) = 5$ and we make the arbitrary choice of vertex $4 \in \mathrm{argmin}_{j \in \bar{S}}\, d(j)$. Step 3 updates $S = \{1, \ldots, 4\}$ and $\bar{S} = \{5, \ldots, 8\}$. Step 4 considers the edge $E_4^+ = \{(4,6)\}$ and checks

$$
j = 6: \quad d(6) = 5; \qquad d(4) + c_{46} = 7 \quad \text{no update.}
$$

We're finished with the fourth iteration and set $\nu = 5$.

Again Step 2 has several choices and we select vertex $5 \in \operatorname{argmin}_{j \in \bar{S}} d(j)$. Step 3 updates $S = \{1, \ldots, 5\}$ and $\bar{S} = \{6, 7, 8\}$. Step 4 considers the edge $E_5^+ = \{(5, 7)\}$ and checks

$$j = 7: \quad d(7) = 5; \qquad d(5) + c_{57} = 6 \quad \text{no update.}$$

We're finished with the fifth iteration and set $\nu = 6$.

In Step 2, we select vertex $6 \in \operatorname{argmin}_{j \in \bar{S}} d(j)$. Step 3 updates $S = \{1, \ldots, 6\}$ and $\bar{S} = \{7, 8\}$. Step 4 considers the edge $E_6^+ = \{(6, 8)\}$ and checks

$$j = 8: \quad d(8) = \infty; \qquad d(6) + c_{68} = 7 \quad \text{set:} \qquad P(8) = 6, \;\; d(8) = 7.$$

We're finished with the sixth iteration and set $\nu = 7$.

In Step 2, we select vertex $7 \in \operatorname{argmin}_{j \in \bar{S}} d(j)$. Step 3 updates $S = \{1, \ldots, 7\}$ and $\bar{S} = \{8\}$. Step 4 considers the edge $E_7^+ = \{(7, 8)\}$ and checks

$$j = 8: \quad d(8) = 7; \qquad d(7) + c_{78} = 13 \quad \text{no update.}$$

We're finished with the seventh iteration and set $\nu = 8$. The next iteration stops in Step 2. Since $d(8) = 7$, the minimum path has length 7. A path furnishing that length is recovered by working backward from vertex 8:

$$P(8) = 6; \quad P(6) = 3; \quad P(3) = 1.$$

An inspection of the various paths (see Example 2.1) confirms that we've obtained the shortest path. □

4.F Stagewise Decision Making and Reinforcement Learning

While Chapter 3 addresses two-stage decision problems, real-world problems typically involve many stages. We attempt to make a good decision today, knowing about the need and opportunity for making future adjustments and decisions. Our objective function involves the sum of today's "cost" and future "costs." If we've complete information, this multi-stage nature is less important and we can simply formulate the problems as a large optimization model involving all the decisions over time. We can do the same in multi-stage problems with a finite number of scenarios and known probabilities; see Section 3.A and Example 3.17 for models involving two stages. However, any such model will be large and probably intractable. There will also be difficulties with capturing incomplete information of more general kinds and situations where our decisions affect the probability of future events.

In this section, we introduce a modeling framework centered on decision making over many stages or time periods. It introduces the concept of a *state*, which records the status at each point in time. We seek a *policy* that for each state prescribes a decision. Our goal is to determine an optimal policy according to some objective function.

Formally, let $t = 0, 1, 2, \ldots, T$ represent time across a discretized planning horizon. For example, $t = 0$ is today, $t = 1$ is tomorrow, etc. Let $\mathcal{S}$ be a finite set of states with $s_t \in \mathcal{S}$ specifying the state at time t of the system under consideration. For example, $\mathcal{S} = \{0, 1, \ldots, S\}$ could be the possible number of items we've in inventory at a warehouse so that $s_t \in \mathcal{S}$ is the inventory at time t.

The set of admissible decisions at time t is $\mathcal{X}_t$, which we also assume is a finite set. For example, $\mathcal{X}_t = \{0, 1\}$ would encode a yes-no decision. (Extensions to continuous decision and state spaces are possible, but involve mathematical details that go beyond our current scope; see, e.g., [4] for further reading.) We also have a description of the system expressed in terms of *transition probabilities*: Given the decision $x_t \in \mathcal{X}_t$,

$$p_{s\bar{s}}(x_t) \text{ is the probability of moving from state } s \in \mathcal{S} \text{ to state } \bar{s} \in \mathcal{S}.$$

Thus, the system is inherently random but we can affect its behavior through the choice of x_t. We refer to this model as a *Markov decision process*.

We capture preferences for different states over others using the *end cost* $g_T(s_T)$ for being in state s_T at time T and the *stagewise cost* $g_t(s_t, x_t)$ for being in state s_t and making decision x_t at time $t = 0, 1, \ldots, T-1$. The *total cost* of being in states $s_0, s_1, \ldots, s_T$ and making decisions $x_0, x_1, \ldots, x_{T-1}$ is then

$$\sum_{t=0}^{T-1} g_t(s_t, x_t) + g_T(s_T).$$

A complication is that we don't directly decide on the state; we can only control the probability of ending up in a particular state. This means that the total cost is random and we shift to minimizing its expected value.

Another complication is that we don't want to decide $x_0, \ldots, x_{T-1}$ *before* time $t = 0$ and then stick to those decisions. Usually, it's better to observe state s_t before settling on the decision at time t. This wait-and-see strategy allows us to *adapt* decisions to the unfolding events. Consequently, we seek a *policy* $\pi = (\pi_0, \ldots, \pi_{T-1})$, with π_t being a mapping that takes a state as input and produces a decision as output. If the state at time t is s_t, then $\pi_t(s_t)$ prescribes an admissible decision from $\mathcal{X}_t$. Assuming that the initial state of the system is s_0, this leads to the *policy optimization problem*

$$\boxed{\operatorname*{minimize}_{\pi \in \Pi} \; g_0\big(s_0, \pi_0(s_0)\big) + \mathbb{E}\Big[\sum_{t=1}^{T-1} g_t\big(\boldsymbol{s}_t, \pi_t(\boldsymbol{s}_t)\big) + g_T(\boldsymbol{s}_T)\Big]} \tag{4.13}$$

where Π is the set of all policies, $\boldsymbol{s}_1, \ldots, \boldsymbol{s}_T$ are random future states (as indicated by the bold face) implicitly given in terms of the transition probabilities, and the policy $\pi = (\pi_0, \ldots, \pi_{T-1})$.

Since $\mathcal{S}$ and $\mathcal{X}_t$ are finite sets, there's a finite but potentially large number of different policies. We can represent a policy in terms of a look-up table with rows and columns corresponding to states and time periods, respectively. The entry in row s and column t of the table would be the decision $x_t = \pi_t(s)$ prescribed for that state and time. The cardinality of $\mathcal{X}_t$ determines the number of possible decisions at time t; the policy optimization problem (4.13) identifies the best one.

We say that a policy $\pi^\star = (\pi_0^\star, \dots, \pi_{T-1}^\star) \in \Pi$ is *optimal* for (4.13) if

$$g_0\big(s_0, \pi_0^\star(s_0)\big) + \mathbb{E}\Big[\sum_{t=1}^{T-1} g_t\big(s_t^\star, \pi_t^\star(s_t^\star)\big) + g_T(s_T^\star)\Big]$$
$$\leq g_0\big(s_0, \pi_0(s_0)\big) + \mathbb{E}\Big[\sum_{t=1}^{T-1} g_t\big(s_t, \pi_t(s_t)\big) + g_T(s_T)\Big] \quad \forall \pi = (\pi_0, \dots, \pi_{T-1}) \in \Pi,$$

where $s_1^\star, \dots, s_T^\star$ are the random states produced by $\pi^\star$ and $s_1, \dots, s_T$ are those produced by π.

Example 4.14 (inventory). Given the planning horizon $t = 0, 1, 2, 3$, we seek an optimal re-stocking policy for a retail store. The store can have up to five items in inventory so the state space is $\mathcal{S} = \{0, 1, \dots, 5\}$. At each time t, the decision is to either re-stock to full capacity ($x_t = 1$) or to do nothing ($x_t = 0$), which means that $\mathcal{X}_t = \{0, 1\}$. If the decision is $x_t = 1$, then the system transitions to state 5 with probability 1:

$$p_{s5}(1) = 1; \quad p_{s\bar{s}}(1) = 0 \quad \text{for } s = 0, 1, \dots, 5, \ \bar{s} = 0, 1, \dots, 4.$$

The other transition probabilities are

$$p_{00}(0) = 1; \qquad \text{and for } s > 0, \ \ p_{s\bar{s}}(0) = \begin{cases} 0.5 & \text{if } \bar{s} = s - 1 \\ 0.5 & \text{if } \bar{s} = s \\ 0 & \text{otherwise.} \end{cases}$$

Thus, under the decision $x_t = 0$, the inventory goes down by 1 with probability 0.5 and stays the same with probability 0.5. If the inventory is zero, then it remains at zero under decision $x_t = 0$.

There's a cost $2s_t$ of having positive inventory s_t, a cost of 25 of having zero inventory, and a cost of 3 to re-stock. This produces the stagewise cost function

$$g_t(s_t, x_t) = \begin{cases} 2s_t + 3x_t & \text{if } s_t > 0 \\ 25 + 3x_t & \text{otherwise.} \end{cases}$$

The initial inventory is $s_0 = 1$ and the end cost is 25 if $s_T = 0$ and zero otherwise.

Detail. We consider two heuristic policies. If we adopt the do-nothing policy $\pi = (\pi_0, \pi_1, \pi_2)$, with $\pi_t(s_t) = 0$ for any s_t, then we face the states

$$
\begin{aligned}
s_0 = 1, s_1 = 1, s_2 = 1, s_3 = 1 &\quad \text{with probability } 0.5^3 \\
s_0 = 1, s_1 = 1, s_2 = 1, s_3 = 0 &\quad \text{with probability } 0.5^3 \\
s_0 = 1, s_1 = 1, s_2 = 0, s_3 = 0 &\quad \text{with probability } 0.5^2 \\
s_0 = 1, s_1 = 0, s_2 = 0, s_3 = 0 &\quad \text{with probability } 0.5^1.
\end{aligned}
$$

The last sequence of states is caused by the inventory dropping to zero for time period 1, which occurs with probability 0.5. Due to the do-nothing policy, the inventory remains at zero with probability 1 for the remaining time periods. The second-to-last sequence is caused by the inventory holding at level 1 for period 1, which occurs with probability 0.5, but then dropping in time period 2, also taking place with probability 0.5. The other rows follow similarly.

The objective function value for this policy becomes

$$
\begin{aligned}
&g_0(s_0, \pi_0(s_0)) + \mathbb{E}\Big[g_1(s_1, \pi_1(s_1)) + g_2(s_2, \pi_2(s_2)) + g_3(s_3)\Big] \\
&= (2 \cdot 1 + 3 \cdot 0) + 0.5^3\big((2 \cdot 1 + 3 \cdot 0) + (2 \cdot 1 + 3 \cdot 0) + 0\big) \\
&+ (2 \cdot 1 + 3 \cdot 0) + 0.5^3\big((2 \cdot 1 + 3 \cdot 0) + (2 \cdot 1 + 3 \cdot 0) + 25\big) \\
&+ (2 \cdot 1 + 3 \cdot 0) + 0.5^2\big((2 \cdot 1 + 3 \cdot 0) + (25 + 3 \cdot 0) + 25\big) \\
&+ (2 \cdot 1 + 3 \cdot 0) + 0.5^1\big((25 + 3 \cdot 0) + (25 + 3 \cdot 0) + 25\big) = 56.625.
\end{aligned}
$$

In contrast, the policy of re-stocking when the inventory reaches 1, i.e., $\pi_t(1) = 1$ and $\pi_t(s_t) = 0$ for $s_t \neq 1$, leads to the following states:

$$
\begin{aligned}
s_0 = 1, s_1 = 5, s_2 = 5, s_3 = 5 &\quad \text{with probability } 0.5^2 \\
s_0 = 1, s_1 = 5, s_2 = 5, s_3 = 4 &\quad \text{with probability } 0.5^2 \\
s_0 = 1, s_1 = 5, s_2 = 4, s_3 = 4 &\quad \text{with probability } 0.5^2 \\
s_0 = 1, s_1 = 5, s_2 = 4, s_3 = 3 &\quad \text{with probability } 0.5^2.
\end{aligned}
$$

The objective function value for this policy is

$$
\begin{aligned}
&g_0(s_0, \pi_0(s_0)) + \mathbb{E}\Big[g_1(s_1, \pi_1(s_1)) + g_2(s_2, \pi_2(s_2)) + g_3(s_3)\Big] \\
&= (2 \cdot 1 + 3 \cdot 1) + 0.5^2\big((2 \cdot 5 + 3 \cdot 0) + (2 \cdot 5 + 3 \cdot 0) + 0\big) \\
&+ (2 \cdot 1 + 3 \cdot 1) + 0.5^2\big((2 \cdot 5 + 3 \cdot 0) + (2 \cdot 5 + 3 \cdot 0) + 0\big) \\
&+ (2 \cdot 1 + 3 \cdot 1) + 0.5^2\big((2 \cdot 5 + 3 \cdot 0) + (2 \cdot 4 + 3 \cdot 0) + 0\big) \\
&+ (2 \cdot 1 + 3 \cdot 1) + 0.5^2\big((2 \cdot 5 + 3 \cdot 0) + (2 \cdot 4 + 3 \cdot 0) + 0\big) = 24.
\end{aligned}
$$

Consequently, the second policy is better because it produces a lower expected cost. □

Finding an optimal policy for a Markov decision process is computationally challenging but a general principle allows us to break the solution process into many simpler optimization problems.

Theorem 4.15 (dynamic programming algorithm). *Given initial state* $s_0 \in \mathcal{S}$, *the value* $V_0(s_0)$ *from the recursive calculations*

$$V_T(s_T) = g_T(s_T) \qquad \forall s_T \in \mathcal{S} \tag{4.14}$$

$$V_t(s_t) = \min_{x_t \in \mathcal{X}_t} g_t(s_t, x_t) + \sum\nolimits_{s \in \mathcal{S}} p_{s_t s}(x_t) V_{t+1}(s), \quad t = 1, \dots, T-1, \ s_t \in \mathcal{S} \tag{4.15}$$

$$V_0(s_0) = \min_{x_0 \in \mathcal{X}_0} g_0(s_0, x_0) + \sum\nolimits_{s \in \mathcal{S}} p_{s_0 s}(x_0) V_1(s) \tag{4.16}$$

is the minimum value of the policy optimization problem (4.13). *Moreover, the policy*

$$\pi^\star = \{\pi_0^\star, \pi_1^\star, \dots, \pi_{T-1}^\star\} \in \Pi$$

is optimal for (4.13) *whenever* $\pi_0^\star(s_0)$ *is a minimizer in* (4.16) *and* $\pi_t^\star(s_t)$ *is a minimizer in* (4.15) *for each* $t = 1, \dots, T-1$, $s_t \in \mathcal{S}$. □

The theorem prescribes an algorithm that works backward. The steps are conceptually straightforward. The *value* $V_T(s_T)$ of being in state s_T at time T is simply the end cost $g_T(s_T)$ by (4.14). The value $V_{T-1}(s_{T-1})$ of being in state s_{T-1} at time $T-1$ is given by the stagewise cost at time $T-1$ plus the expected future value under a best decision x_{T-1}; see (4.15). We compute this for every $s_{T-1} \in \mathcal{S}$ and obtain the necessary input for the computation of $V_{T-2}(s_{T-2})$. We iterate backward in time and eventually compute $V_0(s_0)$ in (4.16), which by the theorem is the minimum value of the policy optimization problem (4.13). The resulting functions $\{V_t, t = 0, 1, \dots, T\}$ are called *value functions* because they specify the optimal objective function values from the various states. It's remarkable that a Markov decision process can be optimized by solving many stagewise subproblems. We refer to this fact as *Bellman's principle*. It produces the *dynamic programming algorithm* of Theorem 4.15; see [4, Section 1.3] for a formal justification.

Theorem 4.15 also specifies an optimal policy for the policy optimization problem (4.13). It's simply given by minimizers of the one-stage subproblems in (4.15) and (4.16). Thus, after computing the value functions $\{V_t, t = 0, 1, \dots, T\}$, decision making is simple. If we're in state s_t at time t, then an optimal decision is

$$x_t^\star \in \operatorname{argmin}_{x_t \in \mathcal{X}_t} g_t(s_t, x_t) + \sum\nolimits_{s \in \mathcal{S}} p_{s_t s}(x_t) V_{t+1}(s). \tag{4.17}$$

This optimization problem specifies a current decision but not future ones. They can't be determined with certainty anyways because future states are unknown. Overall, we've constructed a policy for making decisions to be applied as the circumstances unfold.

Example 4.16 (inventory). Returning to Example 4.14, the dynamic programming algorithm of Theorem 4.15 results in the following calculations.

For $T = 3$, we set $V_3(0) = 25$ and $V_3(s_3) = 0$ for $s_3 > 0$.

At $t = 2$, we obtain

$$V_2(0) = \min_{x_2 \in \{0,1\}} 25 + 3x_2 + 25p_{00}(x_2) = 28$$
$$V_2(1) = \min_{x_2 \in \{0,1\}} 2 + 3x_2 + 25p_{10}(x_2) = 5$$
$$V_2(2) = \min_{x_2 \in \{0,1\}} 4 + 3x_2 = 4$$
$$V_2(3) = \min_{x_2 \in \{0,1\}} 6 + 3x_2 = 6$$
$$V_2(4) = \min_{x_2 \in \{0,1\}} 8 + 3x_2 = 8$$
$$V_2(5) = \min_{x_2 \in \{0,1\}} 10 + 3x_2 = 10.$$

At $t = 1$, we compute

$$V_1(0) = \min_{x_1 \in \{0,1\}} 25 + 3x_1 + 28p_{00}(x_1) + 10p_{05}(x_1) = 38$$
$$V_1(1) = \min_{x_1 \in \{0,1\}} 2 + 3x_1 + 28p_{10}(x_1) + 5p_{11}(x_1) + 10p_{15}(x_1) = 15$$
$$V_1(2) = \min_{x_1 \in \{0,1\}} 4 + 3x_1 + 5p_{21}(x_1) + 4p_{22}(x_1) + 10p_{25}(x_1) = 8.5$$
$$V_1(3) = \min_{x_1 \in \{0,1\}} 6 + 3x_1 + 4p_{32}(x_1) + 6p_{33}(x_1) + 10p_{35}(x_1) = 11$$
$$V_1(4) = \min_{x_1 \in \{0,1\}} 8 + 3x_1 + 6p_{43}(x_1) + 8p_{44}(x_1) + 10p_{45}(x_1) = 15$$
$$V_1(5) = \min_{x_1 \in \{0,1\}} 10 + 3x_1 + 8p_{54}(x_1) + 10p_{55}(x_1) = 19.$$

At $t = 0$, we only need to consider

$$V_0(1) = \min_{x_0 \in \{0,1\}} 2 + 3x_0 + 38p_{10}(x_0) + 15p_{11}(x_0) + 19p_{15}(x_0) = 24.$$

Consequently, an optimal policy produces a minimum value of 24. This coincides with the best policy from Example 4.14. The specific details of the optimal policy follow by (4.17):

$$\pi_0(1) = 1, \quad \pi_1(5) = 0, \quad \pi_2(4) = 0, \quad \pi_2(5) = 0.$$

We've reproduced the policy guessed in Example 4.14. Now, we know it's optimal. We also realize that the inventory will surely be 5 in time period 1, but it can be either 4 or 5 in time period 2. Regardless, the policy prescribes do-nothing in either case. □

While Markov decision processes and Theorem 4.15 provide a versatile framework for modeling decision making over many stages, there are two main challenges. The first one is computational. As the number of states grow, the dynamic programming algorithm of Theorem 4.15 becomes practically intractable. The situation worsens further when the optimization problems in (4.15) and (4.16) become more complex requiring the branch-and-bound algorithm or other nontrivial computations.

There are opportunities for trimming the calculations in the dynamic programming algorithm of Theorem 4.15. Clearly, there's no need to compute $V_t(s_t)$ for a state s_t not reachable at time t from the initial state s_0. Similarly, in Example 4.16, we didn't compute $V_0(s_0)$ for states $s_0 \neq 1$ because the initial state is 1. A more general remedy is to replace V_t by an approximation $\hat{V}_t$, possibly based on neural networks or other function approximations. Then, it might suffice to *estimate* V_t at some states and extrapolate to the other states using the approximation $\hat{V}_t$. We refer to [3, 4, 24] for further details.

A second challenge with Markov decision processes is that we might not know the transition probabilities $p_{s\bar{s}}(x_t)$ in practice. These probabilities model how a system evolves and therefore may have to be learned through observations. We're forced to make decisions while having only incomplete knowledge about their effect on the system. We could *exploit* decisions that appear good according to our current knowledge about the transition probabilities, or *explore* seemingly suboptimal decisions with the hope that they lead us to valuable observations of the system and estimates of $p_{s\bar{s}}(x_t)$ to be exploited later. This brings us to the broad field of *reinforcement learning*.

An approach to reinforcement learning is *Q-learning*. It builds on the definition of a *Q-factor*:

$$Q_t(s_t, x_t) = g_t(s_t, x_t) + \sum\nolimits_{s \in \mathcal{S}} p_{s_t s}(x_t) V_{t+1}(s), \; t = 0, \dots, T-1, \, s_t \in \mathcal{S}, x_t \in \mathcal{X}_t, \quad (4.18)$$

where $V_{t+1}(s)$ is defined by Theorem 4.15. It's immediately clear that

$$V_t(s_t) = \min\nolimits_{x_t \in \mathcal{X}_t} Q_t(s_t, x_t) \quad (4.19)$$

so the definition of Q-factors may seem to have no practical value. They're trivially defined in terms of the value functions from Theorem 4.15. However, such a conclusion would be premature. Q-factors help us to separate the task of estimating an expectation from that of optimizing a decision. It leads to the following algorithm.

Q-Learning Algorithm.

Data. $\lambda^\nu > 0$, $\nu = 0, 1, \dots$; $Q_t^0(s_t, x_t) \in \mathbb{R}$, $t = 0, 1, \dots, T-1$, $s_t \in \mathcal{S}, x_t \in \mathcal{X}_t$.
Step 0. Set $\nu = 0$.
Step 1. Select $t^\nu \in \{0, 1, \dots, T-1\}$, $s^\nu \in \mathcal{S}$, and $x^\nu \in \mathcal{X}_{t^\nu}$.
Step 2. Observe $\bar{s}^\nu$, drawn randomly according to $\{p_{s^\nu s}(x^\nu), s \in \mathcal{S}\}$.
Step 3. Compute

$$\hat{Q}^\nu = g_{t^\nu}(s^\nu, x^\nu) + \min\nolimits_{x \in \mathcal{X}_{t^\nu+1}} Q^\nu_{t^\nu+1}(\bar{s}^\nu, x).$$

Step 4. Update

$$Q^{\nu+1}_{t^\nu}(s^\nu, x^\nu) = (1 - \lambda^\nu) Q^\nu_{t^\nu}(s^\nu, x^\nu) + \lambda^\nu \hat{Q}^\nu.$$

Step 5. For $t = 0, 1, \ldots, T-1$, $s_t \in \mathcal{S}$, $x_t \in \mathcal{X}_t$, $t \neq t^\nu$, $s_t \neq s^\nu$, $x_t \neq x^\nu$, carry over

$$Q_t^{\nu+1}(s_t, x_t) = Q_t^\nu(s_t, x_t).$$

Step 6. Replace ν by $\nu + 1$ and go to Step 1.

The algorithm computes a sequence of estimates $\{Q_t^\nu(s_t, x_t), \nu = 0, 1, \ldots\}$ of the Q-factor $Q_t(s_t, x_t)$ from (4.18). The initial estimate $Q_t^0(s_t, x_t)$ might simply be zero or one could use insight about the problem at hand. Step 1 selects a time, state, and decision for which to update the corresponding Q-factor estimate. This is typically done randomly using a distribution over time, state, and decisions that balance a focus on exploitation of promising states and decisions, and exploring new states and decisions. The theory would require that all times, states, and decisions are sampled infinitely often (see, e.g., [5, Section 5.6]) but there's plenty of room for practical choices. For example, one could roll-forward using the current estimates, i.e., start with

$$t^\nu = 0, \quad s^\nu = s_0, \quad x^\nu \in \operatorname{argmin}_{x \in \mathcal{X}_0} Q_0^\nu(s^\nu, x),$$

and then proceed with

$$t^{\nu+1} = 1, \quad s^{\nu+1} \text{ drawn randomly according to } \{p_{s^\nu s}(x^\nu), s \in \mathcal{S}\},$$
$$x^{\nu+1} \in \operatorname{argmin}_{x \in \mathcal{X}_1} Q_1^{\nu+1}(s^{\nu+1}, x),$$

and so forth. This approach would overly focus on those states and decisions that are promising according to current estimates of the Q-factors. Thus, it should be combined with some exploration of other states and decisions too. We note that the procedure doesn't require us to know the transition probabilities $\{p_{s^\nu s}(x^\nu), s \in \mathcal{S}\}$. It suffices to be able to sample states, conditioned on s^ν and x^ν. For example, the sample may come in the form of an observation of the actual system or by running a complex simulation of the system. The situation remains similar in Step 2, where the transition probabilities may be unknown.

In view of (4.19), the term

$$\min_{x \in \mathcal{X}_{t^\nu+1}} Q_{t^\nu+1}^\nu(\bar{s}^\nu, x) \tag{4.20}$$

is the current estimate of $V_{t^\nu+1}(\bar{s}^\nu)$. Consequently, Step 3 uses a single observation $\bar{s}^\nu$ to compute the estimate $\hat{Q}^\nu$ of $Q_{t^\nu}(s^\nu, x^\nu)$ in (4.18), i.e., the expectation

$$\sum\nolimits_{s \in \mathcal{S}} p_{s^\nu s}(x^\nu) \min_{x \in \mathcal{X}_{t^\nu+1}} Q_{t^\nu+1}^\nu(s, x)$$

is estimate by (4.20). When repeated, such single-sample estimates of expectations suffice just as they did in the case of the stochastic gradient descent method of Section 3.D.

Step 4 averages the estimates using a step size λ^ν, which prevents the noise in $\hat{Q}^\nu$ to grow beyond control. We refer to [5, Section 5.6] for assumptions on the step sizes that ensure convergence of the algorithm in the sense that $Q_t^\nu(s_t, x_t)$ converges to $Q_t(s_t, x_t)$ in (4.18) for all $s_t \in \mathcal{S}$, $x_t \in \mathcal{X}_t$, and $t = 0, 1, \ldots, T-1$ as $\nu \to \infty$. Without knowing the transition probabilities in a Markov decision process, we can therefore still optimize a policy via Q-learning, at least in principle. We refer to [4, 3, 24] for further details about how to overcome the various computational challenges.

4.G Summary and Study Guide

In this chapter, we learned how to formulate and solve optimization problems involving yes-no decisions, integer constraints, and network structure.

- A decision variable might be continuous and thus be free to take any real value, or it could be discrete and then be limited to integer values. A binary variable is restricted to the values 0 and 1, and models yes-no decisions.
- Standard form of notation helps us to communicate complex optimization models by listing indices, sets, data, variables, objective function, and constraints in a systematic manner. It's usually supplemented with an explanation of the constraints.
- The branch-and-bound algorithm is a main approach to solving optimization problems with integer variables. It relies on the concept of relaxation and restriction. Relaxation removes integer restrictions and produces lower bounds on the minimum value of the actual problem. Restriction introduces additional constraints such as fixing some variables and leads to upper bounds on the minimum value of the actual problem. Through the solution of a series of subproblems, organized in a branch-and-bound tree, the algorithm closes the optimality gap, which is the difference between the upper and lower bounds. If a subproblem produces a feasible point, then the corresponding minimum value in the subproblem is an upper bound on the minimum value of the actual problem. The lowest minimum value "across" nodes in a branch-and-bound tree yields a lower bound on the minimum value of the actual problem.
- By considering the convex hull of a feasible set, we can reformulate a nonconvex problem into a convex one. While conceptually elegant, a convex hull is rarely readily available in practice. The best algorithms for solving mixed-integer problems combine the branch-and-bound algorithm with constraint generation in an effort to approximate the convex hull.
- A stronger formulation of a mixed-integer problem is one with a tighter continuous relaxation of the feasible set.
- Networks appear in many applications and introduce specific structure to an optimization model. Shortest-path, maximum-flow, and min-cost flow problems are network optimization problems that can be solved by specialized algorithms. Dijkstra's algorithm is one such algorithm for the shortest-path problem.

- Multi-stage decision making under uncertainty can be modeled using Markov decision processes and solved by the dynamic programming algorithm. While general, the approach often faces computational challenges and tends to be combined with approximations.
- Reinforcement learning such as Q-learning allows us to optimize an unknown system while observing its evolution.

4.H Exercises

Exercise 4.1 (formulation with binary variables). Let x_{ij} be the quantity to be transported from node i to node j in a network. Using the binary variable y_{ij}, write linear inequality constraints that enforce the condition: either nothing is being transported from i to j or the quantity is between 10 and 100.

Exercise 4.2 (formulation with binary variables). In an optimization model, x_k is the amount of money invested in fund k, with $k = 1, 2, \ldots, 10$. (a) Write a constraint ensuring that our total investment doesn't exceed 100. (b) With y_k being a binary variable, explain the meaning of the constraints:

$$\sum_{k=1}^{10} y_k = 3 \tag{4.21}$$

$$0 \le x_k \le 100 y_k, \quad k = 1, \ldots, 10. \tag{4.22}$$

Exercise 4.3 (aircraft allocation with integer constraints). Rewrite the aircraft allocation problem in Example 3.8 using standard form of notation and include the additional constraints that every x_{ij} must be an integer. Select new, intuitive names for indices, sets, data, and variables. Solve the resulting model. Consider both uncertain demand and expected demand. Compare solutions as well as solution times with those from Example 3.8.

Exercise 4.4 (trucking). A trucking company "TransCo" operates a set of trucks and transports goods between various cities. Each truck is allocated to an origin city, from which it can drive to two destination cities with commodities. A truck can be assigned a destination city that's identical to its origin city. Adopt the indices i for city ($i = 1, 2, \ldots, 50$) (can also use j for city), t for truck ($t = 1, 2, \ldots, 100$) and c for commodity ($c = 1, 2, \ldots, 10$). Use the binary variables $\text{ASSIGN}_{it} = 1$ if city i is allocated to be truck t's origin city, and zero otherwise; and $\text{MOVE}_{ijt} = 1$ if truck t drives between origin city i and destination city j, and zero otherwise. ($\text{MOVE}_{iit} = 1$ represents driving within city i.) Also use the variable QUANT_{ijtc} = pounds of commodity c loaded on truck t in city i and delivered to city j, which needs to be nonnegative. We've the following data: maxLoad_t = maximum total load for truck t (pounds) on each of its two trips; demand_{ci} = minimum amount of goods (pounds) of commodity c to be delivered in city i; supply_{ci} = maximum amount of goods (pounds) of commodity c that can be loaded in city i; costMove_{ijtc} = cost of moving one pound of commodity c from city i to city j by truck t; and costAll_{it} = cost of allocating truck t to origin city i.

Write a linear mixed-integer optimization with the total cost of allocating trucks to origin cities and moving the goods between origin and destination cities as objective function. The constraints are: Each truck must be allocated to exactly one origin city and move to at most two destination cities. Each truck can load at most maxLoad_t pounds of goods on each trip. Trucks can only drive from their respective origin cities. Each city must receive enough of all commodities. No city can exceed the stated threshold for the amount loaded of each commodity. At most 10 trucks can drive between any pair of cities. No truck can load in city i and deliver in city j unless it drives between origin city i and destination city j. Use the standard format of notation as laid out in Example 4.5. Determine the number of constraints in each group of constraints.

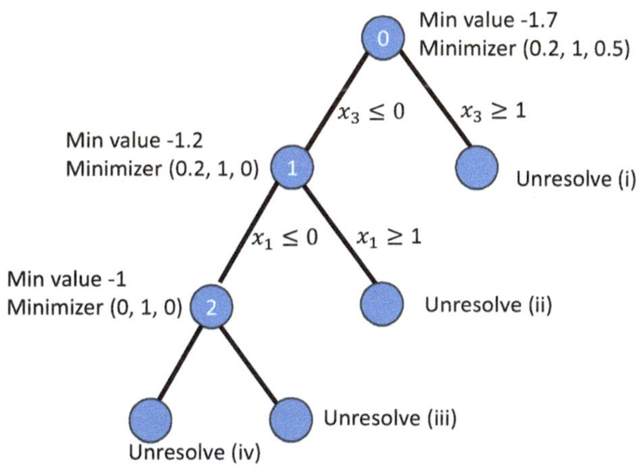

Fig. 4.12: Branch-and-bound tree for Exercise 4.5.

Exercise 4.5 (branch-and-bound algorithm). Consider the branch-and-bound tree in Figure 4.12 for a minimization problem with three binary variables. (a) What's the current upper bound on the minimum value of the actual problem? (b) What's the current lower bound on the minimum value of the actual problem? (c) Suppose that the unsolved node (i) is the next one solved and the minimum value at that node becomes -1.0 with minimizer $(0, 0, 1)$. Now there are three unsolved nodes. Do we need to solve the subproblems corresponding to the unsolved nodes? (d) Given the information in (c), four subproblems have been solved so far during the branch-and-bound algorithm. Discuss how many subproblems remain to be solved before you can conclude that the incumbent solution is a minimizer of the actual problem.

Exercise 4.6 (branch-and-bound algorithm). Suppose that the branch-and-bound algorithm has generated a tree as shown in Figure 4.13. (a) What are the current upper and lower bounds on the minimum value of the actual problem? What's the current optimality gap? (b) Suppose that node 4 is solved with a minimum value of -25.0. The corresponding solution is fractional. Now, what are the current upper and lower bounds? (c) Continuing

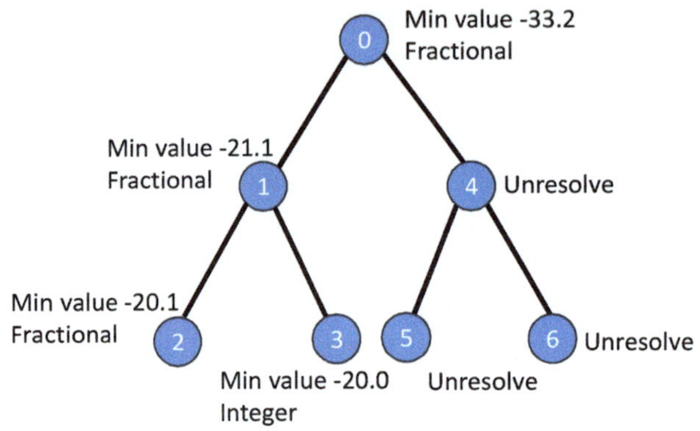

Fig. 4.13: Branch-and-bound tree for Exercises 4.6 and 4.7.

from (b), suppose that node 5 is solved with a minimum value of -21.0. The corresponding minimizer is feasible for the actual problem. Now, what are the current upper and lower bounds? (d) Continuing from (c), suppose that node 6 is found to be infeasible. Now, what are the current upper and lower bounds? How can you retrieve a minimizer of the actual problem at this point in time?

Exercise 4.7 (branch-and-bound algorithm). For the branch-and-bound tree in Figure 4.13, suppose that node 4 is solved and we obtain an integer solution with minimum value of -33.2. Compute the upper bound, the lower bound, and the optimality gap.

Exercise 4.8 (stronger formulation). For a positive scalar α, which of the following two formulations is stronger?

$$\alpha \geq x_1 \geq 0, \quad \alpha \geq x_2 \geq 0, \quad y \in \{0,1\}, \quad x_1 \leq \alpha y, \quad x_2 \leq \alpha y \tag{4.23}$$

$$\alpha \geq x_1 \geq 0, \quad \alpha \geq x_2 \geq 0, \quad y \in \{0,1\}, \quad x_1 + x_2 \leq 2\alpha y. \tag{4.24}$$

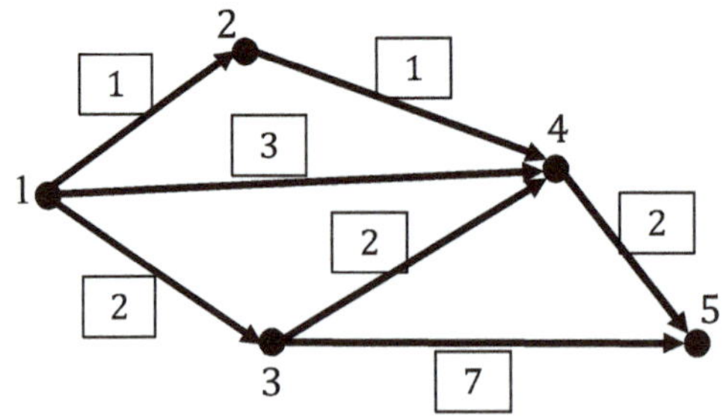

Fig. 4.14: The network for Exercise 4.9.

Exercise 4.9 (shortest paths using Dijkstra's). Use Dijkstra's algorithm to compute by hand the shortest path from vertex 1 to vertex 5 in Figure 4.14. The numbers in squares are edge lengths. Show all your calculations.

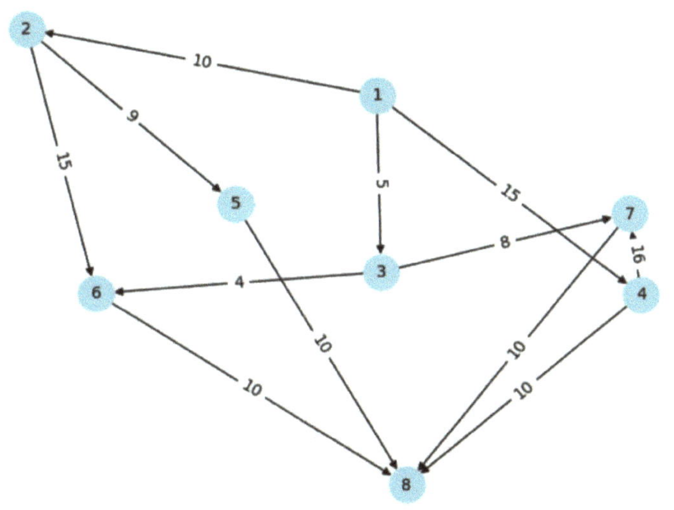

Fig. 4.15: The network for Exercise 4.10. The numbers on the edges are capacities.

Exercise 4.10 (maximum flow). Use a software package such as *networkx* to compute the maximum flow from vertex 1 to vertex 8 in Figure 4.15. The numbers on the edges are edge capacities. Verify by hand calculations that the result makes sense.

Exercise 4.11 (inventory). Consider Example 4.14, but change the data to

$p_{20}(0) = 0.3 \quad p_{21}(0) = 0.4 \quad p_{22}(0) = 0.3$
$p_{30}(0) = 0.25 \quad p_{31}(0) = 0.25 \quad p_{32}(0) = 0.25 \quad p_{33}(0) = 0.25$
$p_{40}(0) = 0.1 \quad p_{41}(0) = 0.2 \quad p_{42}(0) = 0.2 \quad p_{43}(0) = 0.25 \quad p_{44}(0) = 0.25$
$p_{50}(0) = 0 \quad p_{51}(0) = 0.1 \quad p_{52}(0) = 0.2 \quad p_{53}(0) = 0.2 \quad p_{54}(0) = 0.25 \quad p_{55}(0) = 0.25.$

The rest of the data remains unchanged. Use Theorem 4.15 to obtain an optimal policy.

Exercise 4.12 (inventory). The dynamic programming algorithm as applied to Exercise 4.11 results in the following calculations at $t = 2$:

$$V_2(0) = \min_{x_2 \in \{0,1\}} 25 + 3x_2 + 25p_{00}(x_2) = 28$$
$$V_2(1) = \min_{x_2 \in \{0,1\}} 2 + 3x_2 + 25p_{10}(x_2) = 5$$
$$V_2(2) = \min_{x_2 \in \{0,1\}} 4 + 3x_2 + 25p_{20}(x_2) = 7$$
$$V_2(3) = \min_{x_2 \in \{0,1\}} 6 + 3x_2 + 25p_{30}(x_2) = 9$$
$$V_2(4) = \min_{x_2 \in \{0,1\}} 8 + 3x_2 + 25p_{40}(x_2) = 10.5$$
$$V_2(5) = \min_{x_2 \in \{0,1\}} 10 + 3x_2 = 10.$$

(a) How do these six lines of calculations change if the end cost $g_3(0)$ shifts from 25 to 30? (b) How do these six lines of calculations change if the stagewise cost $g_t(s_t, x_t)$ changes from $2s_t + 3x_t$ to $5s_t + 3x_t$ for states $s_t > 0$? (Ignore the change from part (a).)

Chapter 5
Changing It Up: Perturbation and Duality

The solutions of a problem depend on the data used in a particular instance. Poor data would typically produce misleading solutions, poor estimates, and unnecessarily costly decisions. It's therefore important to assess how sensitive a solution is to varying model parameters. Solving a single instance of an optimization problem and presenting the solution as *the* answer is rarely enough. One would need to confirm that the model is suitable and this can, at least in part, be achieved by considering a *family of optimization problems* constructed by perturbing parameters of concern. The resulting sensitivity analysis uncovers troubling situations with unstable solutions and indicates better formulations.

Embedding an actual problem of interest within a family of problems is also a primary path to computationally attractive, alternative problems, which under ideal circumstances, and when properly tuned, may even furnish the minimum value of the actual problem. The tuning of these alternative problems turns out to be intimately tied to finding multipliers in optimality conditions and thus emerges as a main component of several optimization algorithms. In fact, the tuning amounts to solving certain *dual* optimization problems. We'll now turn to the opportunities afforded by these perspectives.

5.A Encoding Perturbations using Rockafellians

Suppose that we've formulated an optimization model to address a particular application. The model involves parameters with values that are somewhat unsettled. The parameters might represent the assumed budget of resources available, the probabilities of various scenarios, or weights of different terms in an objective function. A preliminary study would involve the solution of the resulting problem with nominal values for the parameters. However, a more comprehensive approach considers different values of the parameters and their effect on minimum values and minimizers in an effort to validate the model. To formalize this thinking, let's introduce a function that fully represents the actual objective function as well as parametric perturbations.

J. O. Royset, Optimization Concepts, *Springer Series in Operations Research and Financial Engineering*,
https://doi.org/10.1007/978-3-032-36819-5_5

Definition 5.1 (Rockafellian). *For a function f_0 and the problem*

$$\underset{x\in\mathbb{R}^n}{\text{minimize}}\ f_0(x),$$

we say that the function f, with input vectors u and x, is a Rockafellian *if*

$$f(0,x) = f_0(x) \quad \forall x \in \mathbb{R}^n.$$

□

Suppose that minimizing f_0 represents the actual problem of interest. Its minimization might bring forth a "good" decision, but when examined in isolation f_0 fails to identify the effect of changing parameter values. An associated Rockafellian f explicitly specifies the dependence on a parameter vector u and defines the *family of problems*

$$\left\{ \underset{x\in\mathbb{R}^n}{\text{minimize}}\ f(u,x), \quad u \in \mathbb{R}^m \right\},$$

where we assume that the *perturbation vector u* is m dimensional. We can then examine the minimum values, the sets of minimizers, and other quantities as they vary with u. For example, we may seek to minimize

$$f_0(x) = h(x) + \iota_{(-\infty,0]}\big(g(x) - \alpha\big),$$

with $g(x) \le \alpha$ being a resource constraint. A focus on the resource budget α leads us to a Rockafellian given by

$$f(u,x) = h(x) + \iota_{(-\infty,0]}\big(g(x) - \alpha + u\big),$$

which encodes perturbation of the budget. The problem of minimizing $h(x)$ subject to the constraint $g(x) \le \alpha$ changes to minimizing $h(x)$ subject to $g(x) \le \alpha - u$.

By considering a Rockafellian, we can carry out *sensitivity analysis* and answer the fundamental question: Would we make a significantly different decision (or assessment or prediction) if the parameters were perturbed? If the answer is yes, then it would be prudent to reevaluate model assumptions and parameter values. Any recommendation derived from the solution of the nominal problem should be qualified as being sensitive to assumptions. Of course, a minimization problem has many possible Rockafellians as they represent different ways the problem can be changed and this offers much flexibility in modeling and computations.

At first, one might mix up the present setting with that of stochastic optimization problems (see Chapter 3). There, one has to make a decision *before* a parameter vector ξ of unknown quantities is revealed. The parameters could represent future demand, environmental conditions, and other factors subject to uncertainty. Presently, we're optimizing a decision *after* assuming a value for the perturbation vector u, which then often represents modeling or data *choices* and not "external" randomness. The resulting sensitivity analysis would be useful even for stochastic optimization problems.

For example, we might have a model involving probabilities for various demand scenarios, but the exact values of the probabilities could be unsettled. The perturbation vector u would then represent these probabilities, with the subsequent sensitivity analysis identifying the effect of changing the probabilities.

Example 5.2 (perturbation). The one-dimensional minimization problem

$$\underset{x\in\mathbb{R}}{\text{minimize}}\ x^2+1 \quad \text{subject to} \quad (x-2)(x-4)+1 \le 0$$

has a Rockafellian defined by

$$f(u,x) = x^2 + 1 + \iota_{(-\infty,0]}\big((x-2)(x-4)+1+u\big),$$

which encodes changes to the right-hand side of the inequality constraint. The minimum value of $f(u,x)$ with respect to x doesn't vary continuously in u. Thus, the Rockafellian highlights the sensitivity to changes on the right-hand side of the constraint.

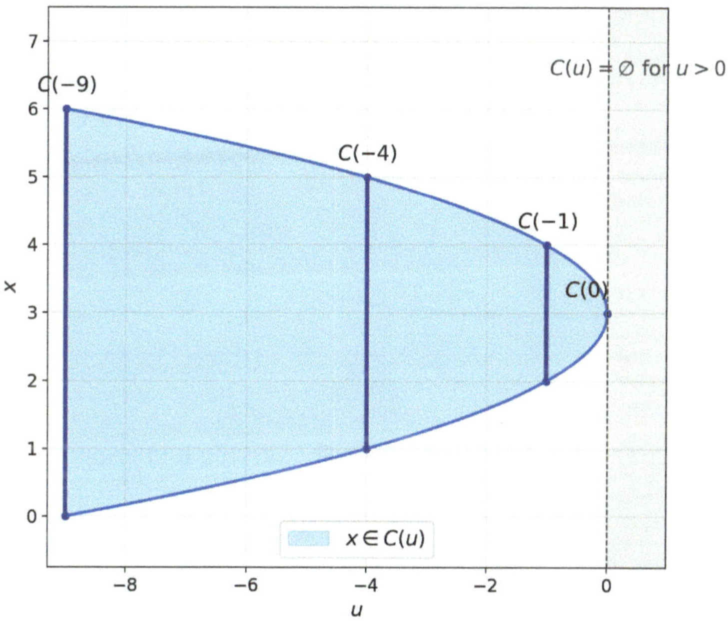

Fig. 5.1: The feasible set $C(u)$ for varying u in Example 5.2.

Detail. The constraint $(x-2)(x-4)+1+u \le 0$ defines the feasible set in Figure 5.1:

$$C(u) = \begin{cases} \big[\,3-\sqrt{-u},\ 3+\sqrt{-u}\,\big] & \text{if } u \le 0 \\ \emptyset & \text{otherwise.} \end{cases}$$

This formula comes from solving the quadratic equation $(x-2)(x-4)+1+u=0$ with respect to x. As functions of u, the minimizers and minimum values become

$$\operatorname{argmin}_x f(u,x) = \begin{cases} \{0\} & \text{if } u < -9 \\ \{3-\sqrt{1-u}\} & \text{if } -9 \le u \le 0 \\ \emptyset & \text{otherwise} \end{cases}$$

$$p(u) = \inf_x f(u,x) = \begin{cases} 1 & \text{if } u < -9 \\ 10-u-6\sqrt{-u} & \text{if } -9 \le u \le 0 \\ \infty & \text{otherwise.} \end{cases}$$

At every $u < 0$, the min-value function p varies continuously as seen from Figure 5.2. However, $p(0) = 10$ and $p(u) = \infty$ for $u > 0$ because the problem becomes infeasible. □

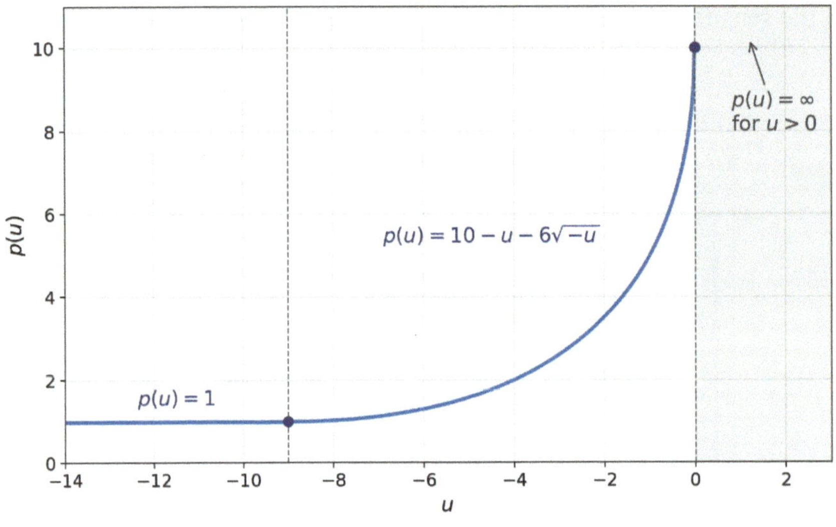

Fig. 5.2: The min-value function p for varying u in Example 5.2.

We may also compute subgradients of min-value functions, which represent first-order estimates of its change. To make precise statements, we'll need some terminology.

Functional Properties. A function g is *proper* if its epigraph is nonempty and $g(x) > -\infty$ for all x. This is a mild assumption for an objective function because $g(x) = -\infty$ in a minimization problem is probably the result of a modeling mistake. A cost of $-\infty$ is too good to be true!

The function is *lower semicontinuous (lsc)* if the epigraph is a closed set in $\mathbb{R}^{n+1}$. For example,

$$g(x) = \begin{cases} 0 & \text{if } x \leq 0 \\ 1 & \text{otherwise} \end{cases} \quad \text{is lsc but} \quad h(x) = \begin{cases} 0 & \text{if } x < 0 \\ 1 & \text{otherwise} \end{cases} \quad \text{isn't lsc.}$$

Informally, epi g includes all its boundary points, while epi h excludes its vertical boundary.

A function f with input vectors u and x is *level-bounded in x locally uniformly in u* if for every vector $\bar{u}$ and scalar α, there exist $\varepsilon > 0$ and a bounded set B such that

$$\{x \mid f(u,x) \leq \alpha\} \subset B \quad \text{for all } u \text{ satisfying } \|u - \bar{u}\|_2 \leq \varepsilon.$$

This property holds, for example, if $f(u,x) = \infty$ for all x outside a bounded set regardless of u. This property tends to hold automatically if f is a Rockafellian for a minimization problem with a bounded feasible set.

The subgradients of a min-value function are given in terms of subgradients of the underlying Rockafellian; see [30, Theorem 5.13] for further details.

Theorem 5.3 (subgradients of min-value function). *For a proper, lsc, and convex function f, which is also level-bounded in x locally uniformly in u, let*

$$p(u) = \inf_x f(u,x)$$

define the min-value function. Then, at any point $\bar{u}$ with $p(\bar{u}) < \infty$, one has

$$\partial p(\bar{u}) = \{y \mid (y,0) \in \partial f(\bar{u},\bar{x})\}$$

for any $\bar{x} \in \operatorname{argmin}_x f(\bar{u},x)$. □

We can view f in the theorem as a Rockafellian for the problem of minimizing $f_0(x) = f(0,x)$. Then, $p(0)$ is the minimum value of the problem and $\partial p(0)$ estimates how it changes under perturbations. The theorem quantifies the rate of change of the min-value function p in terms of the rate of change of the Rockafellian. By Definition 2.7, a subgradient $y \in \partial p(0)$ provides the bound

$$p(u) \geq p(0) + y^\top u \quad \forall u \tag{5.1}$$

because the min-value function p is convex; see [30, Proposition 1.21]. The bound estimates the effect of a particular perturbation without the need for re-optimization. Moreover, we can identify a poorly formulated problem. If the vector y is relatively large in magnitude, then the minimum value tends to change substantially even under small changes to u away from 0 and this calls into question the viability of the formulation. One can also examine the different components of y and see what portion of a perturbation is more influential.

Example 5.4 (perturbation of inequalities). Consider the one-dimensional problem

$$\underset{x\in\mathbb{R}}{\text{minimize}}\ (x-2)^2 \quad \text{subject to} \quad x \geq -1,\ \ x \leq 1.$$

The minimizer is $x^\star = 1$, with minimum value of 1. We're interested in quantifying the change to the minimum value as the second constraint becomes $x \leq 1-u$ for u near zero. Thus, we define a Rockafellian given by

$$f(u,x) = (x-2)^2 + \iota_C(u,x), \quad \text{where } C = \big\{(u,x) \,\big|\, -x \leq 1,\ u+x \leq 1\big\}.$$

We utilize Theorem 5.3 and (5.1) to estimate how the min-value function given by

$$p(u) = \inf_x f(u,x)$$

changes near $u = 0$.

Detail. Since $x^\star = 1$, the subgradients in $\partial p(0)$ are the vectors y satisfying $(y,0) \in \partial f(0,1)$; see Theorem 5.3. We leverage Theorem 2.12 to compute

$$\partial f(0,1) = \big(0,\ 2(1-2)\big) + N_C(0,1) = (0,\ -2) + N_C(0,1).$$

The normal cone $N_C(u,x)$ follows by Proposition 2.22 because the feasible set takes the form

$$C = \big\{w = (u,x) \,\big|\, Dw \leq d\big\}, \text{ where } D = \begin{bmatrix} 0 & -1 \\ 1 & 1 \end{bmatrix} \text{ and } d = \begin{bmatrix} 1 \\ 1 \end{bmatrix}.$$

The constraint $-x \leq 1$ isn't active at $x = 1$. Thus,

$$N_C(0,1) = \{D^\top z \mid z_1 = 0, z_2 \geq 0\} = \big\{(z_2, z_2) \,\big|\, z_2 \geq 0\big\}.$$

We therefore obtain

$$\partial f(0,1) = (0,\ -2) + \big\{(z_2, z_2) \,\big|\, z_2 \geq 0\big\} = \big\{(z_2, z_2 - 2) \,\big|\, z_2 \geq 0\big\}.$$

Following Theorem 5.3, we seek vectors y such that $(y,0) \in \partial f(0,1)$. The first row in this expression says that $y = z_2$. The second row says that $0 = z_2 - 2$. Thus, there's only one possibility: $y = 2$. This means that $\partial p(0) = \{2\}$. Using (5.1), we reach the conclusion that

$$p(u) \geq p(0) + yu = 1 + 2u.$$

To check this result, we compute $p(u)$ directly and find that

$$p(u) = \begin{cases} 0 & \text{if } u < -1 \\ (-u-1)^2 & \text{if } u \in [-1, 2] \\ \infty & \text{if } u > 2. \end{cases}$$

Figure 5.3 shows the min-value function p together with the tangent line given by $1 + 2u$. The tangent line provides the estimates 1.2 and 0.8 for $u = 0.1$ and $u = -0.1$, respectively. These are slightly off compared to the exact values $p(0.1) = 1.21$ and $p(-0.1) = 0.81$. In more complex models, computing the minimum value $p(u)$ for many values of u is computationally costly and it's beneficial to simply estimate the values using $p(0)$, y, and (5.1).

We end by formally verifying that Theorem 5.3 applies. The Rockafellian is proper because its values aren't $-\infty$ and its epigraph isn't empty; the point $((0,0), f(0,0))$ is in the epigraph since $f(0,0) = (0-2)^2 + \iota_C(0,0) = 4 < \infty$. The Rockafellian is convex by Proposition 2.4 and the Hessian rule in Proposition 1.9; note that C is a polyhedron. The Rockafellian is also lsc because C contains all its boundary points and is therefore closed. For any scalars u and α, the level-set

$$\{x \mid f(u,x) \le \alpha\} \subset \{x \mid (x-2)^2 \le \alpha,\ -x \le 1\},$$

which, in turn, is a bounded set. Thus, f is level-bounded in x, locally uniformly in u. □

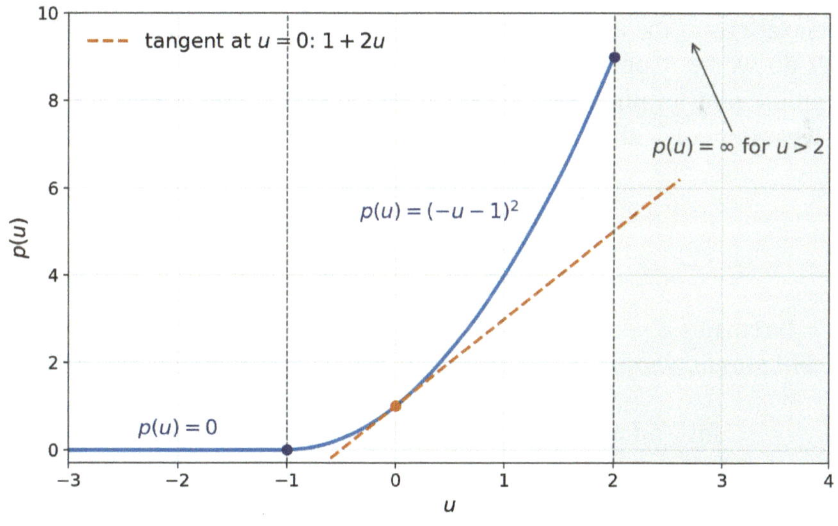

Fig. 5.3: The min-value function p for varying u (blue solid line) in Example 5.4 and its approximation provided by a tangent line (orange dashed line).

5.B Lagrangians and Dual Problems

In addition to their role in sensitivity analysis, Rockafellians for a minimization problem furnish a principal path to constructing *problem relaxations*. These are alternative problems with minimum values no higher than that of the actual one. Relaxations obtained by replacing integer variables by continuous ones are critical for the branch-and-bound algorithm of Section 4.B. Now, we adopt a more general approach to relaxation.

Relaxations based on Rockafellians are computationally useful and, when properly tuned, may even reproduce the minimum value of the actual problem. In contrast, continuous relaxations from Section 4.B tend to produce strict lower bounds. The process of tuning a Rockafellian-based relaxation is in of itself an optimization problem, which we refer to as a *dual problem*.

Let's consider the problem

$$\underset{x\in\mathbb{R}^n}{\text{minimize}}\, f_0(x)$$

and an associated Rockafellian f. The Rockafellian defines a family of problems, among which

$$\underset{x\in\mathbb{R}^n}{\text{minimize}}\, f(0,x)$$

recovers the actual problem of interest because $f(0,x) = f_0(x)$ for all x by Definition 5.1. The choice of Rockafellian could be dictated by a concern about certain model parameters as part of a sensitivity analysis, but just as well by the need for constructing tractable relaxations of the actual problem and this is the goal now.

We can't do worse if permitted to *optimize* the perturbation vector together with x. Thus, the problem of minimizing $f(u,x)$ over both u and x is a relaxation of the actual problem. Regardless of y, this is also the case for the problem

$$\boxed{\underset{u\in\mathbb{R}^m,x\in\mathbb{R}^n}{\text{minimize}}\; f(u,x) - y^\top u}$$

because $u = 0$ remains a possibility. We refer to this problem of optimizing both u and x as a *Rockafellian relaxation*. Consequently, for any y, we obtain the lower bound

$$\begin{aligned}\inf_x f_0(x) &= \inf_x f(0,x) \\ &\geq \inf_{u,x}\{f(u,x) - y^\top u\} = \inf_x l(x,y),\end{aligned} \tag{5.2}$$

where we use the short-hand notation

$$l(x,y) = \inf_u\{f(u,x) - y^\top u\}. \tag{5.3}$$

The function l given by this formula is the *Lagrangian* of f.

A benefit from bringing in a vector y is that we now can tune the relaxation by selecting y appropriately. We interpret the component y_i as the "price" associated with setting u_i nonzero so that $-y_iu_i$ is the additional cost for such u_i to be accrued on top of $f(u,x)$.

Then, $\inf_x l(x, y)$ is the lowest possible cost that can be achieved when one is permitted to "buy" perturbations at prices stipulated by y.

We next examining Lagrangians arising from typical Rockafellians but first introduce terminology that extends maximization in the same manner as infimum generalizes minimization.

Supremum. The *supremum* of a nonempty set $Y \subset [-\infty, \infty]$, denoted by $\sup Y$, is the *lowest* scalar α that satisfies $\alpha \geq \beta$ for all $\beta \in Y$, with $\sup \emptyset$ being set to $-\infty$. When Y contains $\sup Y$, one refers to $\sup Y$ as the *maximum* of Y and denotes it by $\max Y$.

For a function f and a set C, we write

$$\sup_C f = \sup_{x \in C} f(x) = \sup \{ f(x) \mid x \in C \} \quad \text{and} \quad \sup f = \sup_{\mathbb{R}^n} f$$

to denote the supremum of f (on C). If there's a point $x \in C$ with $f(x) > -\infty$ and $f(x) = \sup_C f$, then x is a *maximizer* of f over C and we may replace "sup" by "max."

As an example, the function given by $f(x) = 1 - \exp(-x)$ has no maximizer but $\sup f = 1$.

Example 5.5 (Lagrangian for equalities and inequalities). For functions $f_0, f_1, \ldots, f_m$ and $g_1, \ldots, g_q$ in n dimensions, and a nonempty set $X \subset \mathbb{R}^n$, let's consider the problem

$$\underset{x \in X}{\text{minimize}}\ f_0(x) \ \text{ subject to } \ f_i(x) = 0,\ i = 1, \ldots, m; \ \ g_i(x) \leq 0,\ i = 1, \ldots, q.$$

Among many possibilities, we obtain a Rockafellian by perturbing the right-hand side of the constraints. While such perturbations can be viewed from the angle of sensitivity analysis as changing "budgetary thresholds," they can also be introduced purely for the purpose of constructing a relaxation. Specifically, let $D = \{0\}^m \times (-\infty, 0]^q$ and

$$F(x) = \big(f_1(x), \ldots, f_m(x), g_1(x), \ldots, g_q(x) \big).$$

A Rockafellian centered on right-hand side perturbations of the problem is defined by

$$f(u, x) = \iota_X(x) + f_0(x) + \iota_D\big(F(x) + u\big).$$

The actual problem is then equivalent to minimizing $f(0, x)$ with respect to x. The resulting Lagrangian has

$$l(x, y) = \begin{cases} f_0(x) + F(x)^\top y & \text{if } x \in X, y_{m+1}, \ldots, y_{m+q} \geq 0 \\ -\infty & \text{if } x \in X, y_{m+i} < 0 \text{ for some } i = 1, \ldots, q \\ \infty & \text{otherwise,} \end{cases}$$

where y is an $m+q$-dimensional vector. Moreover, the actual problem is equivalently stated as

$$\underset{x\in\mathbb{R}^n}{\text{minimize}}\ \sup_y l(x,y).$$

Detail. With $x \in X$, the Rockafellian f produces the Lagrangian given by

$$l(x,y) = \inf_u \{f_0(x) + \iota_D(F(x)+u) - y^\top u\};$$

see the definition in (5.3). If there's $y_i < 0$ for some $i \in \{m+1,\dots,m+q\}$, then we can select $u_j = -f_j(x)$ for all $j \in \{1,\dots,m\}$ and $u_j = -g_j(x)$ for all $j \in \{m+1,\dots,m+q\}\setminus\{i\}$ so that $\iota_D(F(x)+u)$ remains zero as $u_i \to -\infty$. But, then

$$f_0(x) + \iota_D(F(x)+u) - y^\top u \to -\infty$$

and $l(x,y) = -\infty$.

If $y_i \geq 0$ for all $i \in \{m+1,\dots,m+q\}$, then $\bar{u}$, with components $\bar{u}_j = -f_j(x)$ for all $j \in \{1,\dots,m\}$ and $\bar{u}_j = -g_j(x)$ for all $j \in \{m+1,\dots,m+q\}$, solves

$$\underset{u\in\mathbb{R}^{m+q}}{\text{minimize}}\ f_0(x) + \iota_D(F(x)+u) - y^\top u$$

and this results in

$$l(x,y) = f_0(x) - y^\top \bar{u} = f_0(x) + y^\top F(x).$$

With the expression for l established, we also see that for $x \in X$,

$$\sup_y l(x,y) = \sup\{f_0(x) + F(x)^\top y \mid y_{m+1},\dots,y_{m+q} \geq 0\} = f(0,x)$$

because if $f_i(x) = 0$ for all $i = 1,\dots,m$ and $g_i(x) \leq 0$ for all $i = 1,\dots,q$, then $y = 0$ is a maximizer. If $g_i(x) > 0$ for some i, then $y_i \to \infty$ brings the supremum to infinity. If $f_i(x) \neq 0$ for some i, then $y_i \to \infty$ or $-\infty$ brings the supremum to infinity again. □

Example 5.6 (Lagrangian for max-function). For a nonempty set $X \subset \mathbb{R}^n$ and functions $f_1,\dots,f_m$ in n dimensions, let's consider the problem

$$\underset{x\in X}{\text{minimize}}\ \max_{i=1,\dots,m} f_i(x)$$

and the associated Rockafellian expressed by

$$f(u,x) = \iota_X(x) + h(F(x)+u),$$

where $F(x) = (f_1(x),\dots,f_m(x))$ and $h(z) = \max\{z_1,\dots,z_m\}$. We recover the actual problem by minimizing $f(0,x)$. The corresponding Lagrangian has

$$l(x,y) = \begin{cases} F(x)^\top y & \text{if } x \in X,\ y \in Y \\ -\infty & \text{if } x \in X,\ y \notin Y \\ \infty & \text{otherwise,} \end{cases}$$

where $Y = \{y \geq 0 \mid \sum_{i=1}^m y_i = 1\}$.

Moreover, the actual problem is equivalently stated as

$$\underset{x \in \mathbb{R}^n}{\text{minimize}}\ \sup_y l(x,y).$$

Detail. With $x \in X$, the Rockafellian f leads to the Lagrangian given by

$$l(x,y) = \inf_u \{h(F(x)+u) - y^\top u\}.$$

If $y_i < 0$ for some i, then we can select $u_j = 0$ for all $j \in \{1,\dots,m\} \setminus \{i\}$ and let $u_i \to -\infty$, which implies that $h(F(x)+u) - y^\top u \to -\infty$ and $l(x,y) = -\infty$.

If $y_i \geq 0$ for all i and $\sum_{i=1}^m y_i = \beta \neq 1$, then $u_i = \alpha$ for all i gives that

$$h(F(x)+u) - y^\top u = \max_{i=1,\dots,m}\{f_i(x)+\alpha\} - \alpha\beta = \max_{i=1,\dots,m} f_i(x) + \alpha(1-\beta).$$

If $\beta > 1$, then $\alpha \to \infty$ makes the previous term approach $-\infty$ and likewise for $\beta < 1$ with $\alpha \to -\infty$. In either case, $l(x,y) = -\infty$.

If $y \in Y$, then $\bar{u} = (-f_1(x),\dots,-f_m(x))$ solves

$$\underset{u \in \mathbb{R}^m}{\text{minimize}}\ h(F(x)+u) - y^\top u$$

because increasing u_i above $\bar{u}_i$ isn't beneficial due to $y_i \leq 1$. Likewise, reducing u_i below $\bar{u}_i$ only benefits $h(F(x)+u)$ if all $u_1,\dots,u_m$ are reduced, but then $-y^\top u$ grows a corresponding amount because $\sum_{i=1}^m y_i = 1$. The minimizer $\bar{u}$ produces the formula for $l(x,y)$.

We see that

$$\sup_y l(x,y) = \max_{i=1,\dots,m} f_i(x) = f(0,x)\ \ \forall x \in \mathbb{R}^n$$

because, generally, $\max_{y \in Y} z^\top y = \max_{i=1,\dots,m} z_i$. □

The Rockafellians in these examples lead to explicit expressions for the corresponding Lagrangians. This is especially useful if they were to be employed within an algorithm for the purpose of computing lower bounds on the minimum values of the actual problems via (5.2). Still, one shouldn't feel confined to these choices; opportunities for innovative constructions abound.

The examples also show that maximizing Lagrangians with respect to y may recover the actual objective function. This fact holds much more generally as we see next; consult [30, Proposition 5.24] for a proof.

Proposition 5.7 (reformulation in terms of Lagrangian). *For the problem of minimizing $f_0(x)$ and a proper Rockafellian f, which for all x is also convex and lsc in u, let l be the corresponding Lagrangian defined by* (5.3). *Then, the actual objective function has the equivalent formulas:*

$$f_0(x) = f(0,x) = \sup_y l(x,y) \ \ \forall x \in \mathbb{R}^n.$$

□

Thus far, a Lagrangian serves in two capacities: Under mild assumptions on the underlying Rockafellian as laid out in Proposition 5.7, its maximization with respect to y recovers the actual objective function and, without any assumptions, its minimization with respect to x furnishes a lower bound on the minimum value of the actual problem via (5.2). This maximization and minimization define

$$\begin{aligned} \forall x \in \mathbb{R}^n: &\quad \varphi(x) = \sup_y l(x,y) \\ \forall y \in \mathbb{R}^m: &\quad \psi(y) = \inf_x l(x,y). \end{aligned}$$

The function φ (Greek "phi") is simply an equivalent expression for the actual objective function under the assumptions of Proposition 5.7 so its minimization coincides in that case with the actual problem. The function ψ (Greek "Psi") quantifies the lower bound in (5.2) so its maximization produces the *best possible* lower bound. We think of this maximization as an effort to tune the relaxations produced by the chosen Rockafellian. A Lagrangian therefore sets up a pair of optimization problems with profound connections.

Specifically, a Rockafellian f associated with the problem of minimizing $f_0(x)$ defines via its Lagrangian l, given by (5.3), a *pair* of *primal* and *dual problems*:

$$\begin{aligned} &\text{(primal)} \quad \underset{x\in\mathbb{R}^n}{\text{minimize}}\ \varphi(x), \quad \text{where} \quad \varphi(x) = \sup_y l(x,y) \\ &\text{(dual)} \quad \underset{y\in\mathbb{R}^m}{\text{maximize}}\ \psi(y), \quad \text{where} \quad \psi(y) = \inf_x l(x,y) \end{aligned}$$

The name "primal problem" is motivated by its equivalence to the actual problem of interest under the assumptions of Proposition 5.7 because then

$$f_0(x) = f(0,x) = \varphi(x) \ \ \forall x \in \mathbb{R}^n.$$

In parallel, we refer to x and y as *primal* and *dual variables*, respectively, and φ and ψ as *primal* and *dual objective functions*, respectively.

Theorem 5.8 (weak duality). *A Lagrangian l, regardless of its expression, produces primal and dual objective functions satisfying*

$$\inf_x \varphi(x) \geq \sup_y \psi(y) \geq \psi(\bar{y}) \quad \forall \bar{y}.$$

□

The theorem establishes that a dual problem is always a relaxation of its primal problem, which reinforces the fact from (5.2) that the minimum value of a Lagrangian bounds from below the minimum value of the actual problem. This can be exploited algorithmically to address difficult optimization problems as we see in the next section. Before turning to the details, however, let's consider a common example of primal and dual problems.

Example 5.9 (dual linear problems). For n-dimensional vector c, m-dimensional vector b, and an $m \times n$-matrix A, the standard linear optimization problem

$$\boxed{\underset{x \in \mathbb{R}^n}{\text{minimize}}\; c^\top x \;\text{ subject to }\; Ax = b,\; x \geq 0}$$

coincides with the primal problem produced by the Rockafellian

$$f(u,x) = c^\top x + \iota_{[0,\infty)^n}(x) + \iota_{\{0\}^m}(b - Ax + u),$$

which then defines the Lagrangian given by

$$l(x,y) = \begin{cases} c^\top x - y^\top (Ax - b) & \text{if } x \geq 0 \\ \infty & \text{otherwise.} \end{cases}$$

The corresponding dual problem is

$$\boxed{\underset{y \in \mathbb{R}^m}{\text{maximize}}\; b^\top y \;\text{ subject to }\; A^\top y \leq c}$$

This is the *standard dual linear optimization problem*.

Detail. In an argument similar to those in Example 5.5, we see that

$$l(x,y) = \inf_u \left\{ f(u,x) - y^\top u \right\} = \begin{cases} b^\top y + (c - A^\top y)^\top x & \text{if } x \geq 0 \\ \infty & \text{otherwise,} \end{cases}$$

which is equivalent to the asserted expression. From Proposition 5.7, we recognize that $f(0,x) = \sup_y l(x,y)$, but this can be seen directly as well. Thus, the actual problem coincides with the primal problem produced by the Rockafellian via this Lagrangian. The dual objective function has

$$\psi(y) = \inf_x l(x, y) = \begin{cases} b^\top y & \text{if } c - A^\top y \geq 0 \\ -\infty & \text{otherwise} \end{cases}$$

and this yields the asserted dual problem.

The process is symmetric. If we start with the problem of minimizing $-b^\top y$ subject to $A^\top y \leq c$ and adopt the Rockafellian

$$\tilde{f}(u, y) = -b^\top y + \iota_{(-\infty,0]^n}(A^\top y - c + u),$$

we obtain the Lagrangian

$$\tilde{l}(y, x) = \begin{cases} -c^\top x + (Ax - b)^\top y & \text{if } x \geq 0 \\ -\infty & \text{otherwise} \end{cases}$$

and dual function

$$\tilde{\psi}(x) = \inf_y \tilde{l}(y, x) = \begin{cases} -c^\top x & \text{if } Ax = b,\ x \geq 0 \\ -\infty & \text{otherwise.} \end{cases}$$

The dual problem to that of minimizing $-b^\top y$ subject to $A^\top y \leq c$ is therefore maximizing $-c^\top x$ subject to $Ax = b$ and $x \geq 0$. After shifting to minimization, we're back at the starting point with the standard linear optimization problem.

The above choice of Rockafellian has the advantage that it leads to an explicit expression for the dual problem. However, there are many alternatives. For example, one might only be concerned about changes to one of the equality constraints. This leads to a different Lagrangian and, in turn, another dual problem. Specifically, let's consider $Ax = b$ *and* $a^\top x = \alpha$, the latter being of concern. This leads to the problem

$$\underset{x \in \mathbb{R}^n}{\text{minimize}}\ c^\top x \ \text{ subject to } \ Ax = b,\ a^\top x = \alpha,\ x \geq 0,$$

which coincides with the primal problem produced by an alternative Rockafellian expressed as

$$f(u, x) = c^\top x + \iota_{[0,\infty)^n}(x) + \iota_{\{0\}^m}(b - Ax) + \iota_{\{0\}}(\alpha - a^\top x + u),$$

where u is a one-dimensional vector. This Rockafellian produces an alternative Lagrangian given by

$$l(x, y) = \inf_u \{f(u, x) - yu\} = \begin{cases} \alpha y + (c - ay)^\top x & \text{if } Ax = b,\ x \geq 0 \\ \infty & \text{otherwise,} \end{cases}$$

which defines a dual problem of maximizing $\psi(y) = \inf_x l(x, y)$. While this dual problem has only a single variable, its objective function isn't generally available in an explicit form; for each y, one needs to solve a linear optimization problem to obtain $\psi(y)$. Still, insight as well as computational benefits may emerge from this approach. □

Interestingly, regardless of convexity in a primal problem and in any of the underlying functions, the dual objective function ψ is concave by [30, Proposition 1.18(a)] and thus its maximization can be achieved by minimizing the convex function $-\psi$. Although the dual problem may only provide a lower bound on the minimum value of the actual problem at hand, it could be more tractable due to this property.

5.C Lagrangian Relaxation

It's important to be able to assess the quality of a candidate solution of a problem and thereby determine if further calculations are needed. Optimality conditions serve this purpose, but relaxations—especially in the form of dual problems—also offer possibilities. For the problem of minimizing $f_0(x)$ and a candidate solution $\bar{x}$, the *optimality gap* of $\bar{x}$ is given by

$$f_0(\bar{x}) - \inf f_0.$$

A relatively small optimality gap indicates that $\bar{x}$ is rather good in the sense that the associated "cost" isn't much higher than the minimum value. This could convince a decision maker to adopt $\bar{x}$, making further efforts to find an even better solution superfluous. An optimality gap is measured in the units of the objective function and is thus well understood by a decision maker in many cases. The meaning of an optimality gap of \$1 million is clear. We're potentially leaving this amount on the table, which might be small or large depending on the circumstances.

Since there are often application-dependent heuristic algorithms for finding a candidate solution $\bar{x}$ of a difficult problem, the challenging part in calculating the optimality gap is to estimate $\inf f_0$. Typically, we prefer a lower bound on $\inf f_0$ as it produces a conservative estimate of the optimality gap and this is where a dual problem can be brought in.

The steps are as follows: First, express the actual objective function f_0 via a Rockafellian f so that $f(0, x) = f_0(x)$ for all x. Second, form the corresponding Lagrangian l according to (5.3). Third, using a vector $\bar{y}$, obtain a lower bound

$$\psi(\bar{y}) = \inf_x l(x, \bar{y}) \leq \inf_x f(0, x) = \inf f_0$$

by minimizing the Lagrangian; see (5.2). These steps are referred to as *Lagrangian relaxation* and are especially productive when the actual problem is difficult, but $l(x, y)$ is relatively simple to minimize with respect to x. The best possible lower bound is achieved by tuning y, i.e., solving the corresponding dual problem to obtain

$$\sup_y \psi(y) \leq \inf f_0$$

and thus bringing $\psi(y)$ as high up as possible.

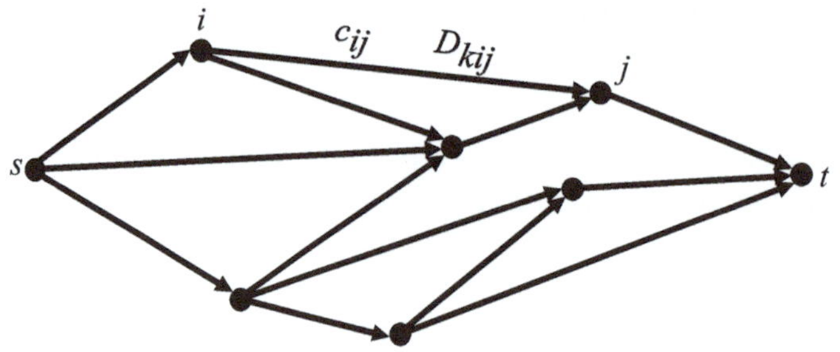

Fig. 5.4: Network with edge lengths c_{ij} and additional weights D_{kij}.

Example 5.10 (constrained shortest-path problem). We extend the shortest-path problem from Example 2.1 by including additional constraints. Let (V, E) be a directed graph with vertex set V and edge set E. Each edge $(i, j) \in E$ connects distinct vertices $i, j \in V$, and it possesses nonnegative length c_{ij} and nonnegative weights D_{kij} for $k = 1, \dots, q$; see Figure 5.4. For nonnegative d_k, $k = 1, \dots, q$, we seek an s-t path along which the sum of the weights D_{kij} doesn't exceed d_k for each k. This is the *constrained shortest-path problem*, which can be addressed by Lagrangian relaxation. In routing of a drone through a discretized 3-dimensional airspace, the weights D_{1ij} might represent fuel consumption along edge (i, j), which can't exceed a capacity d_1. Figure 5.5 illustrates a route satisfying such a fuel constraint while minimizing exposure to enemy radars expressed by c_{ij}; see [27].

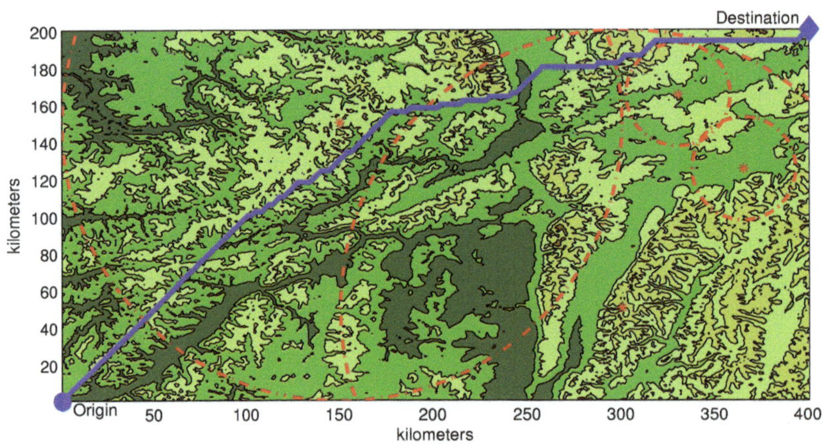

Fig. 5.5: Route for a drone through a 3-dimensional airspace to a destination (blue line) that minimizes exposure to enemy radars (red circles) while satisfying a fuel constraints. Altitude changes to leverage terrain masking aren't shown; see [27].

Detail. We can formulate the constrained shortest-path problem as follows. Suppose that m is the number of vertices in V and n is the number of edges in E. Let A and b be as in

Example 2.1. For each $k = 1, \dots, q$, we place the edge weights $\{D_{kij}, (i, j) \in E\}$ in the vector D_k. The $q \times n$-matrix D has D_k as its kth row. Let $d = (d_1, \dots, d_q)$. With c being the column vector $(c_{ij}, (i, j) \in E)$, a model for the constrained shortest-path problem is then (cf. [1, p. 599]):

$$\boxed{\operatorname*{minimize}_{x\in\{0,1\}^n} c^\top x \quad \text{subject to } Ax = b,\ Dx \le d}$$

A point $\bar{x} \in \{0, 1\}^n$ satisfying $Ax = b$ corresponds to an s-t path with $\bar{x}_{ij} = 1$ if edge (i, j) is on the path and $\bar{x}_{ij} = 0$ otherwise. We assume there's at least one such path. Then, $c^\top \bar{x}$ gives the length of the path and $D_k\bar{x}$ the kth weight of the path. In contrast to Example 2.1, the formulation would be incorrect if $x \in \{0, 1\}^n$ is replaced by $x \in [0, 1]^n$; see also the discussion in Section 4.D.

In the absence of the weight-constraint $Dx \le d$, the model reduces to one for the shortest-path problem as seen in Example 2.1. This opens up an opportunity for applying Lagrangian relaxation via Example 5.5. Let

$$X = \{x \in \{0, 1\}^n \mid Ax = b\},$$

which is nonempty by assumption, and consider the Rockafellian from Example 5.5:

$$f(u, x) = \iota_X(x) + c^\top x + \iota_{(-\infty,0]^q}(Dx - d + u).$$

The actual problem corresponds to minimizing $f(0, x)$. By Example 5.5,

$$l(x, y) = \begin{cases} c^\top x + (Dx - d)^\top y & \text{if } x \in X, y \ge 0 \\ -\infty & \text{if } x \in X, y_k < 0 \text{ for some } k = 1, \dots q \\ \infty & \text{otherwise.} \end{cases}$$

Thus, the chosen Rockafellian results in a Lagrangian without the explicit constraint $Dx \le d$. The corresponding dual problem is

$$\operatorname*{maximize}_{y\in\mathbb{R}^q} \psi(y) = \inf_x l(x, y).$$

For $y \ge 0$, which are the only values of concern because $\psi(y) = -\infty$ otherwise, we obtain explicitly that

$$\psi(y) = -d^\top y + \inf_{x\in X}(c + D^\top y)^\top x.$$

The minimization occurring here is nothing but a shortest-path problem on the directed graph, but with edge lengths changed from c_{ij} to $c_{ij} + \sum_{k=1}^q D_{kij} y_k$, which can be solved efficiently using Dijkstra's algorithm from Section 4.E and related algorithms. The resulting minimum value, modified by $d^\top y$, yields $\psi(y)$, a lower bound on the minimum value of the constrained shortest-path problem as seen from (5.2) and the introductory discussion in this section. The lower bound can be used to assess the optimality gap for any candidate path, for example obtained by a greedy search or an enumeration algorithm; see [10].

Since $y_k \geq 0$, the addition of $\sum_{k=1}^{q} D_{kij} y_k$ to the edge length c_{ij} penalizes edges with high weights. While there's no guarantee that the system of penalties produced by the vector y results in a feasible solution satisfying $Dx \leq d$, it tends to steer a solution toward a path with low weights and hopefully a meaningful lower bound on the minimum value of the constrained shortest-path problem. □

For a Lagrangian l and the corresponding dual objective function ψ, it would certainly be desirable to obtain the best possible lower bound $\sup_y \psi(y)$. Since ψ is concave, the subgradient method of Section 2.G is often viable for this purpose. Returning to the familiar minimization setting, the task at hand is to

$$\operatorname*{minimize}_{y} \ \tilde{\psi}(y) = -\psi(y) = \sup_x -l(x, y).$$

While Section 2.G only describes the subgradient method for real-valued objective functions, it can be extended by use of projections to handle situations involving simple constraints on y [30, Section 2.I]. Thus, minimizing $\tilde{\psi}$ is often possible provided that we can minimize the Lagrangian with respect to x efficiently. This is the case in Example 5.10; it only requires shortest-path calculations.

5.D Strong Duality

Fundamental to a dual problem is its lower-bounding property that we refer to as weak duality; cf. Theorem 5.8. However, the practical usefulness of the property diminishes if the resulting bound is much below the minimum value of the primal problem. A natural question is then: When does the maximum value of the dual problem reach the whole way up to the minimum value of the primal problem? In this section, we examine this question.

Given a Lagrangian l and the corresponding primal and dual objective functions

$$\varphi(x) = \sup_{y \in \mathbb{R}^m} l(x, y) \quad \text{and} \quad \psi(y) = \inf_{x \in \mathbb{R}^n} l(x, y),$$

we say that the pair of optimization problems

$$\left\{ \operatorname*{minimize}_{x \in \mathbb{R}^n} \ \varphi(x), \quad \operatorname*{maximize}_{y \in \mathbb{R}^m} \ \psi(y) \right\},$$

possesses *strong duality* when

$$\inf \varphi = \sup \psi.$$

Example 5.11 (primal-dual relation for linear problems). From Example 5.9, we recall that the problems

$$\underset{x \in \mathbb{R}^n}{\text{minimize}}\ c^\top x \qquad\qquad \underset{y \in \mathbb{R}^m}{\text{maximize}}\ b^\top y$$

$$\text{subject to } Ax = b \qquad\qquad \text{subject to } A^\top y \le c$$

$$x \ge 0$$

are primal-dual pairs under the particular Rockafellian considered there. The two problems are closely connected. If the primal problem has a minimizer, then the dual problem has a maximizer and the minimum value of the primal problem equals the maximum value of the dual problem. If the dual problem has a maximizer, then the primal problem has a minimizer and the minimum value of the primal problem equals the maximum value of the dual problem. Moreover, one has

$$\begin{aligned}
\text{primal problem unbounded} &\implies \text{dual problem infeasible}\\
\text{primal problem infeasible} &\implies \text{dual problem infeasible or unbounded}\\
\text{dual problem unbounded} &\implies \text{primal problem infeasible}\\
\text{dual problem infeasible} &\implies \text{primal problem infeasible or unbounded.}
\end{aligned}$$

We recall that the primal problem being unbounded means that

$$\inf_x\{c^\top x \mid Ax = b, x \ge 0\} = -\infty$$

and being infeasible means that

$$\inf_x\{c^\top x \mid Ax = b, x \ge 0\} = \infty.$$

By analogy, the dual problem is unbounded and infeasible means that

$$\sup_y\{b^\top y \mid A^\top y \le c\}$$

is ∞ and $-\infty$, respectively.

Detail. By the KKT condition of Theorem 2.23, a minimizer of the primal problem is characterized by

$$\begin{aligned}
&\forall j = 1, \dots, n: \quad z_j \ge 0 \text{ if } x_j = 0;\ z_j = 0 \text{ otherwise}\\
&-c = A^\top y - z, \quad Ax = b, \quad x \ge 0.
\end{aligned}$$

After eliminating z and flipping the sign of y, we obtain that

$$\begin{aligned}
&\forall j = 1, \dots, n: \quad \text{if } x_j > 0, \text{ then } y^\top A^j = c_j\\
&A^\top y \le c, \quad Ax = b, \quad x \ge 0,
\end{aligned} \tag{5.5}$$

where A^j is the jth column of A. Again by Theorem 2.23, a maximizer of the dual problem or, equivalently, a minimizer of $-b^\top y$ subject to $A^\top y \le c$, is characterized by

$$\forall j = 1, \ldots, n: \quad x_j \geq 0 \text{ if } y^\top A^j = c_j; \;\; x_j = 0 \text{ otherwise}$$
$$b = Ax, \;\; A^\top y \leq c,$$

where we pretentiously label the multiplier vector by x. We see that the pair $(\bar{x}, \bar{y})$ satisfies these conditions if and only if the pair also satisfies (5.5). In view of Theorem 2.23, this means that $\bar{x}$ is a minimizer of the primal problem with corresponding multiplier vector $\bar{y}$ if and only if $\bar{y}$ is a maximizer of the dual problem with corresponding multiplier vector $\bar{x}$. In such a case, the objective function values become

$$c^\top \bar{x} = \sum\nolimits_{j=1}^{n} c_j \bar{x}_j = \sum\nolimits_{j=1}^{n} \bar{y}^\top A^j \bar{x}_j = \sum\nolimits_{i=1}^{m} A_i \bar{x} \bar{y}_i = b^\top \bar{y},$$

where A_i is the ith row of A. Thus, the assertion about equal objective function values holds. The symmetry between the primal and dual problems implies that one can apply the simplex method or interior-point method to the dual problem and then recover a primal solution by simply recording the corresponding multipliers.

If the primal problem is unbounded, then the dual problem is infeasible by weak duality; see Theorem 5.8. Likewise, if the dual problem is unbounded, then the primal problem must be infeasible. If one of the problems is infeasible, then the other one is either infeasible or unbounded; it can't have a solution as that would have implied one for the first problem as well. A linear optimization problem either has a minimizer, is infeasible, or is unbounded; see [7, Corollary 2.3]. Consequently, we've exhausted the possibilities.

While the multiplier vector y in the KKT condition of Theorem 2.23 appeared secondary to the vector x of decision variables, this example places them on equal footing. The vector y represents the variables of a dual problem intimately connected with the primal one. □

While Example 5.11 furnishes an instance where the maximum value of the dual problem reaches the minimum value of the primal problem, this property isn't automatic. We may have a *duality gap* given by

$$\inf \varphi - \sup \psi.$$

Example 5.12 (failure of strong duality). For the problem of minimizing x^3 subject to $x \geq 0$ and the Rockafellian given by

$$f(u, x) = x^3 + \iota_{(-\infty,0]}(-x + u),$$

we obtain a Lagrangian of the form

$$l(x, y) = \begin{cases} x^3 - xy & \text{if } y \geq 0 \\ -\infty & \text{otherwise;} \end{cases}$$

see Example 5.5. The dual objective function $\psi(y) = -\infty$ for all y, while the minimum value of the corresponding primal problem is zero so the duality gap is ∞.

Detail. In this case, the Lagrangian isn't convex in its first argument. However, strong duality may fail even under convexity. Consider the problem

$$\underset{x\in\mathbb{R}^2}{\text{minimize}}\ e^{-x_1} \ \text{subject to}\ g(x) \le 0, \ \text{where}\ g(x) = \begin{cases} x_1^2/x_2 & \text{if } x_2 > 0 \\ \infty & \text{otherwise,} \end{cases}$$

and a Rockafellian of the form

$$f(u,x) = \begin{cases} e^{-x_1} & \text{if } g(x) + u \le 0 \\ \infty & \text{otherwise.} \end{cases}$$

We note that f is convex. Similar to Example 5.5, this produces a Lagrangian with

$$l(x,y) = \begin{cases} e^{-x_1} + yg(x) & \text{if } x_2 > 0, y \ge 0 \\ \infty & \text{if } x_2 \le 0 \\ -\infty & \text{otherwise.} \end{cases}$$

Consequently, the dual objective function $\psi(y) = 0$ if $y \ge 0$, but $\psi(y) = -\infty$ otherwise. The maximum value of the dual problem is therefore 0. The actual problem coincides with the primal problem of minimizing $\varphi(x) = \sup_y l(x,y)$ with respect to x, which has minimum value of 1. Thus, the duality gap is one even though $l(x,y)$ is convex in x regardless of y. □

Despite this discouraging example, there's a large class of problems beyond linear optimization for which strong duality holds as we see in the next theorem extracted from [25, Theorem 11.39].

Theorem 5.13 (strong duality). *For a proper, lsc, and convex Rockafellian f with corresponding Lagrangian l given by* (5.3)*, the primal and dual problems*

$$\underset{x\in\mathbb{R}^n}{\text{minimize}}\ \varphi(x) = \sup_y l(x,y) \qquad \underset{y\in\mathbb{R}^m}{\text{maximize}}\ \psi(y) = \inf_x l(x,y)$$

satisfy strong duality, i.e.,

$$\inf \varphi = \sup \psi,$$

provided that there's $\varepsilon > 0$ such that the min-value function satisfies

$$p(u) = \inf_x f(u,x) < \infty \quad \forall u \ \text{with} \ \|u\|_2 \le \varepsilon. \tag{5.6}$$

If in addition $p(0) > -\infty$, then

$$\partial p(0) = \operatorname{argmax} \psi,$$

which must be a nonempty and bounded set. □

In addition to convexity of the Rockafellian, the key assumption of the theorem is (5.6). It requires that small perturbations near 0 shouldn't produce an infeasible problem. In Example 5.4, the min-value function has $p(u) = (-u-1)^2$ for $u \in [-1, 2]$ so (5.6) holds and Theorem 5.13 applies. Moreover, the example has $p(0) = 1$ and $\partial p(0) = \{2\}$, which imply that the dual problem has 2 as its unique maximizer, with 1 as the maximum value. We determine all using Theorem 5.13 without having a detailed formulation of the dual problem. (We derive the formulation below in Example 5.14.) The formula for $\partial p(0)$ supplements Theorem 5.3, but most significantly it highlights the profound role played by a dual problem. Under the conditions of the theorem, solving the dual problem furnishes both the minimum value of the actual problem as well as its sensitivity to perturbations as defined by a Rockafellian.

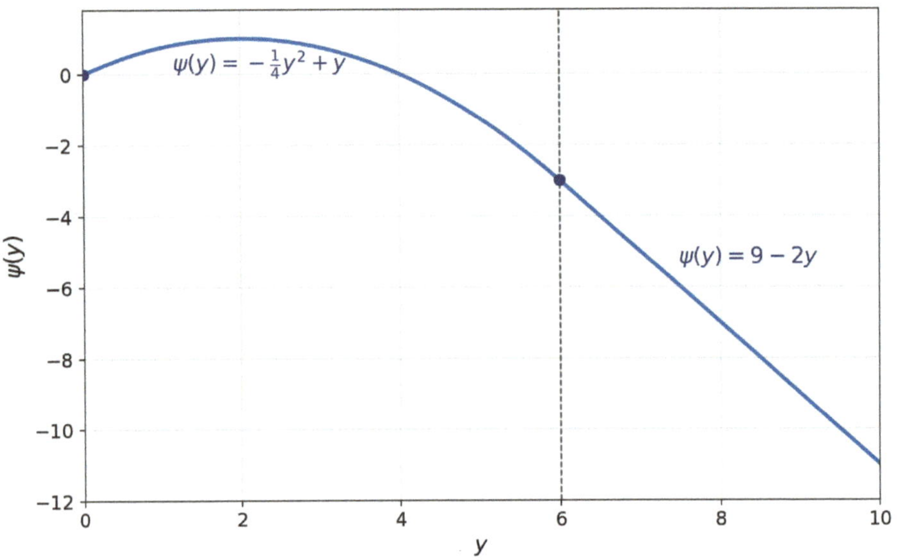

Fig. 5.6: Dual function ψ from Example 5.14.

Example 5.14 (nonlinear dual problem). We return to Example 5.4 and recall the Rockafellian given by

$$f(u,x) = (x-2)^2 + \iota_C(u,x), \quad \text{where } C = \{(u,x) \in \mathbb{R}^2 \mid -x \le 1,\ u+x \le 1\}.$$

This case fits the setting of Example 5.5, with $f_0(x) = (x-2)^2$, $g_1(x) = x - 1$, and $X = \{x \mid -x \le 1\}$. The resulting Lagrangian has the form

$$l(x,y) = (x-2)^2 + y(x-1) \ \text{ for } -x \le 1, y \ge 0.$$

Minimization the Lagrangian with respect to x produces the dual function

$$\psi(y) = \begin{cases} -\frac{1}{4}y^2 + y & \text{if } y \in [0,6] \\ 9 - 2y & \text{if } y > 6. \end{cases}$$

The maximum value of ψ is attained uniquely at $y = 2$; see Figure 5.6. We recall from Example 5.4 that $\partial p(0) = \{2\}$. Thus, we've confirmed $\partial p(0) = \operatorname{argmax} \psi$ as discussed after Theorem 5.13. □

Theorem 5.13 requires that the min-value function doesn't jump to infinity near 0. This holds, for example, when constraint functions leave some "slack," which often is easily verified.

Example 5.15 (Slater constraint qualification). For real-valued convex functions $f_0, g_1, \dots, g_q$ and the problem

$$\underset{x \in \mathbb{R}^n}{\text{minimize}}\ f_0(x) \ \text{ subject to } \ g_i(x) \le 0,\ i = 1, \dots, q,$$

let's consider the Rockafellian given by

$$f(u,x) = f_0(x) + \iota_{(-\infty,0]^q}\big(G(x) + u\big), \ \text{ with } G(x) = \big(g_1(x), \dots, g_q(x)\big).$$

The resulting primal and dual problems (as defined via the Lagrangian given by Example 5.5) satisfy strong duality provided that the following *Slater constraint qualification* holds:

$$\text{there's } \bar{x} \text{ such that } g_i(\bar{x}) < 0,\ \ i = 1, \dots, q.$$

Figure 5.7(left) illustrates a situation with two constraints that satisfy the qualification. If a constraint function doesn't reach below zero as would be the case with $g_1(x) = x^2$, then the qualification fails; see Figure 5.7(right).

Detail. Let $p(u) = \inf_x f(u,x)$. Under the Slater constraint qualification, there exist $\bar{x}$ and $\delta > 0$ such that $g_i(\bar{x}) + u_i \le 0$ when $|u_i| \le \delta$ for all i. Consequently, $p(u) \le f_0(\bar{x})$ is finite for sufficiently small u, which means that p doesn't jump to infinity near 0 as required by Theorem 5.13. □

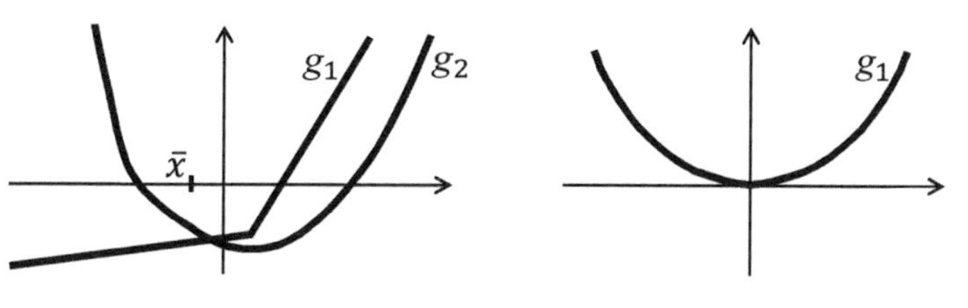

Fig. 5.7: Examples of when the Slater condition holds (left) and fails (right).

Example 5.16 (strong duality without Slater constraint qualification). In the setting of Example 5.15, the Slater constraint qualification is sufficient for strong duality but by no means necessary. For instance, consider the problem of minimizing x subject to $x^2 \leq 0$; see Figure 5.7(right) where $g_1(x) = x^2$. The Slater constraint qualification fails, but strong duality holds.

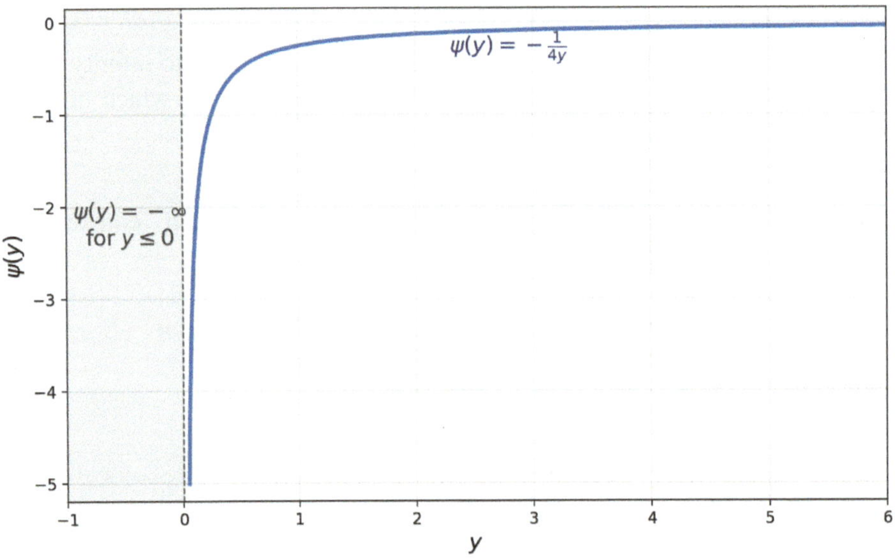

Fig. 5.8: A dual function that achieves strong duality without relying on the Slater condition; see Example 5.16.

Detail. The setting involves the Rockafellian given by

$$f(u,x) = x + \iota_{(-\infty,0]}(x^2 + u).$$

If $u > 0$, then $x^2+u \leq 0$ can't hold for any x and the minimum value $p(u) = \inf_x f(u,x) = \infty$. If $u < 0$, then the constraint $x^2 + u \leq 0$ holds for $x \in [-\sqrt{-u}, \sqrt{-u}]$. In summary, we obtain that the min-value function

$$p(u) = \inf_x f(u,x) = \begin{cases} -\sqrt{-u} & \text{for } u \leq 0 \\ \infty & \text{otherwise} \end{cases}$$

so the requirement of p not jumping to infinity fails and Theorem 5.13 doesn't apply.

Next, let's compute the dual objective function. By Example 5.5, the Lagrangian has

$$l(x,y) = \begin{cases} x + yx^2 & \text{for } y \geq 0 \\ -\infty & \text{otherwise.} \end{cases}$$

The dual objective function at y equals $\inf_x l(x, y)$. We find this value by computing a zero of the partial derivative $l'_x(x, y)$. Specifically, for $y > 0$, $l'_x(x, y) = 1 + 2xy = 0$ has the zero $x = -1/(2y)$. Since the Lagrangian is smooth and convex in x, the optimality condition in Theorem 1.6 establishes that $-1/(2y)$ is a minimizer of the Lagrangian and

$$l(x, y) = x + yx^2 = -\frac{1}{2y} + y\left(\frac{-1}{2y}\right)^2 = -\frac{1}{4y}.$$

Since $\inf_x l(x, 0) = \inf_x x = -\infty$, we conclude that the dual objective function has

$$\psi(y) = \begin{cases} -1/(4y) & \text{for } y > 0 \\ -\infty & \text{otherwise} \end{cases}$$

as seen in Figure 5.8. Consequently, $\sup \psi = 0$. The actual problem has minimum value 0 because the constraint $x^2 \le 0$ dictates $x = 0$. Thus, strong duality holds. □

5.E Reformulations

Strong duality presents numerous possibilities for reformulations. Even for difficult problems without easily obtainable strong duality, there might be parts of a formulation that can be reworked using duality. We'll illustrate the possibilities with two examples involving an infinite number of constraints.

Example 5.17 (linear constraints under uncertainty). Suppose that the inequality constraint in the problem

$$\underset{x \in X}{\text{minimize}}\ f_0(x) \ \text{ subject to } \ a^\top x \le \alpha$$

is unsettled in the sense that the n-dimensional vector a is unknown. Following Chapter 3, we could have modeled a using a random vector with a probability distribution. However, when we lack data from which to build a distribution or when we seek guaranteed feasibility for a range of values of a, a reasonable alternative is to adopt the following simple model of uncertainty: Given a nominal value $\bar{a} = (\bar{a}_1, \dots, \bar{a}_n)$ of a and a nonnegative scaling vector $s = (s_1, \dots, s_n)$, we *assume* that the vector a has the possible values

$$a = (\bar{a}_1 + s_1\xi_1, \dots, \bar{a}_n + s_n\xi_n),$$

where $\xi = (\xi_1, \dots, \xi_n)$ is allowed to vary within a set

$$\Xi = \left\{\xi \,\middle|\, \sum_{j=1}^n |\xi_j| \le \beta,\ |\xi_j| \le 1,\ j = 1, \dots, n\right\}$$

for some $\beta \ge 0$.

Figure 5.9 illustrates the situation when $n = 2$, $\beta = 1.4$, $\bar{a} = (0.2, 0.3)$, and $s = (0.5, 0.5)$. Since β is a number between one and two, the set Ξ becomes an octagon. The vector s shrinks the set by a factor of 0.5 and there's a shift to the center $\bar{a}$. This produces the set of possible a-vectors on the right in the figure.

Generally, any value of a_j between $\bar{a}_j - s_j$ and $\bar{a}_j + s_j$ might be considered, but this is further restricted by the size of β. If $\beta < n$, then not all components of a can be at their extreme values.

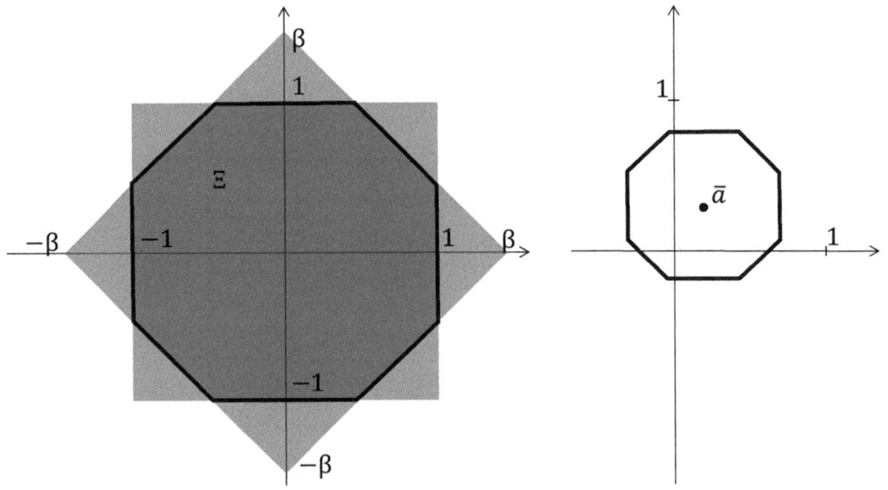

Fig. 5.9: The set Ξ with $\beta = 1.4$ (left) and the possible values of the vector $a = (a_1, a_2)$ centered at $\bar{a} = (0.2, 0.3)$ with $s = (0.5, 0.5)$ (right).

We seek a solution satisfying $a^\top x \leq \alpha$ for every considered value of a and impose the constraints

$$a(\xi)^\top x \leq \alpha \;\; \forall \xi \in \Xi, \quad \text{where } a(\xi) = (\bar{a}_1 + s_1\xi_1, \ldots, \bar{a}_n + s_n\xi_n).$$

Strong duality helps us to overcome the apparent implementation challenge associated with the now infinite number of constraints.

Detail. The infinite collection of constraints is equivalent to the single constraint

$$\sup_{\xi \in \Xi} a(\xi)^\top x \leq \alpha,$$

where the left-hand side can be expressed by

$$\begin{aligned}\sup_{\xi \in \Xi} a(\xi)^\top x &= \sum_{j=1}^n \bar{a}_j x_j + \sup_{\xi \in \Xi} \sum_{j=1}^n s_j x_j \xi_j \\ &= \bar{a}^\top x + \max\Big\{ \sum_{j=1}^n s_j |x_j| \xi_j \;\Big|\; \sum_{j=1}^n \xi_j \leq \beta,\; 0 \leq \xi_j \leq 1,\; j = 1, \ldots, n \Big\}.\end{aligned}$$

The change to only nonnegative ξ_j is permitted due to the fact that $s_j \geq 0$. Let's fix x. The previous expression is the maximum value of a linear optimization problem; a maximizer exists because the maximization takes place over a nonempty and bounded feasible set. This means we can replace "sup" by "max." Moreover, by strong duality from Example 5.11, the maximum value can be computed just as well by solving the corresponding minimization problem, which is referred to as the primal problem in Example 5.11. Specifically, we seek the maximum value of the problem

$$\underset{\xi \in \mathbb{R}^n}{\text{maximize}}\ b^\top \xi \ \text{ subject to } \ A^\top \xi \leq c, \tag{5.7}$$

where

$$b = \big(s_1|x_1|, \dots, s_n|x_n|\big), \qquad c = (\beta, 1, \dots, 1, 0, \dots, 0),$$

and A is an $n \times (1 + 2n)$-matrix with ones in the first column, the $n \times n$ identity matrix occupies the next n columns, and the negative of the $n \times n$ identity matrix fills out the rest. By Example 5.11, the primal problem is to

$$\underset{w \in \mathbb{R}^{1+2n}}{\text{minimize}}\ c^\top w \ \text{ subject to } \ Aw = b,\ w \geq 0.$$

In more detail, this becomes

$$\underset{w \geq 0}{\text{minimize}}\ \beta w_0 + \sum\nolimits_{j=1}^n w_j \ \text{ subject to } \ w_0 + w_j - w_{n+j} = s_j|x_j|, \ \ j = 1, \dots, n.$$

We can eliminate $w_{n+1}, \dots, w_{2n}$ and obtain the equivalent formulation

$$\underset{w_0, \dots, w_n \geq 0}{\text{minimize}}\ \beta w_0 + \sum\nolimits_{j=1}^n w_j \ \text{ subject to } \ w_0 + w_j \geq s_j|x_j|, \ \ j = 1, \dots, n.$$

Since the minimum value here matches the corresponding maximum value in (5.7) by Example 5.11, we obtain that

$$\begin{aligned} &\sup\nolimits_{\xi \in \Xi} a(\xi)^\top x \\ &= \bar{a}^\top x + \inf\Big\{\beta w_0 + \sum\nolimits_{j=1}^n w_j \ \Big|\ w_0 \geq 0,\ w_0 + w_j \geq s_j|x_j|,\ w_j \geq 0,\ j = 1, \dots, n\Big\}. \end{aligned}$$

Thus, finding x such that $\sup_{\xi \in \Xi} a(\xi)^\top x \leq \alpha$ is equivalent to finding x and $w_0, \dots, w_n$ such that

$$\bar{a}^\top x + \beta w_0 + \sum\nolimits_{j=1}^n w_j \leq \alpha, \ \ w_0 \geq 0, \ \ w_0 + w_j \geq s_j|x_j|, \ \ w_j \geq 0, \ \ j = 1, \dots, n.$$

The only remaining nonlinear parts are the $|x_j|$-terms, but these can be reformulated by doubling the number of constraints because $s_j \geq 0$. The problem of minimizing $f_0(x)$ over $x \in X$ subject to $a(\xi)^\top x \leq \alpha$ for all $\xi \in \Xi$ is therefore equivalent to

$$\begin{aligned}\underset{x\in\mathbb{R}^n, w\in\mathbb{R}^{1+n}}{\text{minimize}}\ f_0(x)\ \text{subject to}\ & \bar{a}^\top x + \beta w_0 + \sum\nolimits_{j=1}^n w_j \le \alpha \\ & w_0 + w_j - s_j x_j \ge 0,\ \ j = 1,\dots,n \\ & w_0 + w_j + s_j x_j \ge 0,\ \ j = 1,\dots,n \\ & x \in X,\ \ w_0, w_j \ge 0,\ \ j = 1,\dots,n.\end{aligned}$$

The infinite collection of constraints in n variables is now expressed by $2 + 3n$ constraints using $1 + 2n$ variables. The reformulation remains valid for arbitrary f_0 and X as they didn't enter the derivation. □

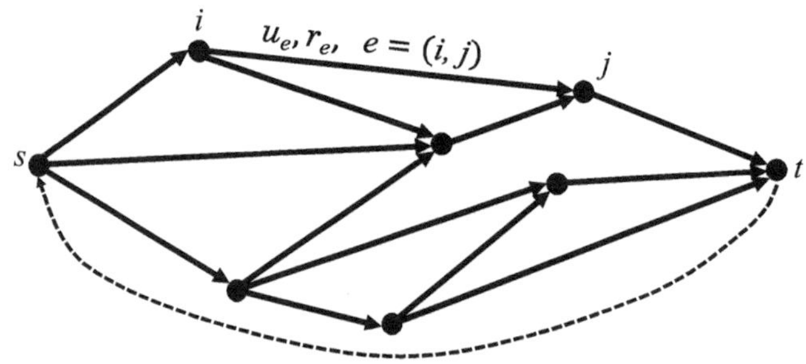

Fig. 5.10: Network for the maximum-flow network-interdiction problem with edge capacities u_e and interdiction costs r_e.

Example 5.18 (maximum-flow interdiction). We return to the maximum-flow problem of Example 4.11 with vertex set V, edge set E, source vertex s, and sink vertex t. A network operator would like to maximize the flow of a commodity from s to t across the edges but is now restricted by an interdictor that brings edge capacities to zero by destroying edges. This setting leads to the *maximum-flow network-interdiction problem*. Using strong duality, we can formulate the problem as a mixed-integer linear optimization problem.

Detail. Each edge $e \in E$ has a capacity $u_e > 0$ and interdiction cost $r_e > 0$; see Figure 5.10. The interdiction cost is the amount of some resource necessary to destroy edge e, i.e., reduce that edge's capacity to zero. We also define the outgoing and incoming edges for vertex i as

$$E_i^+ = \{(i,j) \in E \mid j \in V\} \quad \text{and} \quad E_i^- = \{(j,i) \in E \mid j \in V\},$$

respectively. Let $x_e = 1$ if edge e is interdicted, and let $x_e = 0$ otherwise. The variable y_e denotes the flow on edge $e \in E$ and y_a the flow on an artificial "return edge" $a = (t, s) \notin E$; see Figure 5.10. Let n be the number of edges in E and $y = (y_e, e \in E)$. Then, for a given interdiction plan $x = (x_e, e \in E)$, the maximum flow through the network is

$$f(x) = \sup_{y_a, y} \Big\{ y_a \;\Big|\; \sum_{e \in E_i^+} y_e - \sum_{e \in E_i^-} y_e = \delta_i y_a \;\; \forall i \in V, \\ 0 \le y_e \le u_e(1 - x_e) \;\; \forall e \in E \Big\},$$

with $\delta_s = 1$, $\delta_t = -1$, and $\delta_i = 0$ for all $i \in V \setminus \{s, t\}$; see Example 4.11. The equality constraints ensure flow-balance at each vertex. The last set of inequalities forces the effective capacity of an edge to zero if $x_e = 1$ and otherwise leaves it at the nominal u_e. Equivalently, we retain the original capacities but subtract the excess flow from the total flow and this results in the formulation

$$f(x) = \max_{y_a, y} \Big\{ y_a - \sum_{e \in E} x_e y_e \;\Big|\; \sum_{e \in E_i^+} y_e - \sum_{e \in E_i^-} y_e = \delta_i y_a \;\; \forall i \in V, \\ 0 \le y_e \le u_e \;\; \forall e \in E \Big\}.$$

The maximum is attained because the feasible set is nonempty and the variables $\{y_e, e \in E\}$ are bounded. Thus, it's permissible to replace "sup" by "max." The *maximum-flow network-interdiction problem* then takes the form

$$\boxed{\operatorname*{minimize}_{x \in \mathbb{R}^n} f(x) \;\; \text{subject to} \;\; \sum_{e \in E} r_e x_e \le \alpha, \;\; x \in \{0,1\}^n}$$

where α is the amount of interdiction resource available. Since minimizing f is equivalent to minimizing a scalar τ subject to the constraint $f(x) \le \tau$, the maximum-flow network-interdiction problem can be viewed as resulting in an infinite number of constraints. Using strong duality, we can reformulate the problem in terms of a finite number of linear equalities and inequalities.

Let's fix $x \ge 0$. We see that $-f(x)$ is the minimum value of the linear optimization problem

$$\operatorname*{minimize}_{y_a, y} \; -y_a + \sum_{e \in E} x_e y_e \;\; \text{subject to} \;\; \sum_{e \in E_i^+} y_e - \sum_{e \in E_i^-} y_e = \delta_i y_a \;\; \forall i \in V \\ 0 \le y_e \le u_e \;\; \forall e \in E.$$

We plan to use strong duality from Example 5.11 and just need to place the minimization problem in the right form by eliminating upper bounds on the variables. Introducing the additional variables $y'_e, e \in E$, this leads to the reformulation

$$\operatorname*{minimize}_{y} \; -y_a + \sum_{e \in E} x_e y_e \;\; \text{subject to} \;\; \sum_{e \in E_i^+} y_e - \sum_{e \in E_i^-} y_e = \delta_i y_a \;\; \forall i \in V \\ y_a \ge 0, \;\; y_e + y'_e = u_e, \;\; y_e \ge 0, \;\; y'_e \ge 0 \;\; \forall e \in E,$$

where we redefine y to be the longer vector $(y_a, y_e, e \in E, y'_e, e \in E)$; the additional constraint $y_a \ge 0$ doesn't change the minimum value. The reformulation is therefore of the form:

$$\underset{y}{\text{minimize}}\ c^\top y \quad \text{subject to} \quad Ay = b, \quad y \geq 0,$$

where $c = (-1, x, 0, \ldots, 0)$ is a $(1+2n)$-dimensional vector and $b = (0, \ldots, 0, u_e, e \in E)$ is $(m+n)$-dimensional, with m being the number of vertices. Moreover,

$$A = \begin{bmatrix} -\delta_1 & a_{11} & \cdots & a_{1n} & 0 \cdots 0 \\ -\delta_2 & a_{21} & \cdots & a_{2n} & 0 \cdots 0 \\ \vdots & \vdots & \ddots & \vdots & \vdots \ddots \vdots \\ -\delta_m & a_{m1} & \cdots & a_{mn} & 0 \cdots 0 \\ 0 & 1 & \cdots & 0 & 1 \cdots 0 \\ \vdots & \vdots & \ddots & \vdots & \vdots \ddots \vdots \\ 0 & 0 & \cdots & 1 & 0 \cdots 1 \end{bmatrix}$$

is an $(m+n) \times (1+2n)$-matrix with $a_{ie} = 1$ if $e \in E_i^+$, $a_{ie} = -1$ if $e \in E_i^-$, and zero otherwise. By Example 5.11, the corresponding dual problem is

$$\underset{v \in \mathbb{R}^m, w \in \mathbb{R}^n}{\text{maximize}}\ b^\top \begin{bmatrix} v \\ w \end{bmatrix} \quad \text{subject to} \quad A^\top \begin{bmatrix} v \\ w \end{bmatrix} \leq c.$$

Explicitly, this problem amounts to

$$\underset{v,w}{\text{maximize}} \sum\nolimits_{e \in E} u_e w_e \quad \text{subject to} \quad v_i - v_j + w_e \leq x_e \ \ \forall (i,j) = e \in E$$
$$v_t - v_s \leq -1, \quad w_e \leq 0 \ \ \forall e \in E.$$

Since the maximum value here gives $-f(x)$ by Example 5.11, $f(x)$ coincides with the minimum value of $\sum_{e \in E} -u_e w_e$ under the same constraints. The overall problem can then be formulated as that of minimizing $\sum_{e \in E} -u_e w_e$ over x, v, w subject to these constraints as well as the original constraints on x. We can also replace "≤ -1" by "$= -1$" because of the structure of the problem and this eliminates redundancy. Putting all together, the maximum-flow network-interdiction problem can equivalently be stated as

$$\underset{x,v,w}{\text{minimize}} \sum\nolimits_{e \in E} -u_e w_e \quad \text{subject to} \quad v_i - v_j + w_e \leq x_e \ \ \forall (i,j) = e \in E$$
$$v_t - v_s = -1$$
$$\sum\nolimits_{e \in E} r_e x_e \leq \alpha$$
$$w_e \leq 0, \quad x_e \in \{0,1\} \ \ \forall e \in E.$$

We conclude that an objective function involving a supremum over an uncountable number of possibilities can in this case be reformulated into a mixed-integer linear optimization problem.

As a concrete instance, let's consider the evacuation of Washington, D.C., after a terrorist attack and the amount of traffic that can "flow" out of Washington. A robust plan needs to consider the effect of destroyed roadways and bridges. The maximum-flow network-interdiction problem informs decision makers about evacuation routes as well as points of vulnerability. A data set involving all roads with speed limits of 30 miles per hour or higher in Maryland, Virginia, and Washington, D.C., gives a network of vertices and edges. There are 152 vertices in the Baltimore-Washington region representing points of departure and they're connected to an artificial source vertex s. At the periphery, 35 vertices represent evacuation centers and they're connected to an artificial sink vertex t. The resulting network has 3672 vertices and 10031 edges. If $\alpha = 4$ so that only four edges can be interdicted, we obtain a minimizer of the maximum-flow network interdiction problem visualized in Figure 5.11; four marked locations have been interdicted and the flow out of Washington to the evacuation centers is reduced from a nominal capacity of 194 to 146 units. The network is moderately resilient: the reduction from 194 to 146 is significant but not catastrophic. Moreover, the points of interdiction could be proposed as locations where hardening of the infrastructure might be especially beneficial; cf. [31] for further information. □

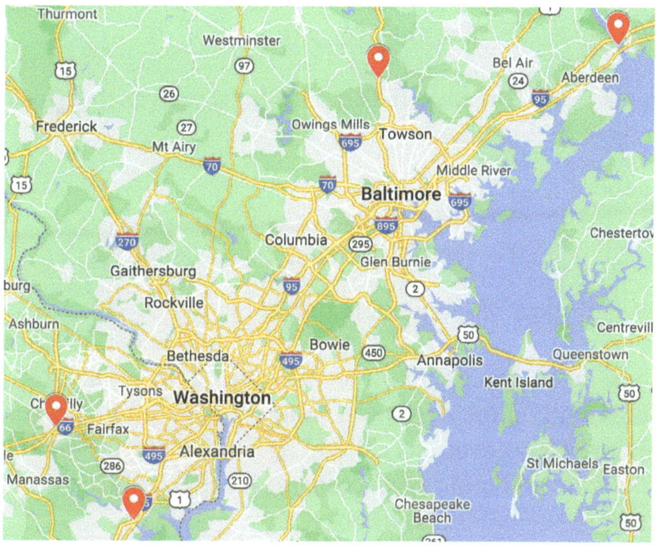

Fig. 5.11: Optimal interdiction (marked) in a maximum-flow network-interdiction problem involving 3672 vertices and 10031 edges; see [30, Section 5.D].

5.F Summary and Study Guide

In this chapter, we learned how to carry out sensitivity analysis and to reformulate difficult problems using duality theory. It also became clear that multipliers in the KKT condition are variables in dual optimization problems.

- A Rockafellian function f encodes changes to an actual problem of minimizing $f_0(x)$. It has two input vectors u and x, with u representing the change relative to f_0 so that $f(u,x) = f_0(x)$ when $u = 0$.
- Sensitivity analysis aims to determine how much the min-value function $p(u) = \inf_x f(u,x)$ changes as u varies near 0; recall that $p(0)$ is the minimum value of the actual problem. Specifically, we seek subgradients of the min-value function p at 0, which quantify the rate of change of p at 0 and these can be identified by computing subgradients of the Rockafellian f. Under convexity, we've that

$$p(u) \geq p(0) + y^\top u \quad \forall u \text{ and } y \in \partial p(0).$$

- Given a vector y, the problem of minimizing $f(u,x) - y^\top u$ with respect to u and x is a Rockafellian relaxation of the actual problem. It produces a lower bound on the minimum value of the actual problem.
- A Rockafellian defines a Lagrangian

$$l(x,y) = \inf_u f(u,x) - y^\top u,$$

which takes simple forms for common Rockafellians such as those encoding right-hand side perturbations of inequalities.

- A Lagrangian defines a

$$\text{primal objective function } \varphi(x) = \sup_y l(x,y)$$
$$\text{dual objective function } \psi(y) = \inf_x l(x,y).$$

The former tends to be identical to the actual objective function $f_0(x)$. The latter provides a lower bound on the minimum value $\inf_x \varphi(x)$; this property is called weak duality. Minimizing $l(x,y)$ over x is called Lagrangian relaxation. It can be a practical approach to a difficult problem, but in general it only provides a lower bound on the minimum value of the actual problem.

- Minimizing the primal objective function is a primal problem. Maximizing the dual objective function is a dual problem.
- Strong duality holds if maximizing the dual objective function produces the same value as minimizing the primal objective function. This holds generally for Rockafellians that are convex and also satisfy some technical properties.
- For the linear optimization problem

$$\underset{x}{\text{minimize}}\ c^\top x \text{ subject to } Ax = b,\ x \geq 0,$$

and a standard Rockafellian encoding right-hand side perturbation of constraints, we obtain a primal problem that's equivalent to this problem and a dual problem that simplifies to

$$\underset{y}{\text{maximize}}\ b^\top y \ \text{ subject to } \ A^\top y \leq c.$$

This primal-dual pair satisfies strong duality except, possibly, when the primal problem is infeasible.

- Strong duality allows us to reformulate problems with an infinite number of constraints or with objective functions that are defined in terms of maximization problems. The latter case enables us to convert a min-max problem into a min-min problem, with computational benefits.

5.G Exercises

Exercise 5.1 (Rockafellians). For nonnegative p_i and θ, consider the following Rockafellians:

(a) $$f(u,x) = \sum_{i=1}^m \big((p_i + u_i) f_i(x) + \iota_{[0,\infty)}(p_i + u_i)\big)$$

(b) $$f(u,x) = f_0(x) + \sum_{i=1}^q \iota_{(-\infty,0]}\big(g_i(x) + u_i\big)$$

(c) $$f(u,x) = f_0(x) + \sum_{i=1}^m \Big(\iota_{\{0\}}\big(f_i(x) + u_i\big) + \tfrac{1}{2}\theta u_i^2\Big).$$

In each case, interpret the perturbation encoded by the Rockafellian and write the unperturbed problem without the use of indicator functions.

Exercise 5.2 (sensitivity analysis). Modify the derivations in Example 5.4 to determine the sensitivity in the minimum value of the problem

$$\underset{x\in\mathbb{R}}{\text{minimize}}\ \exp(-x) \ \text{ subject to } \ x \geq -1, \ \ x \leq 1$$

to changes in the second constraints as encoded by the Rockafellian

$$f(u,x) = \exp(-x) + \iota_C(u,x), \quad \text{where } C = \{(u,x) \mid -x \leq 1, \ u + x \leq 1\}.$$

Compare the estimated changes with the actual minimum values as u varies near from 0.

Exercise 5.3 (Lagrangian relaxation). Consider the one-dimensional problem of minimizing $(x-1)^2$ subject to $x = 0$ and the Rockafellian

$$f(u,x) = (x-1)^2 + \iota_{\{0\}}(x+u) + \tfrac{1}{2}\theta u^2,$$

where θ is a nonnegative parameter. Solve the Rockafellian relaxation for different multipliers y and parameters θ. Discuss the results.

Exercise 5.4 (Lagrangian). In the setting of Example 5.5 with $f_0(x) = \exp(x)$ and $g_1(x) = -x$, write the formula for the Lagrangian, plot the Lagrangian as a function of x for many different y, and compare it with the function $f_0 + \iota_{[0,\infty)}$. Confirm that (5.2) holds.

Exercise 5.5 (dual problem). Consider the linear optimization problem

$$\operatorname*{minimize}_{x\in\mathbb{R}^2}\ x_1 - 2x_2 \ \text{ subject to } \ 2x_1 + x_2 = 3,\ x_2 \geq 0,\ x_1 \geq 0,$$

which has minimizer $(0,3)$. (a) For the Rockafellian in Example 5.9, write the corresponding dual problem. (b) Compute the maximum value of the dual problem and compare it with the minimum value of the actual problem.

Exercise 5.6 (primal-dual pairs). Consider the linear optimization problem

$$\operatorname*{minimize}_{x\in\mathbb{R}^3}\ x_1 + x_2 + x_3 \ \text{ subject to } \ x_2 = 1,\ x_1 \geq 0,\ x_2 \geq 0,\ x_3 \geq 0.$$

(a) For this specific instance, identify A, b, and c in the standard form: minimize $c^\top x$ subject to $Ax = b$ and $x \geq 0$. (b) Write the dual problem from Example 5.9 explicitly in this case. (c) Find the maximum value of the dual problem. (d) For any y and $x \geq 0$, the Lagrangian in this case becomes

$$l(x,y) = c^\top x + y^\top(b - Ax).$$

Given $y = 1$, compute the minimum value of the Lagrangian with respect to $x \geq 0$ and compare it with the maximum value of the dual problem.

Exercise 5.7 (primal-dual pairs). (a) Show that the linear optimization problem of minimizing $c^\top x$ subject to $Ax = b$ and $x \geq 0$, with A being an $m \times n$-matrix, is convex. (b) For $A = [1 \ \ 1]$, $b = 1$, and $c = (0,1)$, find the minimum value. (c) Write the dual problem from Example 5.9 explicitly in this case. (d) Find the maximum value of the dual problem in (c). (e) For any y and $x \geq 0$, the Lagrangian in this case becomes

$$l(x,y) = c^\top x + y^\top(b - Ax).$$

Given $y = 0$, compute the minimum value of the Lagrangian with respect to $x \geq 0$ and compare it with the maximum value of the dual problem.

Exercise 5.8 (infeasible primal-dual pairs). In the setting of Example 5.11, construct a primal-dual pair with both problems being infeasible.

Exercise 5.9 (primal-dual pairs). Given the functions

$$f_0(x) = 2x_1^2 + x_2^2 + 3x_3^2 \quad \text{and} \quad g(x) = -2x_1 - 3x_2 - x_3 + 1,$$

consider the problem of minimizing $f_0(x)$ subject to $g(x) \leq 0$ and the Rockafellian defined by $f(u,x) = f_0(x) + \iota_{(-\infty,0]}(g(x) + u)$. Determine the corresponding Lagrangian as well as the primal and dual problems. Solve the problems and verify strong duality. Check the assumptions in Theorem 5.13.

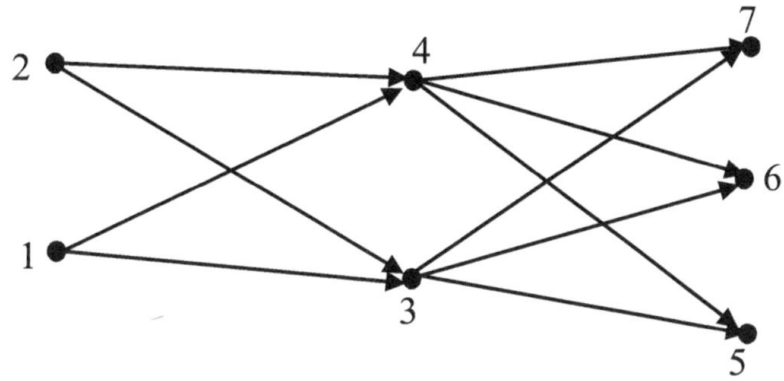

Fig. 5.12: Supply network for a retailer in Exercise 5.10.

Exercise 5.10 (supply chain). Suppose that the supply chain of a retailer consists of two factories, two warehouse, and three stores. In Figure 5.12, each facility is represented by a vertex and the movement of a commodity is represented by an edge. Vertices 1 and 2 represent factories, vertices 3 and 4 represent warehouses, and vertices 5, 6, and 7 represent stores. We assume there's just one type of commodity. Let x_{ij} be a continuous nonnegative variable that measures the amount of commodity that moves between vertex i and vertex j, where $i, j = 1, 2, \dots, 7$. If a pair of vertices isn't connected with an edge in Figure 5.12, no movement can take place directly between that pair of vertices. Use E for the set of edges (i, j).

(a) Let c_{ij} be the per-unit transportation cost between vertex i and vertex j. Formulate an objective function that represents the total transportation cost for the retailer.

(b) Assume that there's a supply of 5 and 4 units of the commodity at vertices 1 and 2, respectively. Also, assume that each store (vertices 5, 6, and 7) requires 3 units of the commodity. Using the objective function from (a), formulate a minimum transportation-cost model with flow-balance constraints at each vertex. There are no other constraints.

(c) The retailer's supply chain is vulnerable to terrorist attacks. Let w_{ij} be a binary variable that's 1 if edge (i, j) is attacked and 0 otherwise. If an edge is attacked, the cost increases from c_{ij} to $10c_{ij}$. For a given w_{ij}, $i, j = 1, 2, \dots, 7$, modify the minimum transportation-cost model to account for these attacks.

(d) The retailer would like to know about the worst possible attacks that can take place. Use the model in (c) and formulate a max-min problem where it's assumed that at most α attacks can take place.

(e) For the Rockafellian in Example 5.9, write the corresponding dual problem to the model in (c). Combine the resulting optimization problem with the max-min problem in (d) to construct a max-max problem. Solve the maximization problem using $c_{13} = 1.0$, $c_{14} = 1.5$, $c_{23} = 0.8$, $c_{24} = 1.3$, $c_{35} = 2.1$, $c_{36} = 1.8$, $c_{37} = 0.1$, $c_{45} = 2.5$, $c_{46} = 3.5$, $c_{47} = 1.7$, and discuss the results for $\alpha = 0, 1, 2$, and 3. The discussion should include a quantitative assessment of the best responses to the worst attacks under different α.

Appendix A
Solutions to Exercises

Exercise 1.1 Solution. Derivatives are given by $f'(x) = x^2 - 1$; see Figure A.1 for plots. The Fermat rule states that $x^2 - 1 = 0$ is a necessary condition for x to be a minimizer. So no x satisfying $x^2 - 1 \neq 0$ can be a minimizer of f. These are all $x \in (-\infty, -1) \cup (-1, 1) \cup (1, \infty)$, i.e., the whole real line excluding -1 and 1.

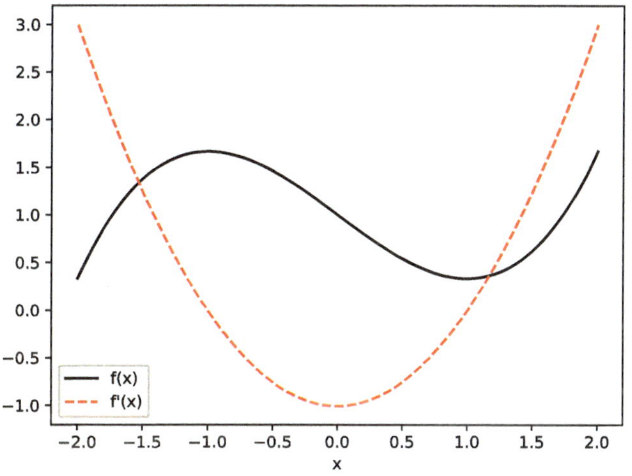

Fig. A.1: Function f (solid black line) and its derivative f' (dashed red line) in Exercise 1.1.

Exercise 1.2 Solution. Derivatives are given by $f'(x) = 6x + 3x^2 = x(6+3x)$. The Fermat rule states that $x(6 + 3x) = 0$ is a necessary condition for x to be a minimizer. So no x except $x = 0$ and $x = -2$ can be a minimizer of f.

J. O. Royset, Optimization Concepts, *Springer Series in Operations Research and Financial Engineering*,
https://doi.org/10.1007/978-3-032-36819-5

Exercise 1.3 Solution. The gradients of f are given by

$$\nabla f(x) = \begin{bmatrix} f'_{x_1}(x) \\ f'_{x_2}(x) \end{bmatrix} = \begin{bmatrix} 4x_1 \\ 2(x_2 - 1) \end{bmatrix}.$$

Figures A.2, A.3, and A.4 plot f, f'_{x_1}, and f'_{x_2}, respectively.

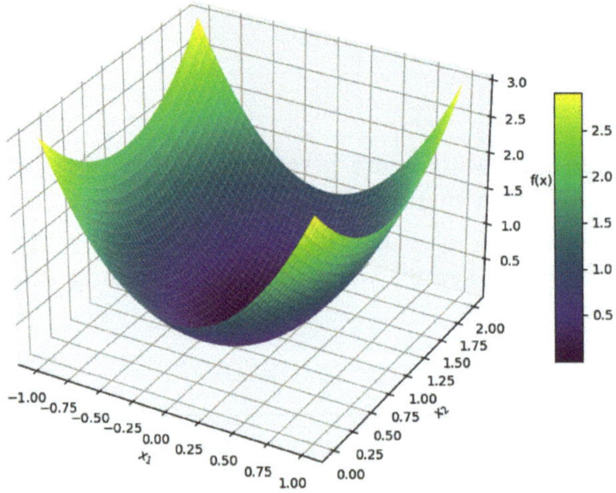

Fig. A.2: Function f in Exercise 1.3.

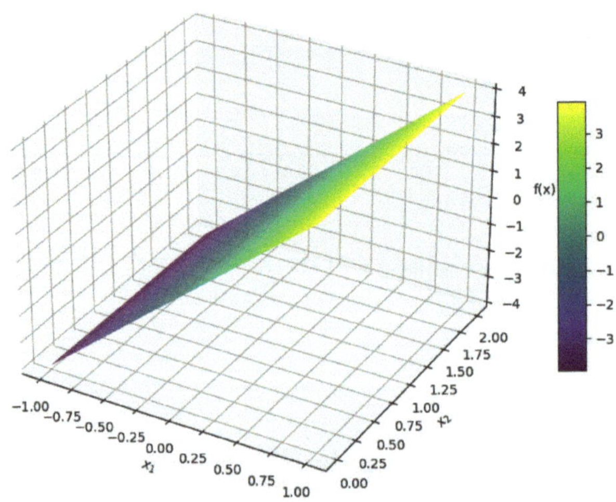

Fig. A.3: The partial derivative f'_{x_1} in Exercise 1.3.

The Fermat rule states that

$$\begin{bmatrix} 4x_1 \\ 2(x_2 - 1) \end{bmatrix} = \begin{bmatrix} 0 \\ 0 \end{bmatrix}$$

is a necessary condition for $x = (x_1, x_2)$ to be a minimizer. So no x except $(0, 1)$ can be a minimizer of f.

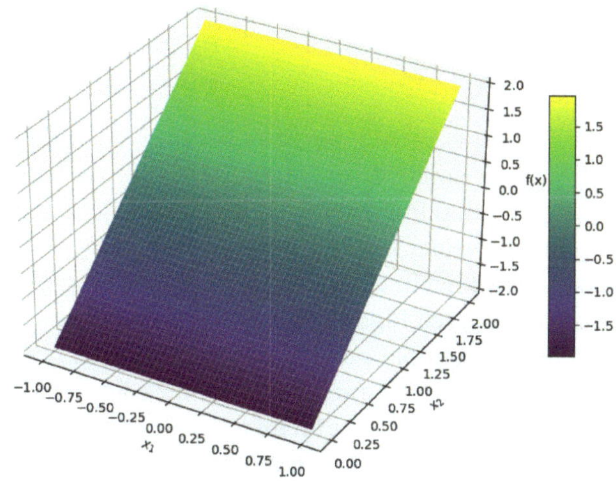

Fig. A.4: The partial derivative f'_{x_2} in Exercise 1.3.

Exercise 1.4 Solution. The gradients of f are given by

$$\nabla f(x) = \begin{bmatrix} f'_{x_1}(x) \\ f'_{x_2}(x) \end{bmatrix} = \begin{bmatrix} 6x_1 + x_2 \\ 1 + x_1 \end{bmatrix}.$$

This vector vanishes when $6x_1 + x_2 = 0$ and $1 + x_1 = 0$, which means that $x_1 = -1$ and $x_2 = 6$. By the Fermat rule, we conclude that no x except $(-1, 6)$ can be a minimizer of f.

Exercise 1.5 Solution. One can use the Python package numpy.linalg.solve to solve the system of equations $D^\top Dc = D^\top y$. For different polynomial degree n, the minimum values are:

polynomial degree n	minimum value
1	0.284949
2	0.277311
3	0.277096
4	0.272061
5	0.252653
6	0.241712
7	0.232915
8	0.114840
9	0.114774
10	0.000144

We know that the last row should be zero because a 10-degree polynomial can fit 11 points perfectly. The small error of 0.000144 is due to the imprecision in numpy.linalg.solve; keep in mind that the matrix $D^\top D$ is getting rather ill-conditioned.

In terms of the minimum value, larger n is better with $n = 10$ being "perfect" in the sense that the obtained polynomial interpolates the data. However, plots of the polynomials reveal a more nuanced picture; see Figures A.5, A.6, and A.7.

The case $n = 2$ produces a simple polynomial clearly indicating the general trend of the data. If used for the purpose of extrapolation between and beyond the data, a simple polynomial of this kind might be more useful than a high-degree polynomial that tends to be overly responsive to spurious variations in the data.

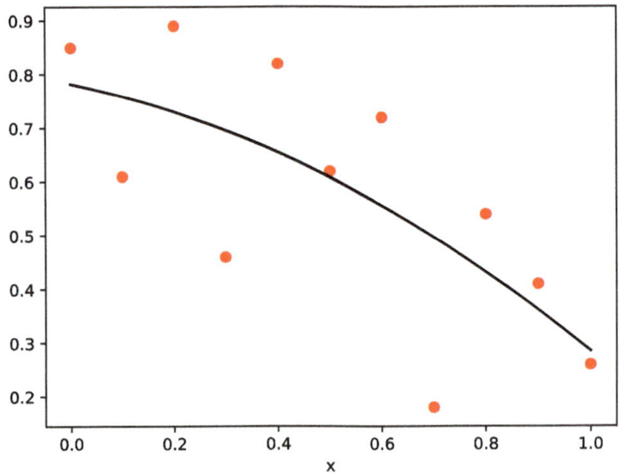

Fig. A.5: Fitted polynomial of degree $n = 2$ in Exercise 1.5.

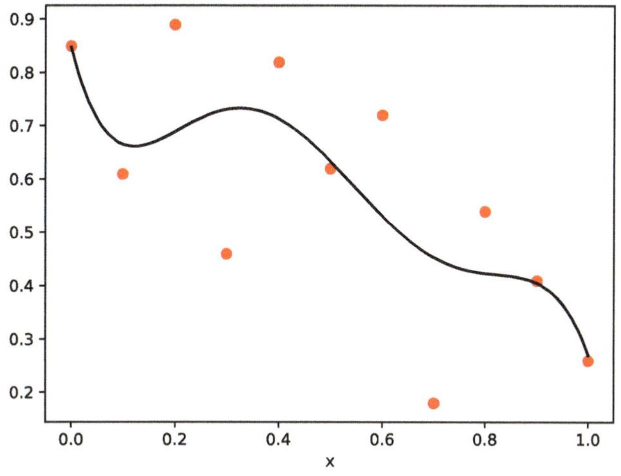

Fig. A.6: Fitted polynomial of degree $n = 5$ in Exercise 1.5.

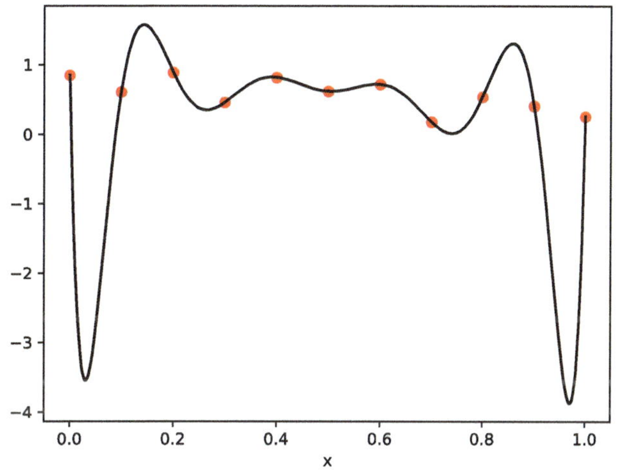

Fig. A.7: Fitted polynomial of degree $n = 10$ in Exercise 1.5.

Exercise 1.6 Solution. We apply the fact that $P(x^\star) = 1 - o/s = 4/7$; see Example 1.3. The assumed uniform distribution has $P(\xi) = 0$ for $\xi < 10$, $P(\xi) = \xi/10 - 1$ for $\xi \in [10, 20]$ and $P(\xi) = 1$ for $\xi > 20$; see Figure A.8. Thus, we need to solve the equation $x^\star/10 - 1 = 4/7$. It produces $x^\star = 110/7 \approx 15.71$. The expected demand $\mathbb{E}[\xi] = 15$, and $P(15) = 1/2$. Thus, the expected demand can't be optimal because $P(15) \neq 4/7$.

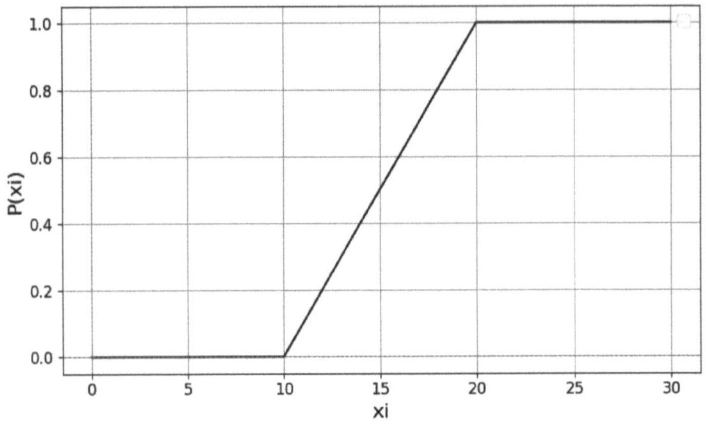

Fig. A.8: Distribution function P in Exercise 1.6.

Exercise 1.7 Solution. Since $f(x) = 2x_1^2 + (x_2 - 1)^2$, it's easy to compute the Hessian matrix and use Proposition 1.9. Specifically,

$$\nabla f(x) = \begin{bmatrix} 4x_1 \\ 2(x_2 - 1) \end{bmatrix}, \qquad \nabla^2 f(x) = \begin{bmatrix} 4 & 0 \\ 0 & 2 \end{bmatrix}.$$

The Hessian matrix is positive semidefinite because $4 \geq 0$, $2 \geq 0$, and $4 \cdot 2 - 0 \cdot 0 \geq 0$; see the gray box before Proposition 1.9. Since this holds for all x, it follows from Proposition 1.9 that f is convex.

Since f is convex, we know from Theorem 1.6 that x is a minimizer of f if and only if $\nabla f(x) = 0$. Thus, we solve the equations

$$4x_1 = 0, \quad 2(x_2 - 1) = 0$$

to obtain $x = (0, 1)$. This is the unique solution to the equations so it also the unique minimizer of f.

Exercise 1.8 Solution. We consider the Hessian rule in Proposition 1.9 and compute

$$\nabla f(x) = \begin{bmatrix} 2(x_1 - 3) \\ \vdots \\ 2(x_n - 3) \end{bmatrix}$$

so that $\nabla^2 f(x)$ becomes the $n \times n$-matrix with 2 on the diagonal and zero elsewhere. The Hessian matrix is positive semidefinite because 2 is its only eigenvalue; see the gray box before Proposition 1.9. Since this holds for all x, it follows from Proposition 1.9 that f is convex.

Exercise 1.9 Solution. We proceed in four steps. First, we conclude that the function given by $f(x) = 2x_1^2 + 3x_2^2 - 1$ is convex. This holds by the Hessian rule in Proposition 1.9. Specifically,

$$\nabla f(x) = \begin{bmatrix} 4x_1 \\ 6x_2 \end{bmatrix}, \qquad \nabla^2 f(x) = \begin{bmatrix} 4 & 0 \\ 0 & 6 \end{bmatrix},$$

which is positive semidefinite because $4 \geq 0$, $6 \geq 0$, and $4 \cdot 6 - 0 \cdot 0 \geq 0$. Second, we invoke the level-set rule (Proposition 1.7(b)) to conclude that the set

$$\{x \mid 2x_1^2 + 3x_2^2 - 1 \leq 0\}$$

is convex. Third, we realize that $\{x \mid x_1 + x_2 = 0.1\}$ is convex by using Proposition 1.7(a). Fourth, we invoke the intersection rule (Proposition 1.7(c)) to reach the conclusion that

$$\{x \mid 2x_1^2 + 3x_2^2 - 1 \leq 0\} \cap \{x \mid x_1 + x_2 = 0.1\} = \{x \mid 2x_1^2 + 3x_2^2 - 1 \leq 0,\ x_1 + x_2 = 0.1\}$$

is convex. It might be helpful to draw these sets.

Exercise 1.10 Solution. By Proposition 1.8(a,b,c), we just need to show that the functions given by

$$g_{ik}(x) = \exp(c_{ik}^\top x + \beta_{ik})$$

are convex for each i and k. We can write g_{ik} as a composition of the exponential function and an affine function, and thus it's convex by Proposition 1.8(d). Specifically,

$$g_{ik}(x) = h(c_{ik}^\top x + \beta_{ik}), \quad \text{where } h(z) = \exp(z).$$

It's clear that h is convex by the Hessian rule in Proposition 1.9: take the derivative twice and conclude that it's nonnegative.

Exercise 1.11 Solution. (a) By Proposition 1.8(a,b), we just need to show that the functions given by

$$g_i(x) = \exp(a_i^\top x)$$

are convex for each i. We can write g_i as a composition of the exponential function and a linear function, and thus it's convex by Proposition 1.8(d). Specifically,

$$g_i(x) = h(a_i^\top x), \quad \text{where } h(z) = \exp(z).$$

It's clear that h is convex by the Hessian rule in Proposition 1.9: take the derivative twice and conclude that it's nonnegative.

(b) In combination with (a), the level-set rule in Proposition 1.7(b) ensures that the set is convex.

Exercise 1.12 Solution. The function f is defined as a sum of terms with nonnegative weights. By Proposition 1.8(a,b), we can therefore focus on the functions given by

$$g_j(x) = \exp(-swx_j/a_j).$$

We can write g_j as a composition of the exponential function and a linear function, and thus it's convex by Proposition 1.8(d). Specifically,

$$g_j(x) = h(-swx_j/a_j), \quad \text{where } h(z) = \exp(z).$$

It's clear that h is convex by the Hessian rule in Proposition 1.9: take the derivative twice and conclude that it's nonnegative.

The intersection rule (Proposition 1.7(c)) allows us to concentrate on the sets

$$\left\{x \,\middle|\, \sum_{j=1}^{n} x_j = t\right\} \text{ and } \{x \mid x_1 \geq 0\}, \ldots, \{x \mid x_n \geq 0\}.$$

The first set is an affine set and thus convex by Proposition 1.7(a). Each set in the second group is convex by Definition 1.4: If $x^0, x^1 \in \{x \mid x_j \geq 0\}$, then

$$(1 - \lambda)x^0 + \lambda x^1 \in \{x \mid x_j \geq 0\}$$

as well regardless of $\lambda \in [0, 1]$.

Exercise 1.13 Solution. By Proposition 1.15, w is a descent direction of f at $\bar{x} = (1, 2)$ if

$$\nabla f(\bar{x})^\top w = 2(\bar{x}_1 - 1)w_1 + 4\bar{x}_2 w_2 = 8w_2 < 0.$$

Thus, any direction $w = (w_1, w_2)$ with $w_2 < 0$ is a descent direction of f at $\bar{x}$. For example, $(0, -1)$ and $(5, -5)$ are descent directions.

Exercise 1.14 Solution. We compute the gradient

$$\nabla f(x) = \begin{bmatrix} 2x_1 \\ 2x_2 \end{bmatrix}.$$

The Lipschitz condition

$$\left\|\nabla f(x) - \nabla f(\bar{x})\right\|_2 \leq \kappa \|x - \bar{x}\|_2$$

is equivalent to having

$$\left\|\nabla f(x) - \nabla f(\bar{x})\right\|_2^2 \leq \kappa^2 \|x - \bar{x}\|_2^2.$$

Writing this out in our specific instance, we find obtain

$$(2x_1 - 2\bar{x}_1)^2 + (2x_2 - 2\bar{x}_2)^2 \leq \kappa^2\big((x_1 - \bar{x}_1)^2 + (x_2 - \bar{x}_2)^2\big).$$

This simplifies to

$$4\big((x_1 - \bar{x}_1)^2 + (x_2 - \bar{x}_2)^2\big) \leq \kappa^2\big((x_1 - \bar{x}_1)^2 + (x_2 - \bar{x}_2)^2\big)$$

and we see that $\kappa = 2$ works. We reach the same result by recalling that κ can be set to the largest eigenvalue of $\nabla^2 f(x)$ across all x. By (1.12), this means that a step size $\lambda = 1/\kappa = 1/2$ is reasonable.

Exercise 1.15 Solution. Since $\sigma = 2$, the function is given by the formula $f(x) = x_1^2 + 2(x_2 - 1)^2$. This leads to the gradient

$$\nabla f(x) = \begin{bmatrix} 2x_1 \\ 4x_2 - 4 \end{bmatrix}.$$

Following the directions leading up to (1.12), we need to compute κ such that

$$\left\|\nabla f(x) - \nabla f(\bar{x})\right\|_2 \leq \kappa \|x - \bar{x}\|_2 \quad \forall x, \bar{x}.$$

For the given function, we find that

$$\begin{aligned}\left\|\nabla f(x) - \nabla f(\bar{x})\right\|_2^2 &= \left\|(2x_1,\ 4x_2 - 4) - (2\bar{x}_1,\ 4\bar{x}_2 - 4)\right\|_2^2 \\ &= (2x_1 - 2\bar{x}_1)^2 + (4x_2 - 4 - 4\bar{x}_2 + 4)^2 = 4(x_1 - \bar{x}_1)^2 + 16(x_2 - \bar{x}_2)^2 \\ &\leq 16(x_1 - \bar{x}_1)^2 + 16(x_2 - \bar{x}_2)^2 = 16\|x - \bar{x}\|_2^2.\end{aligned}$$

We conclude that

$$\left\|\nabla f(x) - \nabla f(\bar{x})\right\|_2 \leq 4\|x - \bar{x}\|_2 \quad \forall x, \bar{x}.$$

We reach the same conclusion that $\kappa = 4$ by looking at the largest eigenvalue of $\nabla^2 f(x)$ across all x. Consequently, we select $\lambda = 1/4$. A positive number smaller than that will also work according to the theory. But, generally, we prefer longer steps so we pick the largest value.

The iterations are

$$\begin{aligned}x^1 &= x^0 - \lambda\nabla f(x^0) = (2,2) - \tfrac{1}{4}(4,4) = (1,1) \\ x^2 &= x^1 - \lambda\nabla f(x^1) = (1,1) - \tfrac{1}{4}(2,0) = (1/2,1) \\ x^3 &= x^2 - \lambda\nabla f(x^2) = (1/2,1) - \tfrac{1}{4}(1,0) = (1/4,1).\end{aligned}$$

We can also compute the function values

$$\begin{aligned}f(x^0) &= 2^2 + 2(2-1)^2 = 6 \\ f(x^1) &= 1^2 + 2(1-1)^2 = 1 \\ f(x^2) &= (1/2)^2 + 2(1-1)^2 = 1/4 \\ f(x^3) &= (1/4)^2 + 2(1-1)^2 = 1/16.\end{aligned}$$

This confirms the fact that the gradient descent method indeed behaves as a descent method, i.e., the function values decrease.

Compared to (1.12), we note that $\beta = \|x^0 - x^\star\|_2^2 = 5$ so that $\beta/(2\lambda\nu) = 10/\nu$. Thus,

$$f(x^\nu) - \min f \le 10/\nu.$$

The right-hand side is therefore $10/1 = 10$, $10/2 = 5$, and $10/3 \approx 3.3$ for $\nu = 1, 2, 3$. These values are conservative bounds on the actual progress.

We see in Figure A.9 that the algorithm indeed makes progress toward the minimizer $x^\star = (0, 1)$ and the minimum value of 0.

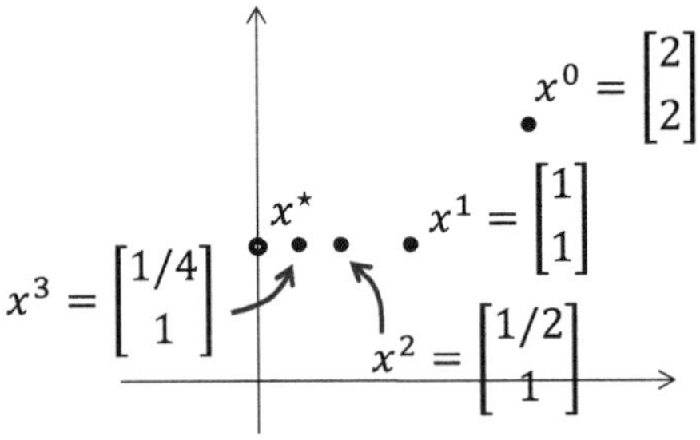

Fig. A.9: Progress of the gradient descent method in Exercise 1.15.

Exercise 1.16 Solution. Since $\sigma = 2$, $f(x) = x_1^2 + 2(x_2 - 1)^2$. This leads to the gradient and Hessian:

$$\nabla f(x) = \begin{bmatrix} 2x_1 \\ 4x_2 - 4 \end{bmatrix}, \quad \nabla^2 f(x) = \begin{bmatrix} 2 & 0 \\ 0 & 4 \end{bmatrix}.$$

The iteration is

$$x^1 = x^0 - \lambda\big(\nabla^2 f(x^0)\big)^{-1}\nabla f(x^0) = \begin{bmatrix} 2 \\ 2 \end{bmatrix} - \begin{bmatrix} 1/2 & 0 \\ 0 & 1/4 \end{bmatrix}\begin{bmatrix} 4 \\ 4 \end{bmatrix} = \begin{bmatrix} 0 \\ 1 \end{bmatrix}.$$

We move directly to the minimizer in one iteration!

Exercise 1.17 Solution. Newton's method starting at $(-1.2, 1)$ takes only six iterations to end near $(1, 1)$, but has a volatile trajectory where it goes "off the map" before it returns and converges quickly; see Figure A.10.

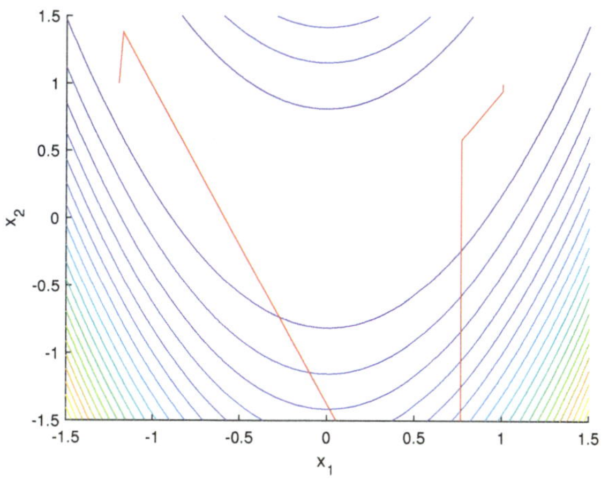

Fig. A.10: Progress of Newton's method in Exercise 1.17 across six iterations starting at $(-1.2, 1)$ and ending near $(1, 1)$.

Exercise 2.1 Solution. The feasible set C is given in Figure A.11. Since the coefficients in front of x_1 and x_2 are positive, the objective function prefers low values for x_1 and x_2. Thus, within C, the best we can do is $(0, 0)$.

The minimizer shifts to $(4, 0)$ if we prefer high x_1 and low x_2, for example, with the objective function $-x_1 + x_2$.

The minimizer shifts to $(4, 6)$ if we prefer high x_1 and high x_2, for example, with the objective function $-x_1 - x_2$.

The argmin-set becomes $\{4\} \times [0, 6]$ if we prefer high x_1 and x_2 doesn't matter, for example, with the objective function $-x_1$.

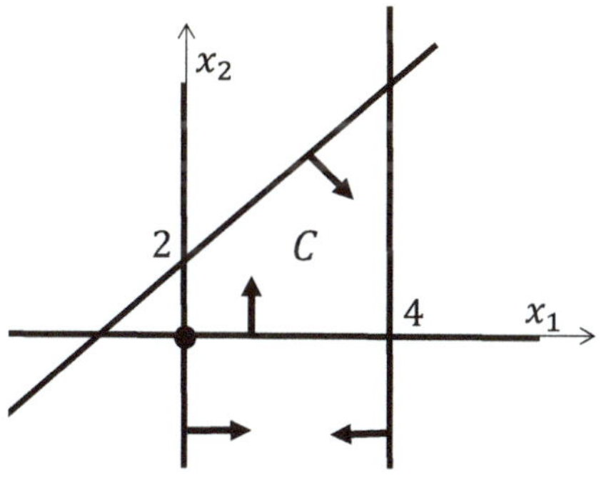

Fig. A.11: Feasible set in Exercise 2.1 with the minimizer at $(0, 0)$ marked with a dot.

Exercise 2.2 Solution. Figure A.12 shows epi f, argmin f, and min f. Proposition 2.4 helps us to conclude that f is convex. We know that f_0 is convex (e.g., use the Hessian rule) and C is convex because it's a line segment. The proposition then gives the conclusion.

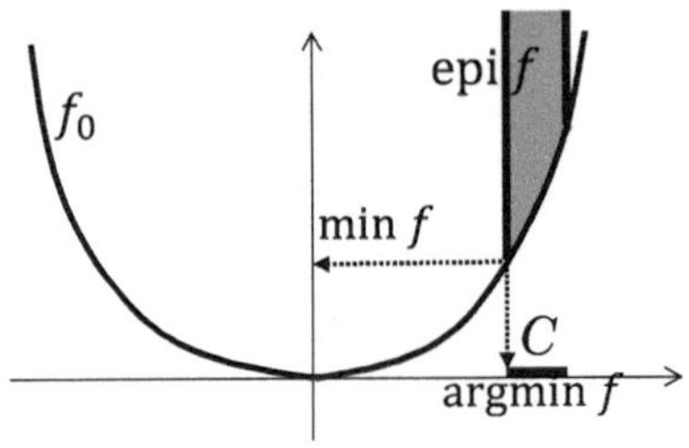

Fig. A.12: Solution of Exercise 2.2.

Exercise 2.3 Solution. We set $c = (0, 3)$, $d = (2, 4, 0, 0)$, and

$$Q = \begin{bmatrix} 4 & 0 \\ 0 & 0 \end{bmatrix}, \quad D = \begin{bmatrix} -1 & 1 \\ 1 & 0 \\ -1 & 0 \\ 0 & -1 \end{bmatrix}.$$

A and b are void. The matrix Q is positive semidefinite because $4 \geq 0$ and the other entries are zero.

Exercise 2.4 Solution. (a) The quadratic function h is convex by the Hessian rule in Proposition 1.9. We can then use the composition rule in Proposition 1.8(d) to conclude that f_0 is convex. (b) Since C is convex, Proposition 2.4 and part (a) ensure that f is convex.

Exercise 2.5 Solution. (a) $\sum_{i=1}^{5} x_{ijt} \geq r_{it}$ for all $j = 1, \ldots, 4$, $t = 1, 2, 3$. (b) $\sum_{j=1}^{4} x_{ijt} \leq a_{it}$ for all $i = 1, \ldots, 5$; $t = 1, 2, 3$. (c) 12 and 15.

Exercise 2.6 Solution. (a) $x_{pwq} + a_{pwq} = s_{pwq} + x_{pwq'}$ for all $p, w, q, q' = q + 1$ with the convention that $q = 5$ is the same as $q = 1$. (b) $0.6x_{pwq} \leq x_{pwq'}$ for all $p, w, q, q' = q + 1$ with the convention that $q = 5$ is the same as $q = 1$. (c) $x_{3,1,q} \leq 0.4x_{2,1,q}$ for all q.

Exercise 2.7 Solution. (a) $\sum_i x_{ci} = s_c$ for all c; $\sum_c x_{ci} = \sum_d y_{id}$ for all i; $\sum_i y_{id} = t_d$ for all d. (b) $\sum_c x_{ci} \leq K_i$ for all i. (c) $\sum_{c,i} \alpha_{ci} x_{ci} + \sum_{i,d} \beta_{id} y_{id}$.

Exercise 2.8 Solution. An equivalent formulation of the lasso regression problem from Example 2.6 is

$$\underset{c \in \mathbb{R}^{1+n}, z \in \mathbb{R}^{1+n}}{\text{minimize}} \ \|Dc - y\|_2^2 + \theta \sum\nolimits_{j=0}^{n} z_j \quad \text{subject to} \quad c_j \leq z_j,\ -c_j \leq z_j, \quad j = 0, 1, \ldots, n.$$

This is a quadratic optimization problem. To see that it fits the canonical form

$$\underset{x}{\text{minimize}} \ \tfrac{1}{2} x^\top Q x + a^\top x \quad \text{subject to} \quad Ax = b, Dx \leq d$$

note that

$$\|Dc - y\|_2^2 = c^\top D^\top Dc - (2D^\top y)^\top c + \|y\|_2^2.$$

We can drop the constant term $\|y\|_2^2$ since it doesn't affect the minimizer. We see that

$$Q = \begin{bmatrix} 2D^\top D & 0 \\ 0 & 0 \end{bmatrix}$$

is a $2(1+n) \times 2(1+n)$-matrix, $a = [-(2D^\top y)^\top \;\; \theta \;\; \ldots \;\; \theta]^\top$ is $2(1+n)$ dimensional, and the decision vector $x = (c, z)$ is of the same size.

Results for $n = 30$ and a range of θ follow in the table below.

θ	$\|Dc - y\|_2^2$	$\theta\|c\|_1$	number of zero terms
0.0001	0.2008	0.0114	0
0.001	0.2511	0.0134	11
0.01	0.2747	0.0141	23
0.1	0.2832	0.1158	28
1	0.5786	0.5327	30

The results are similar for smaller n. We note that the penalty term $\theta\|c\|_1$ allows us to pick a large n and then let the optimization algorithm "select" the terms to include in the polynomial being fitted. Those terms that don't improve the fit significantly are set to zero.

The last column of the table lists the number of c_j with absolute value less than 10^{-12}. We note that there's a trade-off between the two terms in the objective function and their relative importance are determined by θ. As θ increases, it becomes more important to pick a small c-vector at the expense of a poorer fit in the sense of the squared error $\|Dc - y\|_2^2$. For large θ, we obtain a "simple" polynomial involving just a few terms due to many zero coefficients.

Let's also take a look at the resulting fits. For $\theta = 0.0001$, we obtain the polynomial in Figure A.13, which seems to overfit the data. All the coefficients are nonzero.

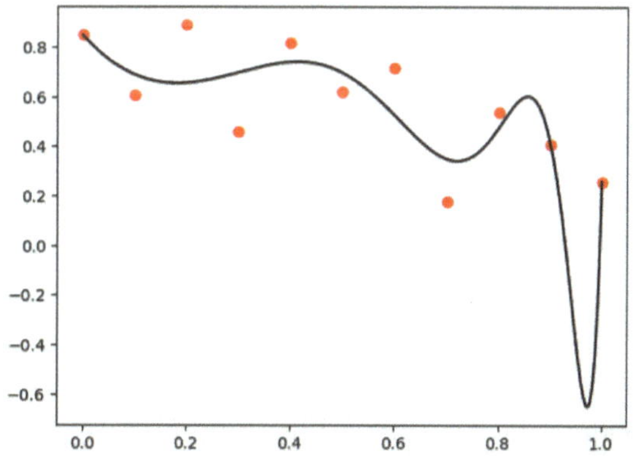

Fig. A.13: Solution of Exercise 2.8 for $\theta = 0.0001$.

For $\theta = 0.001$, we obtain the better polynomial in Figure A.14. We see how the optimization makes some of the coefficients (nearly) zero in an attempt to find a simple polynomial that also fits the data well. Eleven coefficients are essentially zero.

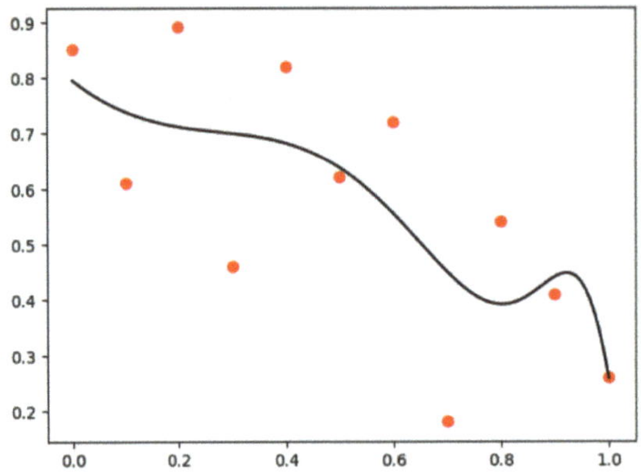

Fig. A.14: Solution of Exercise 2.8 for $\theta = 0.001$.

For $\theta = 0.01$, we obtain the polynomial in Figure A.15. This case seems to hit a good middle ground. The resulting polynomial represents the data quite well without being overly complex. After rounding of coefficients, it has the formula

$$p(x) = 0.780 - 0.213x - 0.276x^2 + 0.029x^8 + 0.033x^9 - 0.008x^{28} - 0.014x^{29} - 0.065x^{30}.$$

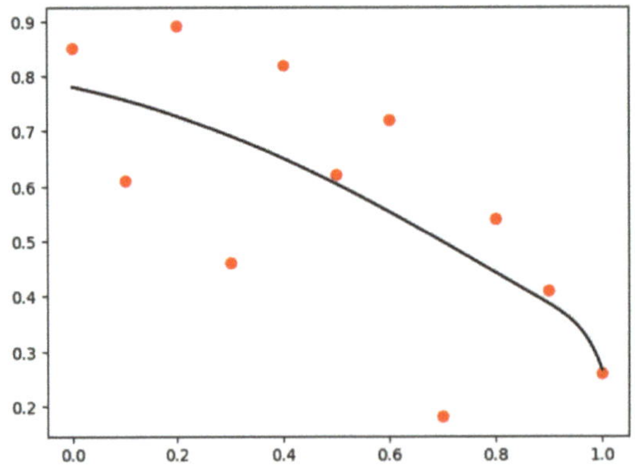

Fig. A.15: Solution of Exercise 2.8 for $\theta = 0.01$.

For $\theta = 0.1$, we obtain the polynomial in Figure A.16. This is an even simpler polynomial, and it might be the best one for making predictions:

$$p(x) = 0.730 - 0.043x - 0.385x^2.$$

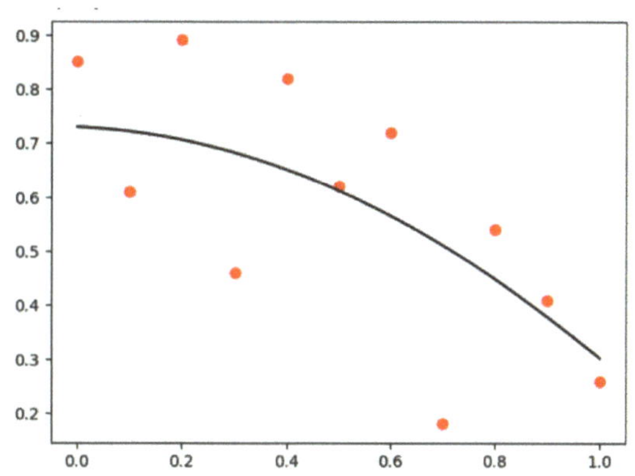

Fig. A.16: Solution of Exercise 2.8 for $\theta = 0.1$.

For $\theta = 1$, we obtain a flat line; θ is too large.

Exercise 2.9 Solution. (a) Since $f(x) = |3x|$ only has a kink at 0, we can concentrate on that point with the other points being taken care of by the usual derivative rules. By definition, a subgradient of f at 0 is any v satisfying

$$f(x) \geq f(0) + vx \quad \forall x \quad \text{or, equivalently,} \quad |3x| \geq vx \quad \forall x.$$

We see that $v \in [-3, 3]$. Thus $\partial f(0) = [-3, 3]$. Also computing subgradients at other points than 0, we find that

$$\partial f(x) = \begin{cases} \{-3\} & \text{if } x < 0 \\ [-3, 3] & \text{if } x = 0 \\ \{3\} & \text{if } x > 0. \end{cases}$$

Figure A.17(left) illustrates the situation.

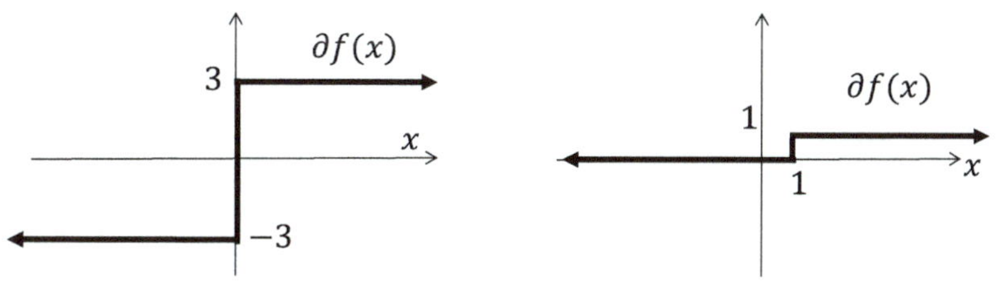

Fig. A.17: Solution of Exercise 2.9; part (a) to the left and part (b) to the right.

(b) Since $f(x) = \max\{x + 1, 2\}$ has a kink only at 1, we concentrate on that point and check for v satisfying

$$f(x) \geq f(1) + v(x - 1) \quad \forall x \quad \text{or, equivalently,} \quad \max\{x + 1, 2\} \geq v(x - 1) \quad \forall x.$$

We consider two cases. If $x > 1$, then this inequality becomes

$$x + 1 \geq v(x - 1),$$

which means that $v \leq 1$. If $x \leq 1$, then the inequality becomes

$$2 \geq v(x - 1)$$

and we'll need $v \geq 0$. We conclude that $\partial f(1) = [0, 1]$. In summary,

$$\partial f(x) = \begin{cases} \{0\} & \text{if } x < 1 \\ [0, 1] & \text{if } x = 1 \\ \{1\} & \text{if } x > 1. \end{cases}$$

Figure A.17(right) illustrates the situation.

Exercise 2.10 Solution. (a) See Figure A.18 for the epigraph of f. (b) Graphically from the figure, we see that

$$\partial f(x) = \begin{cases} \{0\} & \text{if } x < 0 \\ [0, 3] & \text{if } x = 0 \\ \{3\} & \text{if } x > 0. \end{cases}$$

The extended Fermat rule in Theorem 2.9 then confirms that every x in $(-\infty, 0]$ is a minimizer because at such point the function has a zero subgradient.

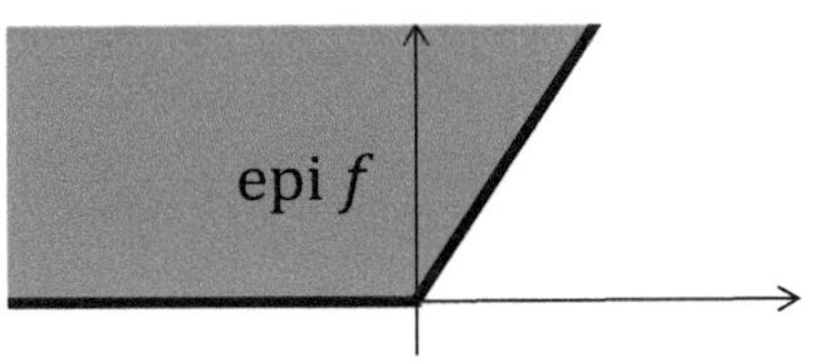

Fig. A.18: Epigraph of f in Exercise 2.10.

Exercise 2.11 Solution. (a) Figure A.19 shows the set C and its normal cone at $(0, 1)$. (b) Since

$$\nabla f(x) = \begin{bmatrix} 2x_1 \\ 6x_2 \end{bmatrix},$$

the gradient at the point $(0, 1)$ is $(0, 6)$. This vector points straight up and can then be "cancelled" by the normal vector $(0, -6)$, which points straight down. We've satisfied the condition $-\nabla f(x) \in N_C(x)$ at the point $(0, 1)$. Consequently, $(0, 1)$ is a minimizer by Theorem 2.13.

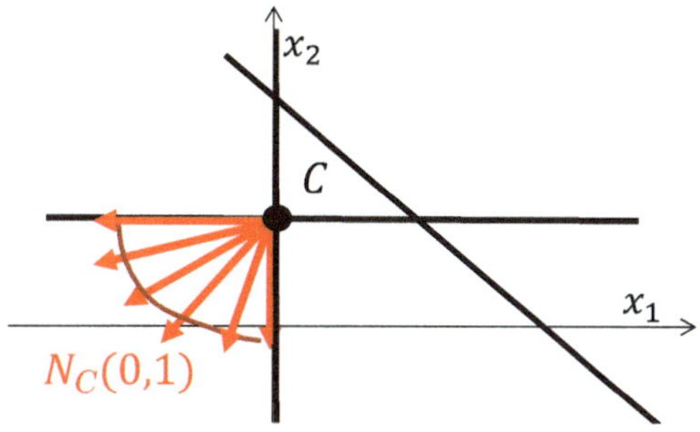

Fig. A.19: Solution of Exercise 2.11.

Exercise 2.12 Solution. We obtain the gradient

$$\nabla f(\bar{x}) = \left[-\frac{swp_1}{a_1} e^{-sw\bar{x}_1/a_1} \quad \dots \quad -\frac{swp_n}{a_n} e^{-sw\bar{x}_n/a_n} \right]^\top.$$

For any $\bar{x}$, this expression and the definition of subgradients produce the approximation

$$f(x) \geq f(\bar{x}) - \sum_{j=1}^{n} \frac{swp_j}{a_j} e^{-sw\bar{x}_j/a_j}(x_j - \bar{x}_j).$$

We can construct such lower bounding approximations at any number of points, say at $\bar{x}^1, \dots, \bar{x}^m$. The maximum of these lower bounding approximations remains a lower bounding approximation. Thus,

$$f(x) \geq \max_{i=1,\dots m} f(\bar{x}^i) - \sum_{j=1}^{n} \frac{swp_j}{a_j} e^{-sw\bar{x}^i_j/a_j}(x_j - \bar{x}^i_j).$$

We've constructed an approximation of the desired form with

$$c_i = \left[-\frac{swp_1}{a_1} e^{-sw\bar{x}^i_1/a_1} \quad \dots \quad -\frac{swp_n}{a_n} e^{-sw\bar{x}^i_n/a_n} \right]^\top$$

$$b_i = \sum_{j=1}^{n} p_j e^{-sw\bar{x}^i_j/a_j} + \sum_{j=1}^{n} \frac{swp_j}{a_j} \bar{x}^i_j e^{-sw\bar{x}^i_j/a_j}.$$

The resulting approximating problem

$$\underset{x \in C}{\text{minimize}} \max_{i=1,\dots,m} c_i^\top x + b_i$$

has a minimum value that's no higher than that of the actual problem because we're using a lower bounding approximation of the objective function. As m increases, the lower bounding approximation remains the same or becomes higher because the maximization takes places over more terms.

The reformulation to a linear optimization problem can be accomplished by introducing an auxiliary decision variable γ. The approximating problem is equivalent to

$$\underset{x\in C,\gamma\in\mathbb{R}}{\text{minimize}}\ \gamma \quad \text{subject to} \quad \max_{i=1,\ldots,m} c_i^\top x + b_i = \gamma.$$

In light of the goal of minimizing γ, we can replace the equality by $\leq$ because the process of minimization forces γ to satisfy the inequality with equality. Thus, we've the equivalent problem

$$\underset{x\in C,\gamma\in\mathbb{R}}{\text{minimize}}\ \gamma \quad \text{subject to} \quad \max_{i=1,\ldots,m} c_i^\top x + b_i \leq \gamma.$$

We seek to have the larger of the terms $c_i^\top x + b_i, i = 1, \ldots, m$ less than or equal to γ. This is equivalent to having *all* the terms less than or equal to γ. We've reached the final equivalent problem

$$\underset{x\in C,\gamma\in\mathbb{R}}{\text{minimize}}\ \gamma \quad \text{subject to} \quad c_i^\top x + b_i \leq \gamma, \quad i = 1, \ldots, m.$$

The new constraints are linear inequalities. Thus, the feasible set remains polyhedral. Since the objective function is linear, the problem is therefore a linear optimization problem.

Exercise 2.13 Solution. For (a), let I be the $(n+1)\times(n+1)$ identity matrix. Then,

$$\|\tilde{A}x - \tilde{b}\|_2^2 = \|Ax - b\|_2^2 + \theta\|x\|_2^2, \quad \text{where } \tilde{A} = \begin{bmatrix} A \\ \sqrt{\theta} I \end{bmatrix}, \quad \tilde{b} = \begin{bmatrix} b \\ 0 \end{bmatrix}.$$

Consequently, the regularized least-squares problem can be expressed as minimizing $\|\tilde{A}x - \tilde{b}\|_2^2$. For (b), similarly set $\tilde{A} = WA$ and $\tilde{b} = Wb$ and obtain

$$\|\tilde{A}x - \tilde{b}\|_2^2 = \left\|W(Ax - b)\right\|_2^2.$$

For (c), we obtain the following results using the Python package numpy.linalg.solve and $n = 5$.

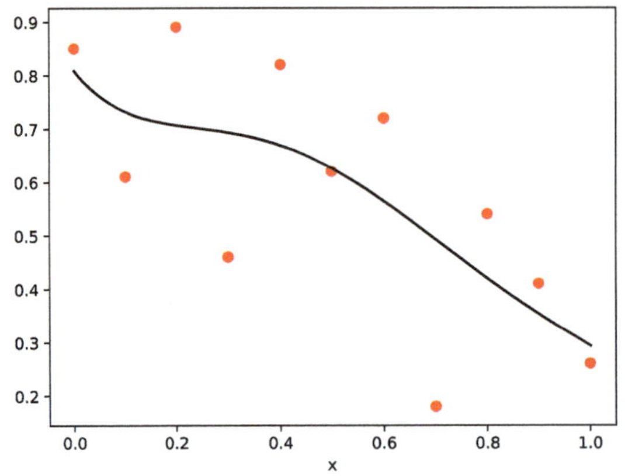

Fig. A.20: Solution of Exercise 2.13(c) with $\theta = 0.00001$.

In the case of the regularized least-squares problem with $\theta = 0.00001$, we obtain the polynomial in Figure A.20. Compared to the result in Figure 1.12, we see that the regularization dampens the oscillations. If $\theta = 0.001$, then this effect is strengthen further as seen in Figure A.21.

The weighted least-squares problem allows us to place more emphasis on some data points. For example, with W set to the diagonal matrix with diagonal values $(1, 1, 1, 1, 1, 1, 1, 20, 20, 20, 20)$, we obtain the fit in Figure A.22. It's clear that the last four data points are emphasized; cf. the results in Figure 1.12.

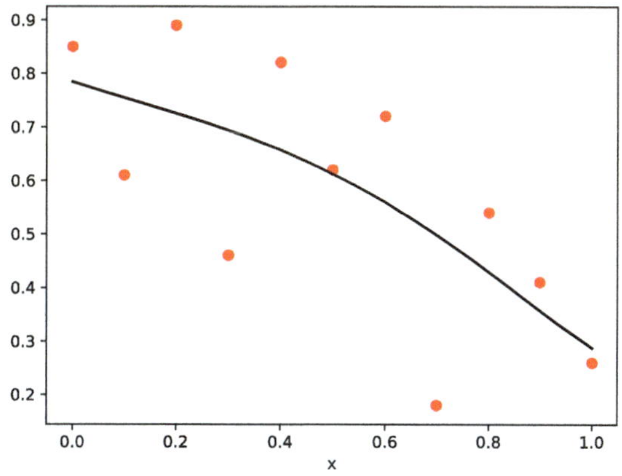

Fig. A.21: Solution of Exercise 2.13(c) with $\theta = 0.001$.

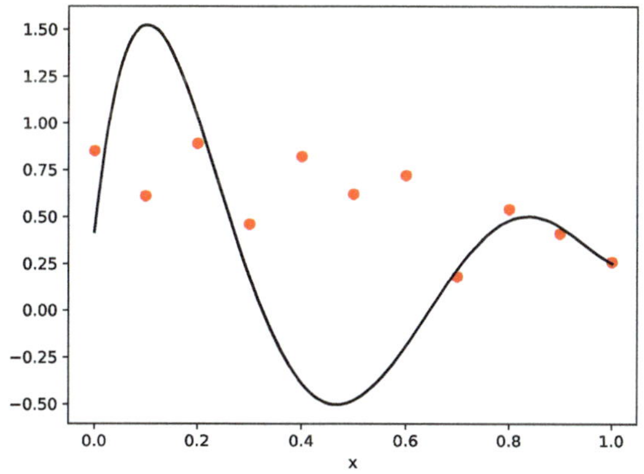

Fig. A.22: Solution of Exercise 2.13(c) for weighted least-squares problem.

Exercise 2.14 Solution. We obtain the gradient and Hessian

$$\nabla f(x) = \begin{bmatrix} 4(x_1 - 1) \\ 6(x_2 + 1) \\ 8x_3 \end{bmatrix}, \qquad \nabla^2 f(x) = \begin{bmatrix} 4 & 0 & 0 \\ 0 & 6 & 0 \\ 0 & 0 & 8 \end{bmatrix}.$$

The eigenvalues of the Hessian are 4, 6, and 8. Since they're nonnegative, it follows that the Hessian is positive semidefinite. This holds for all x and f is therefore convex by Proposition 1.9. By Example 2.18, we conclude that $\operatorname{argmin}_C f$ consists of the points (x_1, x_2, x_3) satisfying

$$-4(x_1 - 1) \in \begin{cases} \{0\} & \text{if } 1 < x_1 < 2 \\ [0, \infty) & \text{if } x_1 = 2 \\ (-\infty, 0] & \text{if } x_1 = 1 \end{cases}$$

$$-6(x_2 + 1) \in \begin{cases} \{0\} & \text{if } 0 < x_2 < \infty \\ (-\infty, 0] & \text{if } x_2 = 0 \end{cases}$$

$$-8x_3 \in \begin{cases} \{0\} & \text{if } -1 < x_3 < 1 \\ [0, \infty) & \text{if } x_3 = 1 \\ (-\infty, 0] & \text{if } x_3 = -1. \end{cases}$$

The first inclusion has one solution $x_1 = 1$ because $-4(x_1 - 1)$ can't be zero while $1 < x_1 < 2$ and it isn't in the interval $[0, \infty)$ when $x_1 = 2$, but it's in the interval $(-\infty, 0]$ when $x_1 = 1$.

The second inclusion holds for $x_2 = 0$. The third inclusion holds for $x_3 = 0$. Thus we've the solution, which is a minimizer: $\operatorname{argmin}_C f = \{(1, 0, 0)\}$.

Exercise 2.15 Solution. The normal cones to C are given by Proposition 2.22. In the notation there, we've $b = 1$, $d = (0, 0)$, and

$$A = [1 \;\; 1] \qquad D = \begin{bmatrix} -1 & 0 \\ 0 & -1 \end{bmatrix}.$$

This means that a normal vector v at a point $x \in C$ must be of the form

$$v = \begin{bmatrix} 1 \\ 1 \end{bmatrix} y + \begin{bmatrix} -1 & 0 \\ 0 & -1 \end{bmatrix} \begin{bmatrix} z_1 \\ z_2 \end{bmatrix}$$

for some scalar y with $z_i \geq 0$ if $x_i = 0$, and $z_i = 0$ otherwise, $i = 1, 2$.

While also accounting for the feasible set C, this presents three cases:

Case 1: If $x_i > 0$ for $i = 1, 2$, then

$$v = \begin{bmatrix} y \\ y \end{bmatrix}$$

for some scalar y as illustrated at the point $\bar{x}$ in Figure A.23.

Case 2: If $x_1 = 0$ and $x_2 = 1$, then

$$v = \begin{bmatrix} y \\ y \end{bmatrix} + \begin{bmatrix} -z_1 \\ 0 \end{bmatrix}$$

for some scalar y and $z_1 \geq 0$ as illustrated at the point x' in the figure.

Case 3: If $x_1 = 1$ and $x_2 = 0$, then

$$v = \begin{bmatrix} y \\ y \end{bmatrix} + \begin{bmatrix} 0 \\ -z_2 \end{bmatrix}$$

for some scalar y and $z_2 \geq 0$ (not illustrated in the figure).

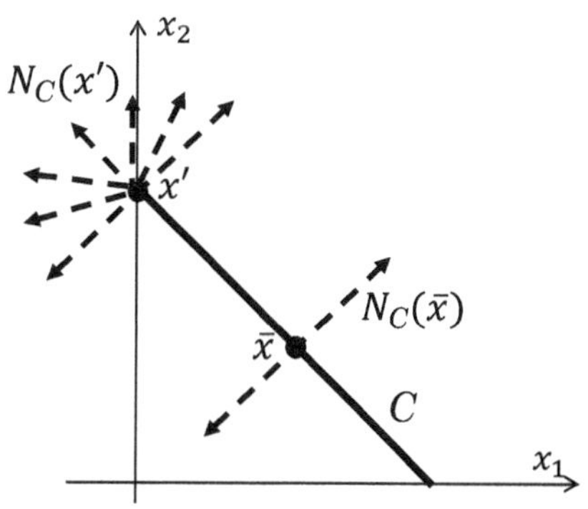

Fig. A.23: Normal cones in Exercise 2.15.

Next, let's invoke the KKT condition of Theorem 2.23. The gradient of f is

$$f(x) = \begin{bmatrix} 2(x_1 - 1) \\ 1 \end{bmatrix}.$$

Since we focus on the point $x = (x_1, x_2) = (1, 0)$, the KKT condition specializes to

$$\begin{bmatrix} -2(1 - 1) \\ -1 \end{bmatrix} = \begin{bmatrix} y \\ y \end{bmatrix} + \begin{bmatrix} 0 \\ -z_2 \end{bmatrix}$$

because the first constraint isn't active and thus $z_1 = 0$. The condition holds with the first equation giving $y = 0$. The second equation specifies $z_2 = 1$, which indeed is nonnegative. Thus, the KKT condition holds at $x = (1, 0)$ and the point must be a minimizer.

Exercise 2.16 Solution. The objective function is convex by the Hessian rule and the constraint is linear. Let $f(x) = (x - 1)^2$, $D = [1]$, and $d = 0$. By the KKT condition of Theorem 2.23, x is a minimizer if and only if

$$-\nabla f(x) = D^\top z \quad \text{or} \quad -2(x - 1) = z.$$

Here, $z \geq 0$ and is actually equal to zero if the constraint isn't active. In our case, the constraint is active at $x^\star = 0$ and the condition simplifies to $-2(0-1) = z$. Consequently, $z = 2$, which is nonnegative. So $x^\star$ satisfies the KKT condition. This is the only solution of the KKT condition and thus the only minimizer.

Exercise 2.17 Solution. The objective function given by $f(x) = (x_1 - 1)^2 + x_2^2$ is convex by the Hessian rule. The constraint is of the form $Dx \leq d$, with $D = [1 \;\; 0]$ and $d = 0$. The KKT condition then fully characterizes minimizers. The condition is

$$-\nabla f(x) = D^\top z \quad \text{or, equivalently,} \quad -(2x_1 - 2) = 1z, \quad -2x_2 = 0z,$$

where $z \geq 0$ and actually zero when the constraint isn't active. At $x^\star$, the constraint is active and the condition holds with $z = 2$. This is the only solution because if the constraint isn't active, then $x_1 < 0$ and $-(2x_1 - 2) = 0$, and this isn't possible.

Exercise 2.18 Solution. In this case $A = [1 \;\; 1]$ and thus the check $Aw = 0$ becomes $w_1 + w_2 = 0$ and this holds for $w = (1, -1)$. We also need to check that the nonnegativity constraints aren't violated. The first constraint is active but $w_1 \geq 0$ so we'll not decrease x_1. The second constraint isn't active, so there are no further restrictions to check.

Exercise 2.19 Solution. (a) We set $A = [2 \;\; 2 \;\; -1]$ and $b = 4$. (b) In this case $m = 1$ and $n = 3$. There are three choices:

(i) x_1 is basic and x_2, x_3 are nonbasic. Then $B = [2]$ and $2x_1 = 4$. Thus, $x_1 = 2$. As always, the nonbasic variables are zero.

(ii) x_2 is basic and x_1, x_3 are nonbasic. Then $B = [2]$ and $2x_2 = 4$. Thus, $x_2 = 2$.

(iii) x_3 is basic and x_1, x_2 are nonbasic. Then $B = [-1]$ and $-x_3 = 4$. Thus, $x_3 = -4$ but this violates the nonnegativity constraint.

Thus, there are two basic *feasible* points: $(2, 0, 0)$ and $(0, 2, 0)$.

(c) A feasible direction w at $(2, 0, 0)$ must satisfy $Aw = 0$, i.e., $2w_1 + 2w_2 - w_3 = 0$. We also need that $w_2, w_3 \geq 0$ to avoid violating $x_2, x_3 \geq 0$. There's no concern about w_1 because $x_1 = 2$ and this variable can change in both directions. For example, one can set $w_2 = 1$ and $w_3 = 0$, and this forces $w_1 = -1$.

Exercise 2.20 Solution. (a) We need to check $Aw = 0$. Since $A = [2 \;\; 1]$ in this case, $Aw = 2w_1 + w_2 = 0$. We also check potential violation of the nonnegativity constraints. It's only $x_1 \geq 0$ that's active, but $w_1 \geq 0$. Consequently, the inequality can't be violated in this direction. (b) $A = [2 \;\; 1]$, $b = 1$, $c = (3, 2)$, and $x = (x_1, x_2)$. (c) There are two possibilities: (i) x_1 is basic and x_2 is nonbasic. Then, $x_2 = 0$ and $x_1 = B^{-1}b = (1/2)1 = 1/2$. (ii) x_2 is basic and x_1 is nonbasic. Then, $x_1 = 0$ and $x_2 = B^{-1}b = 1 \cdot 1 = 1$.

Exercise 2.21 Solution. (a) Since the algorithm has stopped, we know that the reduced cost

$$c_4 - c_B^\top B^{-1} A^4 \geq 0.$$

Any increase in c_4 will maintain nonnegativity of the right-hand side, and hence optimality. (b) We're concerned about satisfying $x_B = B^{-1}b \geq 0$, which becomes

$$\begin{bmatrix} 0 & 1 & 0 \\ 1 & 0 & 1 \\ 1 & 2 & 0 \end{bmatrix} \begin{bmatrix} 1 \\ 1 \\ b_3 \end{bmatrix} \geq 0.$$

This means that $1 + b_3 \geq 0$. Consequently, $b_3 \geq -1$ maintains the feasibility of (x_B, x_N).

Exercise 2.22 Solution. We need to make sure that $x + \lambda w \geq 0$, i.e.,

$$(0,2,0,0,2,3) + \lambda(1,4,0,0,-1,2) \geq 0.$$

We find that $\lambda = \min_j\{-x_j/w_j \mid w_j < 0\} = -2/(-1) = 2$ provides the step size.

Exercise 2.23 Solution. We need to make sure that $x + \lambda w \geq 0$. Since w is nonnegative, one can select any λ and the problem is unbounded.

Exercise 2.24 Solution. (a) We obtain $x = (x_1, \ldots, x_4)$, $b = (1,1)$ and

$$A = \begin{bmatrix} 1 & 1 & 1 & 0 \\ 0 & 1 & 0 & 1 \end{bmatrix}.$$

(b) We've the basic variables

$$x_B = B^{-1}b = \begin{bmatrix} 1 & 0 \\ 0 & 1 \end{bmatrix} \begin{bmatrix} 1 \\ 1 \end{bmatrix} = \begin{bmatrix} 1 \\ 1 \end{bmatrix}$$

and $x = (0,0,1,1)$, which is feasible because it's a nonnegative vector. (c) To ensure feasibility, $x_3 = x_4 = 0$. We're at the point $x = (0,1,0,0)$. One possibility is to let x_1, x_3 be nonbasic and x_2, x_4 be basic. This is permissible because the second and fourth columns of A are linearly independent and the nonbasic variables are zero. Indeed,

$$x_B = B^{-1}b = \begin{bmatrix} 1 & 0 \\ 1 & 1 \end{bmatrix}^{-1} \begin{bmatrix} 1 \\ 1 \end{bmatrix} = \begin{bmatrix} 1 & 0 \\ -1 & 1 \end{bmatrix} \begin{bmatrix} 1 \\ 1 \end{bmatrix} = \begin{bmatrix} 1 \\ 0 \end{bmatrix}$$

produces the correct values for the basic variables.

There are two other possibilities at the point $x = (0,1,0,0)$ for selecting two nonbasic variables (which must be zero) and two basic variables: x_1, x_4 being nonbasic, or x_3, x_4 being nonbasic.

Exercise 2.25 Solution. Figure A.24 answers (a). For (b), there are three extreme points $(0,1)$, $(1,0)$, and $(2,0)$. For (c), we find that

$$A = \begin{bmatrix} 1 & 1 & -1 & 0 \\ 1/2 & 1 & 0 & 1 \end{bmatrix} \quad \text{and} \quad b = \begin{bmatrix} 1 \\ 1 \end{bmatrix}.$$

(d) If x_1 and x_2 are nonbasic and x_3 and x_4 are basic, then

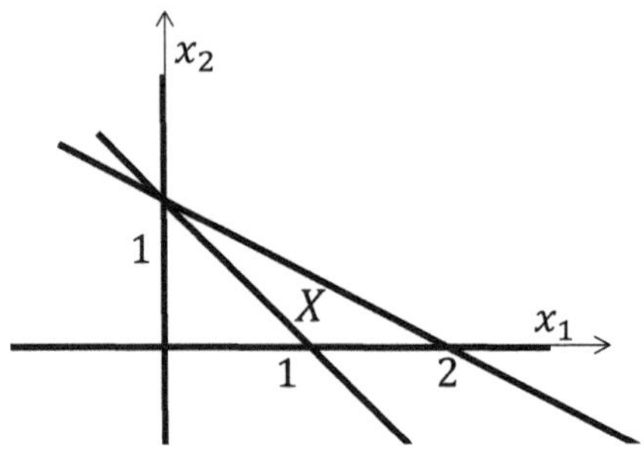

Fig. A.24: Feasible set X in Exercise 2.25.

$$x_B = B^{-1}b = \begin{bmatrix} -1 & 0 \\ 0 & 1 \end{bmatrix}^{-1} \begin{bmatrix} 1 \\ 1 \end{bmatrix} = \begin{bmatrix} -1 & 0 \\ 0 & 1 \end{bmatrix} \begin{bmatrix} 1 \\ 1 \end{bmatrix} = \begin{bmatrix} -1 \\ 1 \end{bmatrix}.$$

So $x_3 = -1$ and $x_4 = 1$, which is infeasible.

(e) Extreme point $(x_1, x_2) = (0, 1)$ requires $x_3 = x_4 = 0$ to ensure feasibility. There are now three representations: x_1, x_3 nonbasic, x_2, x_4 basic; x_1, x_4 nonbasic, x_2, x_3 basic; x_3, x_4 nonbasic, x_1, x_2 basic. Extreme point $(x_1, x_2) = (1, 0)$ requires $x_3 = 0$ and $x_4 = 1/2$ for feasibility. There's one representation: x_2, x_3 nonbasic, x_1, x_4 basic. Extreme point $(x_1, x_2) = (2, 0)$ requires $x_3 = 1$ and $x_4 = 0$ for feasibility. There's one representation: x_2, x_4 nonbasic, x_1, x_3 basic. So there are five basic representations at the three extreme points.

(f) If x_2 and x_4 are nonbasic, then x_1 and x_3 are basic. Consequently,

$$x_B = B^{-1}b = \begin{bmatrix} 1 & -1 \\ 1/2 & 0 \end{bmatrix}^{-1} \begin{bmatrix} 1 \\ 1 \end{bmatrix} = \begin{bmatrix} 0 & 2 \\ -1 & 2 \end{bmatrix} \begin{bmatrix} 1 \\ 1 \end{bmatrix} = \begin{bmatrix} 2 \\ 1 \end{bmatrix},$$

which means that $x_1 = 2$, $x_3 = 1$. The point $x = (2, 0, 1, 0)$ is feasible.

Exercise 2.26 Solution. First, none of the reduced costs are improving, so the current point must be a minimizer. But $3 < 4$ which can't happen. Second, $\bar{c}_4 = 0$ which indicates multiple minimizers but only one extreme point has value 4.

Exercise 2.27 Solution. Let's introduce the auxiliary decision variables $v = (v_1, \ldots, v_m)$ to represent the max-term in the objective function. Then, we obtain a reformulation in terms of a quadratic problem:

$$\begin{aligned} \underset{(a,\alpha)\in\mathbb{R}^{n+1}, v\in\mathbb{R}^m}{\text{minimize}} \quad & \frac{1}{m} \sum_{i=1}^{m} v_i + \theta \|a\|_2^2 \\ \text{subject to} \quad & 1 - y_i(a^\top x^i + \alpha) \le v_i, \ \ i = 1, \ldots, m \\ & 0 \le v_i, \ \ i = 1, \ldots, m. \end{aligned}$$

Exercise 2.28 Solution. There are several possible data set online. We consider the spambase data set from https://archive.ics.uci.edu/dataset/94/spambase and obtain the following results using the simplex method and an implementation of the subgradient method that starts with step size $\lambda = 0.02$ and reduces it gradually over five stages using a reduction factor of 0.9. The maximum number of iterations is set to 500, with the algorithm also stopping if the change in solution is less than 10^{-6}.

	simplex method				subgradient method			
θ	hinge train	hinge test	correct (%) test	time (sec.)	hinge train	hinge test	correct (%) test	time (sec.)
0.000	0.1858	0.2214	92.6	0.2	0.2262	0.2119	92.9	0.8
0.001	0.1933	0.2042	92.8	0.1	0.2295	0.2146	93.0	0.9
0.010	0.2501	0.2405	91.8	0.1	0.2625	0.2477	91.7	0.8
0.020	0.2987	0.2877	89.4	0.1	0.3020	0.2895	90.1	0.8
0.050	0.3991	0.3928	84.9	0.1	0.4005	0.3947	84.5	0.8
0.100	0.4943	0.4787	80.9	0.1	0.4939	0.4786	81.0	0.8

It might be possible to tune the parameters in the subgradient method. For example, with $\theta = 0.001$, if the initial step size is changed from 0.02 to 0.2, then computing time is slightly reduced to 0.8 seconds. More significantly, the objective function value is reduced from 0.2436 to 0.2208. The lower value is quite close to the minimum value obtained by the simplex method: 0.2178. The test accuracy also improves to 93.4% after this change. We note that a lower objective function value doesn't necessarily mean better test accuracy.

With $\theta = 0.1$, we obtain an affine classifier with nonzero coefficients in front of columns:

5 (word_freq_our)
16 (word_freq_free)
19 (word_freq_you)
21 (word_freq_your)
25 (word_freq_hp)
27 (word_freq_george)
46 (word_freq_edu)
52 (char_freq_)
55 (capital_run_length_average)

These features appear to be the more important ones.

Exercise 2.29 Solution. At $x = 1$, the three parts of the formula for f have values $-x - 1 = -2$, $x - 1 = 0$, and $2x - 2 = 0$. Thus, the second and third terms are the ones defining f near $x = 1$. (A picture of the epigraph might help you realize this.) The slope corresponding to these two terms at $x = 1$ are 1 and 2. Thus, $\partial f(1) = [1, 2]$. Repeating such reasoning, we compute the subgradients

$$\partial f(x) = \begin{cases} \{-1\} & \text{if } x < 0 \\ [-1, 1] & \text{if } x = 0 \\ \{1\} & \text{if } x \in (0, 1) \\ [1, 2] & \text{if } x = 1 \\ \{2\} & \text{if } x > 1. \end{cases}$$

We're now ready to start the subgradient method from $x^0 = 1$. We compute

$$x^1 = x^0 - \tfrac{1}{2}w^0 = 1 - \tfrac{1}{2} \cdot 1 = \tfrac{1}{2}$$
$$x^2 = x^1 - \tfrac{1}{2}w^1 = \tfrac{1}{2} - \tfrac{1}{2} \cdot 1 = 0.$$

Since $x = 0$ is the minimizer by Theorem 2.9, we stop.

Exercise 3.1 Solution. The number of possible outcomes $\nu = 16$ so in addition to the nonnegativity constraints, the stochastic optimization problem has 32 constraints and a minimizer $x^\star = (1072.6,\ 0,\ 251.4,\ 0)$ with minimum value $-\$15900$. A solution of the deterministic version suggests manufacturing the same type of boards but in significantly different quantities.

Exercise 3.2 Solution. The 0.8-quantile is 1 as obtained by drawing the distribution function. Since we know that superquantiles are no smaller than the corresponding quantile and no larger than the largest possible outcome, the 0.8-superquantile is 1.

Exercise 3.3 Solution. The following table gives the mean, standard deviation, and α-quantiles as well as α-superquantiles for $\xi_0, \xi_1, \xi_2, \xi_3$.

			quantile/superquantile				
	mean	std	0	0.5	0.75	0.9	1
ξ_0	2.08	0.15	*/2.08	2.00/2.20	2.20/2.27	2.30/2.30	*/2.30
ξ_1	1.00	0.00	*/1.00	1.00/1.00	1.00/1.00	1.00/1.00	*/1.00
ξ_2	1.16	0.31	*/1.16	1.20/1.42	1.40/1.50	1.50/1.58	*/1.60
ξ_3	1.06	0.12	*/1.06	1.10/1.15	1.10/1.20	1.20/1.20	*/1.20

The α-quantile isn't defined for $\alpha = 0$ and $\alpha = 1$, and we write "*" in the table. A 1-superquantile of a random variable always corresponds to the highest possible outcome and the 0-superquantile coincides with the mean. We notice that in general the 0.5-quantile (median) can be different than the mean of a random variable. An α-quantile of a random variable is never larger than the corresponding α-superquantile.

The next table reports the same quantities for each of the random variables $\xi_0 - 2\xi_1$, $\xi_0 - 2\xi_2$, and $\xi_0 - 2\xi_3$.

			quantile/superquantile				
	mean	std	0	0.5	0.75	0.9	1
$\xi_0 - 2\xi_1$	0.08	0.15	*/0.08	0.00/0.20	0.20/0.27	0.30/0.30	*/0.30
$\xi_0 - 2\xi_2$	−0.24	0.62	*/−0.24	−0.50/0.25	0.20/0.63	0.50/0.83	*/0.90
$\xi_0 - 2\xi_3$	−0.04	0.29	*/−0.04	−0.10/0.18	0.10/0.36	0.40/0.48	*/0.50

These numbers represent how well we'll do when investing everything in one asset alone, without any diversification. If we put all the money in the bank, we'll have a shortfall of 0.08, i.e., \$80,000, on average. In the worst case, the shortfall is \$300,000. If we invest all in the first fund corresponding to ξ_2, then we'll avoid a shortfall on average but in the worst case the shortfall can be \$900,000. The situation is similar for investments in the second fund corresponding to ξ_3, but then the worst-case shortfall is less—but still very bad!

We solve (3.4), with $C = \{x \mid x_1 + x_2 + x_3 = 2,\ x_i \geq 0, i = 1, 2, 3\}$ and obtain the following results using different α:

	α						
	0	0.2	0.4	0.6	0.8	0.9	0.99
x_1	0.00	0.00	0.00	0.00	0.00	0.00	0.00
x_2	2.00	1.50	1.00	0.70	0.73	0.50	0.50
x_3	0.00	0.50	1.00	1.30	1.27	1.50	1.50
γ	−0.90	−0.55	−0.20	−0.02	−0.01	0.15	0.15
min-value	−0.24	−0.09	−0.01	0.04	0.10	0.15	0.15
prob. broke	1/3	1/3	1/4	1/6	1/12	1/4	1/4

We recall that the minimum values and optimal γ reported are the corresponding α-superquantiles and α-quantiles of the shortfall $\xi_0 - \xi_1 x_1 - \xi_2 x_2 - \xi_3 x_3$, respectively.

While the probability of going broke actually increases as we pass from the plan obtained by using $\alpha = 0.8$ to the one under $\alpha = 0.9$, the latter plan has the advantage of lower worst-case costs. In detail, we find that the two plans perform as follows across the 12 scenarios, where we also include the three heuristic plans:

	heuristics			optimal using α	
scenario	(2,0,0)	(0,2,0)	(0,0,2)	0.8	0.9
1	0.3	−0.9	0.5	−0.01	0.15
2	0.2	−0.8	0.4	−0.04	0.10
3	−0.1	−0.9	−0.1	−0.39	−0.30
4	0	−0.8	0.2	−0.16	−0.05
5	0.1	−0.5	0.1	−0.12	−0.05
6	0.2	−0.4	0	−0.15	−0.10
7	−0.1	−0.5	−0.3	−0.37	−0.35
8	0	−0.2	−0.4	−0.33	−0.35
9	0	0.2	−0.2	−0.05	−0.10
10	0.1	0.5	−0.3	−0.01	−0.10
11	−0.1	0.5	−0.3	−0.01	−0.10
12	0.3	0.9	−0.1	0.26	0.15
prob. broke	1/6	1/3	1/3	1/12	1/4

Thus, $\alpha = 0.8$ produces only one "broke-scenario." But, when that happens, the shortfall is somewhat large: 0.26. In contrast, $\alpha = 0.9$ produces three such scenarios but all with moderate shortfalls: 0.15, 0.10, 0.15. Comparing the 0.9-superquantiles of the shortfall for the two plans, we obtain that the former ($\alpha = 0.8$) has 0.22 while the latter

($\alpha = 0.9$) has 0.15. Both optimal plans vastly outperform the three heuristic plans in the sense of most meaningful metrics.

Exercise 3.4 Solution. In addition to the "standard" constraints of nonnegativity and integral-equals-one, continuity and convexity constraints result in the model

$$\begin{aligned}
&\underset{a,b\in\mathbb{R}^q}{\text{minimize}}\ -\frac{1}{\nu}\sum\nolimits_{i=1}^{\nu} \ln(a_{j_i}\xi^i + b_{j_i})\\
&\text{subject to}\quad a_j\tau_j + b_j \geq 0,\ \ j = 1,\dots,q\\
&\qquad a_j\tau_{j+1} + b_j \geq 0,\ \ j = 1,\dots,q\\
&\sum\nolimits_{j=1}^{q} \tfrac{1}{2}a_j(\tau_{j+1}^2 - \tau_j^2) + b_j(\tau_{j+1} - \tau_j) = 1\\
&\qquad a_{j-1}\tau_j + b_{j-1} = a_j\tau_j + b_j,\ \ j = 2,\dots,q\\
&\qquad a_j \geq a_{j-1},\ \ j = 2,\dots,q.
\end{aligned}$$

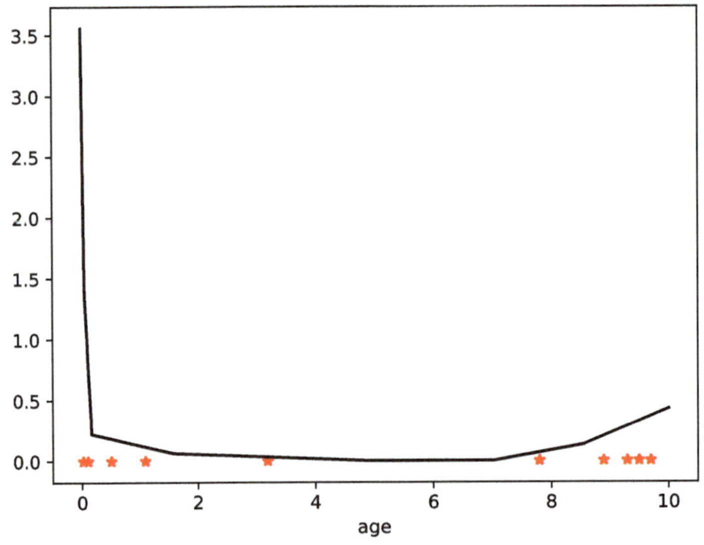

Fig. A.25: Density function using $q = 1000$ in Exercise 3.4.

Figure A.25 shows the resulting density function when $q = 1000$. The density function has some kinks, which can be made more gradual by also including the constraints

$$a_j - 0.01 \leq a_{j-1},\ \ j = 2,\dots,q.$$

This results in the density function in Figure A.26. These additional constraints worsen the minimum value from 1.43 to 1.54 but the "look" is smoother.

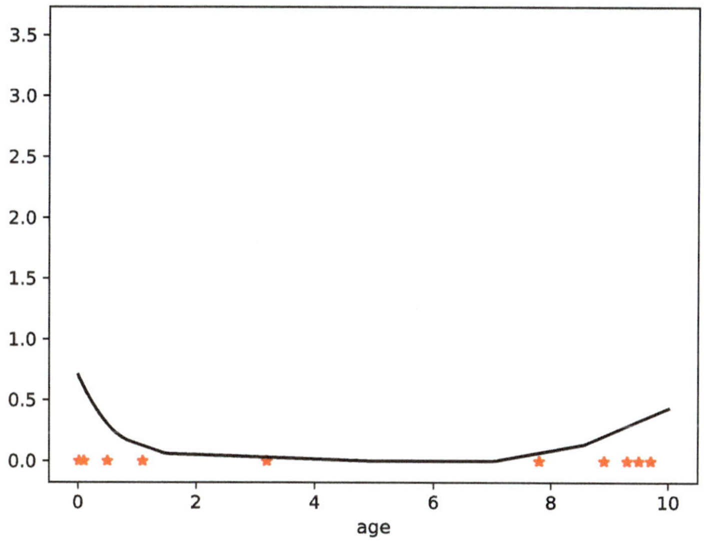

Fig. A.26: Density function in Exercise 3.4 when also using "smoothing" constraints.

Exercise 3.5 Solution. Since the objective function

$$f(x) = \frac{1}{m}\sum_{i=1}^{m} f_i(x), \quad \text{with } f_i(x) = (x-i)^2,$$

the stochastic gradient descent method selects one i in each iteration, computes the gradient of f_i at the current point, and then updates

$$x^{\nu+1} = x^{\nu} - \lambda^{\nu}\nabla f_i(x^{\nu}), \quad \text{where } \nabla f_i(x^{\nu}) = 2(x^{\nu} - i).$$

Exercise 3.6 Solution. We start by identifying a minimizer by using the Fermat rule and Theorem 1.6. After solving

$$\nabla f(x) = \begin{bmatrix} 6x_1 + 0.1x_2 + 2(x_1 - 1) \\ 2x_2 + 0.1x_1 + 4(x_2 - 2) \end{bmatrix} = \begin{bmatrix} 0 \\ 0 \end{bmatrix},$$

we obtain $x^\star = (0.2334, 1.3294)$ (rounding to four digits), which is a minimizer because f is convex; the Hessian

$$\nabla^2 f(x) = \begin{bmatrix} 8 & 0.1 \\ 0.1 & 6 \end{bmatrix}$$

is positive semidefinite and we can apply Proposition 1.9. The largest eigenvalue of this matrix is 8.005 and we can conclude that the Lipschitz constant of the objective function is $\kappa = 8.005$, i.e.,

$$\left\|\nabla f(x) - \nabla f(\bar{x})\right\|_2 \leq 8.005\|x - \bar{x}\|_2 \quad \forall x, \bar{x} \in \mathbb{R}^2.$$

Thus, a step size of less that 1/8.005 = 0.125 is theoretically advised in the gradient descent method. Using this method with step size $\lambda = 0.1$, we obtain after five iterations $x = (0.2334, 1.3294)$. The trajectory is given in Figure A.27. The gradient descent method is rather effective at obtaining the minimizer.

Final point: [0.23338195 1.32944363]

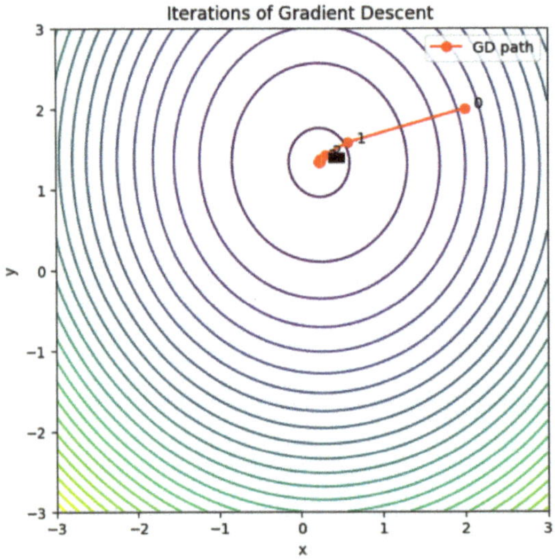

Fig. A.27: Trajectory of gradient descent method over five iterations using $\lambda = 0.1$.

The stochastic gradient descent method also applies because we can interpret the objective function to be the average of five terms:

$$f(x) = \tfrac{1}{5} \sum\nolimits_{i=1}^{5} f_i(x),$$

with $f_1(x) = 15x_1^2$, $f_2(x) = 5x_2^2$, $f_3(x) = 0.5x_1x_2$, $f_4(x) = 5(x_1 - 1)^2$, and $f_5(x) = 10(x_2 - 2)^2$. Here, the term 1/5 triggers multiplication of each function with a factor of five to match the format in (3.7). The stochastic gradient descent method is stochastic(!) so different runs tend to be different; see Figures A.28 and A.29 for examples over 100 iterations.

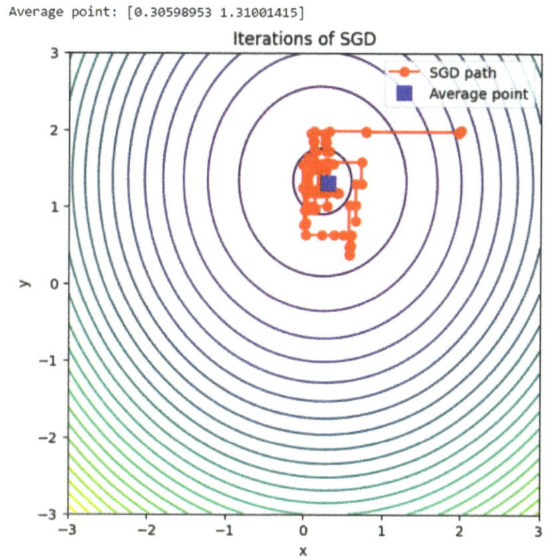

Fig. A.28: Trajectory of stochastic gradient descent method and the resulting average point over 100 iterations using $\lambda = 0.02$.

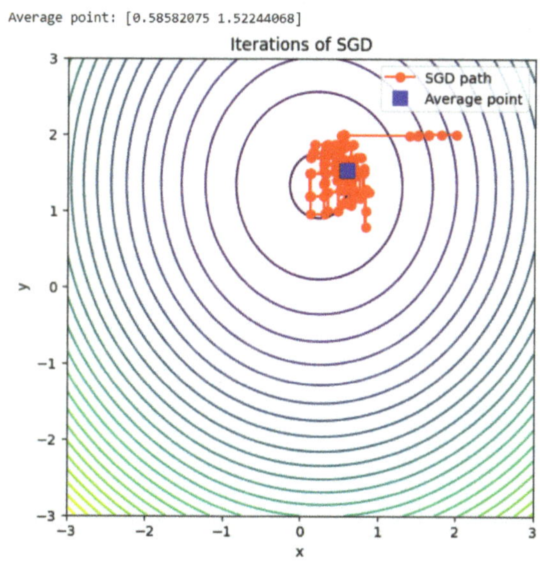

Fig. A.29: Another trajectory of stochastic gradient descent method and the resulting average point over 100 iterations using $\lambda = 0.02$.

When using step size $\lambda = 0.01$, the variability reduces as we see in Figures A.30 and A.31. If we increase the iteration limit to 1000, then five different runs with $\lambda = 0.01$ give the solutions: (0.2759, 1.3509), (0.2597, 1.3793), (0.2152, 1.3250), (0.2714, 1.3191), (0.2254, 1.3624). We observe that the stochastic gradient method has a hard time producing highly accurate solutions. The inherent randomness, however, is helpful when solving nonconvex problems as the method might not get stuck in a local minimizer—a major problem for the gradient descent method.

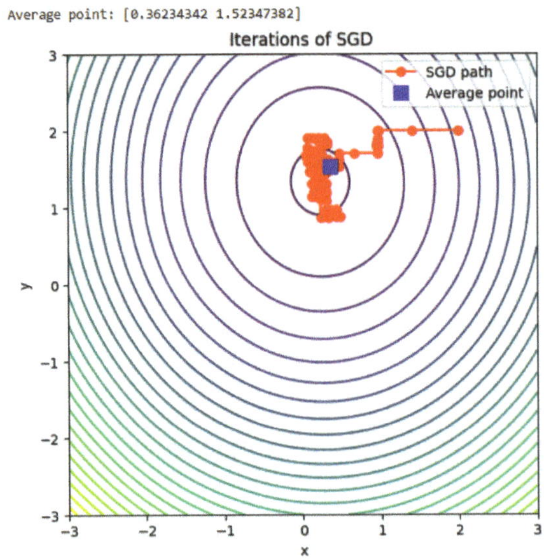

Fig. A.30: Another trajectory of stochastic gradient descent method and the resulting average point over 100 iterations using $\lambda = 0.01$.

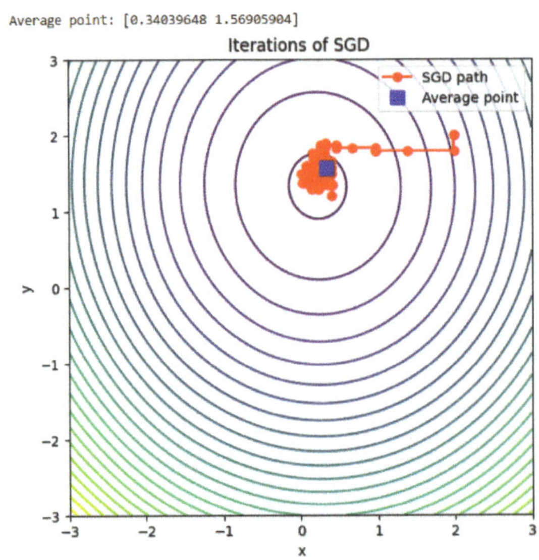

Fig. A.31: Another trajectory of stochastic gradient descent method and the resulting average point over 100 iterations using $\lambda = 0.01$.

Exercise 3.7 Solution. The results depend on the data set. One can check the implementation of the stochastic gradient descent method on a small data set by solving an equivalent linear optimization problem; see Section 2.F.

Exercise 3.8 Solution. The model is reasonable because it reflects that a decision about allocation of aircraft to routes needs to be done months in advance, well before the airline knows the passenger demand. Thus, it adopts a two-stage decision framework. The model allows for fractional solutions, while the number of aircraft and seats are obviously integers. This could be satisfactory during such planning because the data might be somewhat imprecise. Regardless, the resulting problem is convex, which has computational benefits.

A minimizer of the last formulation in Example 3.15 is $x_{11}^\star = 10$, $x_{22}^\star = 12.85$, $x_{23}^\star = 0.82$, $x_{24}^\star = 5.33$, $x_{32}^\star = 4.31$, $x_{35}^\star = 20.69$, $x_{41}^\star = 7.34$ and $x_{43}^\star = 7.66$, with other variables being zero.

Exercise 3.9 Solution. Define the second-stage decision vector $y = (y_1^+, y_1^-, \dots, y_{m_2}^+, y_{m_2}^-)$ and write

$$f(\xi, x) = \inf_y \left\{ \sum_{i=1}^{m_2} \beta_i y_i^+ - \alpha_i y_i^- \;\middle|\; y_i^+ - y_i^- = \xi_i - T_i x,\ y_i^+ \geq 0,\ y_i^- \geq 0,\ i = 1, \dots, m_2 \right\}$$

from which one can recognize a, W, d, and T in (3.16).

Exercise 3.10 Solution. (a) The per-item profits are:

Desks: $60 - 8 \cdot 2 - 2 \cdot 5.2 - 4 \cdot 4 = 17.6$ dollars.

Tables: $40 - 6 \cdot 2 - 1.5 \cdot 5.2 - 2 \cdot 4 = 12.2$ dollars.

Chairs: $10 - 1 \cdot 2 - 0.5 \cdot 5.2 - 1.5 \cdot 4 = -0.6$ dollars.

Chairs aren't profitable on a per-item basis.

(b) Using x_l for acquisition of lumber, x_c and x_f for carpentry and finishing hours, y_d, y_t, and y_c for production quantities for desks, tables, and chairs, and s_d, s_t, and s_c for sales quantities, we obtain the model

$$\begin{aligned}
\text{minimize } & 2x_l + 5.2x_c + 4x_f - 60s_d - 40s_t - 10s_c \\
\text{subject to } & 8y_d + 6y_t + y_c \leq x_l \\
& 2y_d + 1.5y_t + 0.5y_c \leq x_c \\
& 4y_d + 2y_t + 1.5y_c \leq x_f \\
& s_d \leq y_d, s_t \leq y_t, s_c \leq y_c \\
& s_d \leq 150, s_t \leq 125, s_c \leq 300 \\
& x_l, x_c, x_f, y_d, y_t, y_c, s_d, s_t, s_c \geq 0.
\end{aligned}$$

A minimizer is $x_l = 1950.0$, $x_c = 487.5$, $x_f = 850.0$, $s_d = 150.0$, $s_t = 125.0$, $s_c = 0.0$, $y_d = 150.0$, $y_t = 125.0$, $y_c = 0.0$, with minimum value -4165.0.

(c) Rerunning the model for each scenario, we obtain:

	Low Scenario	Likely Scenario	High Scenario
Lumber used	520.0	1860.0	3500.0
Carpentry hours used	130.0	465.0	875.0
Finishing hours used	240.0	820.0	1500.0
Desks sold	50.0	150.0	250.0
Tables sold	20.0	110.0	250.0
Chairs sold	0.0	0.0	0.0
Desks produced	50.0	150.0	250.0
Tables produced	20.0	110.0	250.0
Chairs produced	0.0	0.0	0.0
Minimum value	−1124.0	−3982.0	−7450.0

In all three cases, the results indicate that we shouldn't produce chairs.

(d) While the x- and y-decisions must be made before knowing the demand, the actual sales are evaluated after we learn the demand. This produces the model

$$\begin{aligned}
\text{minimize } & 2x_l + 5.2x_c + 4x_f - 0.3(60s_d^{\text{low}} + 40s_t^{\text{low}} + 10s_c^{\text{low}}) \\
& -0.4(60s_d^{\text{like}} + 40s_t^{\text{like}} + 10s_c^{\text{like}}) - 0.3(60s_d^{\text{high}} + 40s_t^{\text{high}} + 10s_c^{\text{high}}) \\
\text{subject to } & 8y_d + 6y_t + y_c \le x_l \\
& 2y_d + 1.5y_t + 0.5y_c \le x_c \\
& 4y_d + 2y_t + 1.5y_c \le x_f \\
& s_d^{\text{low}} \le y_d,\ s_t^{\text{low}} \le y_t,\ s_c^{\text{low}} \le y_c \\
& s_d^{\text{like}} \le y_d,\ s_t^{\text{like}} \le y_t,\ s_c^{\text{like}} \le y_c \\
& s_d^{\text{high}} \le y_d,\ s_t^{\text{high}} \le y_t,\ s_c^{\text{high}} \le y_c \\
& s_d^{\text{low}} \le 50,\ s_t^{\text{low}} \le 20,\ s_c^{\text{low}} \le 200 \\
& s_d^{\text{like}} \le 150,\ s_t^{\text{like}} \le 110,\ s_c^{\text{like}} \le 250 \\
& s_d^{\text{high}} \le 250,\ s_t^{\text{high}} \le 250,\ s_c^{\text{high}} \le 500
\end{aligned}$$

with all variables being nonnegative. A minimizer is summarized below:

Quantity	Value
Lumber used	1060.0
Carpentry hours used	265.0
Finishing hours used	420.0
Desks produced	50.0
Tables produced	110.0
Chairs produced	0.0
Desks sold (low/likely/high)	50.0 / 50.0 / 50.0
Tables sold (low/likely/high)	20.0 / 110.0 / 110.0
Chairs sold (low/likely/high)	0.0 / 0.0 / 0.0

The corresponding minimum value is −1142.0. Again, we see that there are no chairs being produced. We also note that the uncertainty about the demand makes us conservative. We only produce 50 desks and 110 tables, which is tailored to the low scenario for desks and the likely scenario for tables. Thus, if the high scenario occurs, we're unable to satisfy the demand. Also, if the low scenario occurs, we produce 110 tables but sell only 20.

The profit of \$1142 is low compared to (c), where we obtain \$1124, \$3982, and \$7450, in the low, likely, and high scenarios, respectively. In (c), we assume knowledge of the demand when acquiring material and setting production numbers. This might be overly optimistic.

(e) While the x-decision must be made before knowing the demand, production and sales are determined after we learn the demand. This produces the model

$$
\begin{aligned}
\text{minimize } & 2x_l + 5.2x_c + 4x_f - 0.3(60s_d^{\text{low}} + 40s_t^{\text{low}} + 10s_c^{\text{low}}) \\
& -0.4(60s_d^{\text{like}} + 40s_t^{\text{like}} + 10s_c^{\text{like}}) - 0.3(60s_d^{\text{high}} + 40s_t^{\text{high}} + 10s_c^{\text{high}}) \\
\text{subject to } & 8y_d^{\text{low}} + 6y_t^{\text{low}} + y_c^{\text{low}} \leq x_l \\
& 2y_d^{\text{low}} + 1.5y_t^{\text{low}} + 0.5y_c^{\text{low}} \leq x_c \\
& 4y_d^{\text{low}} + 2y_t^{\text{low}} + 1.5y_c^{\text{low}} \leq x_f \\
& 8y_d^{\text{like}} + 6y_t^{\text{like}} + y_c^{\text{like}} \leq x_l \\
& 2y_d^{\text{like}} + 1.5y_t^{\text{like}} + 0.5y_c^{\text{like}} \leq x_c \\
& 4y_d^{\text{like}} + 2y_t^{\text{like}} + 1.5y_c^{\text{like}} \leq x_f \\
& 8y_d^{\text{high}} + 6y_t^{\text{high}} + y_c^{\text{high}} \leq x_l \\
& 2y_d^{\text{high}} + 1.5y_t^{\text{high}} + 0.5y_c^{\text{high}} \leq x_c \\
& 4y_d^{\text{high}} + 2y_t^{\text{high}} + 1.5y_c^{\text{high}} \leq x_f \\
& s_d^{\text{low}} \leq y_d^{\text{low}},\ s_t^{\text{low}} \leq y_t^{\text{low}},\ s_c^{\text{low}} \leq y_c^{\text{low}} \\
& s_d^{\text{like}} \leq y_d^{\text{like}},\ s_t^{\text{like}} \leq y_t^{\text{like}},\ s_c^{\text{like}} \leq y_c^{\text{like}} \\
& s_d^{\text{high}} \leq y_d^{\text{high}},\ s_t^{\text{high}} \leq y_t^{\text{high}},\ s_c^{\text{high}} \leq y_c^{\text{high}} \\
& s_d^{\text{low}} \leq 50,\ s_t^{\text{low}} \leq 20,\ s_c^{\text{low}} \leq 200 \\
& s_d^{\text{like}} \leq 150,\ s_t^{\text{like}} \leq 110,\ s_c^{\text{like}} \leq 250 \\
& s_d^{\text{high}} \leq 250,\ s_t^{\text{high}} \leq 250,\ s_c^{\text{high}} \leq 500
\end{aligned}
$$

with all variables nonnegative. A minimizer, with minimum value −1730.0, is given by $x_l = 1300.0$, $x_c = 325.0$, $x_f = 540.0$, and the following scenario dependent numbers:

	Low Scenario	Likely Scenario	High Scenario
Desks produced	50.0	80.0	80.0
Tables produced	20.0	110.0	110.0
Chairs produced	**200.0**	0.0	0.0
Desks sold	50.0	80.0	80.0
Tables sold	20.0	110.0	110.0
Chairs sold	200.0	0.0	0.0

This model produces new insight. In the low scenario, we now produce chairs! Chairs become a useful outlet for excess lumber and labor hours when the demand is low. Thus, it might be a good idea to keep the production line for chairs.

The profit is \$1730, which is better than \$1142 from part (d). The flexibility afforded by a delayed production decision is quite beneficial.

Exercise 3.11 Solution. (a) $\inf f = -\infty$. (b) $\inf f = 0$. (c) $\inf f = 0$.

Exercise 3.12 Solution. There are no explicit constraints on x because there's no "subject to" or other restriction indicated under "minimize" in the problem formulation. For the induced constraints, we find that

$$f_1(x) = \inf_y\{y \mid 2 - x \le y,\ y \le 0\} = \begin{cases} 2 - x & \text{if } x \ge 2 \\ \infty & \text{otherwise.} \end{cases}$$

$$f_2(x) = \inf_y\{y \mid 4 - x \le y,\ y \le 0\} = \begin{cases} 4 - x & \text{if } x \ge 4 \\ \infty & \text{otherwise.} \end{cases}$$

Thus, the expectation $(1/2)f_1(x) + (1/2)f_2(x)$ imposes the induced constraint $x \ge 2$.

Exercise 4.1 Solution. We need the two constraints: $10y_{ij} \le x_{ij}$ and $x_{ij} \le 100y_{ij}$.

Exercise 4.2 Solution. (a) The constraint is

$$\sum_{k=1}^{10} x_k \le 100.$$

(b) The constraint (4.21) specifies that exactly three binary variables take the value 1, while the remaining ones take the value 0. This means that we'll invest in exactly three funds. The ten constraints in (4.22) ensure that if fund k isn't selected (i.e., $y_k = 0$), then the investment amount x_k in that fund must be zero. Otherwise, if fund k is selected, then the investment amount can be up to 100.

Exercise 4.3 Solution. The standard form formulation of the model becomes:

Indices

t	aircraft type	
r	airline route	
s	scenario	

Sets

$\mathcal{T}$	set of aircraft types	$\mathcal{T} = \{1, \ldots, m_1\}$
$\mathcal{R}$	set of airline routes	$\mathcal{R} = \{1, \ldots, m_2\}$
$\mathcal{S}$	set of scenarios	$\mathcal{S} = \{1, \ldots, S\}$

Data

opCost_{tr}	operating cost of type t on route r	(dollar)
cap_{tr}	passenger capacity for type t on route r	(count)
rev_r	revenue per passenger on route r	(dollar)
avail_t	number of aircraft of type t	(count)
demand_{rs}	demand on route r in scenario s	(count)
prob_s	probability of scenario s	(-)

Variables

ASSIGN_{tr}	number of aircraft of type t on route r
EMPTY_{rs}	number of empty seats on route r in scenario s

Objective

$$\text{minimize} \sum_{t \in \mathcal{T}} \sum_{r \in \mathcal{R}} \left(- \text{cap}_{tr} \text{rev}_r + \text{opCost}_{tr} \right) \text{ASSIGN}_{tr} + \sum_{s \in \mathcal{S}} \text{prob}_s \sum_{r \in \mathcal{R}} \text{rev}_r \text{EMPTY}_{rs}$$

Constraints

$$\sum_{r \in \mathcal{R}} \text{ASSIGN}_{tr} \le \text{avail}_t, \quad t \in \mathcal{T} \tag{A.1}$$

$$\sum_{t \in \mathcal{T}} \text{cap}_{tr} \text{ASSIGN}_{tr} - \text{EMPTY}_{rs} \le \text{demand}_{rs}, \quad r \in \mathcal{R}, \ s \in \mathcal{S} \tag{A.2}$$

$$\text{ASSIGN}_{tr} \in \{0, 1, 2, \ldots\}, \quad t \in \mathcal{T}, r \in \mathcal{R} \tag{A.3}$$

$$\text{EMPTY}_{rs} \ge 0, \quad r \in \mathcal{R}, s \in \mathcal{S}. \tag{A.4}$$

Description. The objective function computes the cost minus revenue for full aircraft plus the expected revenue corresponding to empty seats. The constraints (A.1) ensure that we don't assign more aircraft of each type than what's available. The constraints (A.2) compute the number of empty seats in each scenario by comparing the demand in the scenario with the total capacity of all the assigned aircrafts.

Currently, there are no numerical results available; see Exercise 3.8 for a solution of the continuous relaxation.

Exercise 4.4 Solution. The standard form formulation of the model becomes:

Indices

i, j	city	(~ 50)
t	truck	(~ 100)
c	commodity	(~ 10)

Sets

$\mathcal{I}$	set of cities	$\mathcal{I} = \{1, \ldots, 50\}$
$\mathcal{T}$	set of trucks	$\mathcal{T} = \{1, \ldots, 100\}$
C	set of commodities	$C = \{1, \ldots, 10\}$

Data

maxLoad_t	max total load for truck t	(pounds)
demand_{ci}	min amount of c to delivered in i	(pounds)
supply_{ci}	max amount of c that can be loaded in i	(pounds)
costMove_{ijtc}	cost of moving one pound of c from i to j by truck t	(dollar)
costAll_{it}	cost of allocating truck t to origin i	(dollar)

Variables

ASSIGN_{it}	1 if city i is t's origin city; 0 otherwise	
MOVE_{ijt}	1 if truck t drives between i and j; 0 otherwise	
QUANT_{ijtc}	amount of c loaded on t in i and delivered to j	(pounds)

Objective

$$\text{minimize} \sum_{i\in\mathcal{I}} \sum_{t\in\mathcal{T}} \text{costAll}_{it}\text{ASSIGN}_{it} + \sum_{i\in\mathcal{I}} \sum_{j\in\mathcal{I}} \sum_{t\in\mathcal{T}} \sum_{c\in C} \text{costMove}_{ijtc}\text{QUANT}_{ijtc}$$

Constraints

$$\sum_{i\in\mathcal{I}} \text{ASSIGN}_{it} = 1 \qquad \forall t \in \mathcal{T} \quad \text{(A.5)}$$

$$\sum_{j\in\mathcal{I}} \text{MOVE}_{ijt} \leq 2\,\text{ASSIGN}_{it} \qquad \forall i \in \mathcal{I}, t \in \mathcal{T} \quad \text{(A.6)}$$

$$\sum_{t\in\mathcal{T}} \text{MOVE}_{ijt} \leq 10 \qquad \forall i \in \mathcal{I}, j \in \mathcal{I} \quad \text{(A.7)}$$

$$\sum_{c\in C} \text{QUANT}_{ijtc} \leq \text{maxLoad}_t \text{MOVE}_{ijt} \qquad \forall t \in \mathcal{T}, i \in \mathcal{I}, j \in \mathcal{I} \quad \text{(A.8)}$$

$$\sum_{j\in\mathcal{I}} \sum_{t\in\mathcal{T}} \text{QUANT}_{jitc} \geq \text{demand}_{ci} \qquad \forall i \in \mathcal{I}, c \in C \quad \text{(A.9)}$$

$$\sum_{j\in\mathcal{I}} \sum_{t\in\mathcal{T}} \text{QUANT}_{ijtc} \leq \text{supply}_{ci} \qquad \forall i \in \mathcal{I}, c \in C \quad \text{(A.10)}$$

$$\text{QUANT}_{ijtc} \geq 0 \qquad \forall i \in \mathcal{I}, j \in \mathcal{I}, t \in \mathcal{T}, c \in C$$

$$\text{ASSIGN}_{it} \in \{0, 1\} \qquad \forall i \in \mathcal{I}, t \in \mathcal{T}$$

$$\text{MOVE}_{ijt} \in \{0, 1\} \qquad \forall i \in \mathcal{I}, j \in \mathcal{I}, t \in \mathcal{T}$$

Description. The objective function combines the cost of moving commodities and assigning trucks to cities. Constraints (A.5) assign each truck to a city (100 constraints). Constraints (A.6) ensure that a truck can only move from city i to two cities if it's assign to city i (5000 constraints). Constraints (A.7) limit the number of trucks between any pair of cities to 10 (2500 constraints). Constraints (A.8) ensure that the load on truck t is within the allowable limit and actually zero if not moving (250000 constraints). Constraints (A.9) make sure that enough commodities are delivered to each city (500 constraints). Constraints (A.10) enforce supply restrictions at the origin cities (500 constraints). The remaining constraints ensure that the variables are of the right kind.

Exercise 4.5 Solution. (a) Node (2) gives a feasible solution and value -1. (b) The root note has value -1.7 and thus is a lower bound. Since no subproblem has been solved on the right-hand side of the tree, the root node provides the only lower bound. (c) There's no need to solve below node (2) because of integrality at that node. But we need to solve node (ii) since it can't be bounded away. (d) We only need to solve (ii) and possibly its two branches. There's no need for further branching after that; with three binary variables the tree can't get any deeper. The other branches are terminated due to integrality. Thus, there are three subproblems left at most.

Exercise 4.6 Solution. (a) The lower bound is -33.2 and the upper bound is -20.0, with a gap of 13.2. (b) The lower bound becomes $\min\{-20.1, -20.0, -25.0\} = -25.0$, while the upper bound remains at -20.0. (c) The lower bound remains at -25.0 but the upper bound improves to -21.0. (d) The lower bound becomes $\min\{-20.1, -20.0, -21.0, \infty\} = -21.0$ and the upper bound is still -21.0. Hence the solution providing the current upper bound (from node 5) is optimal.

Exercise 4.7 Solution. The best upper bound is $\min\{-20.0, -33.2\} = -33.2$. We compute lower bounds by looking across the tree and retain the lowest value: $\min\{-20.1, -20.0, -33.2\} = -33.2$. The gap is 0.

Exercise 4.8 Solution. Consider the continuous relaxations of the sets defined by (4.23) and (4.24):

$$S' = \left\{(x_1, x_2, y) \,\middle|\, \alpha \geq x_1 \geq 0,\ \ \alpha \geq x_2 \geq 0,\ \ y \in [0,1],\ \ x_1 \leq \alpha y,\ \ x_2 \leq \alpha y\right\}$$
$$S = \left\{(x_1, x_2, y) \,\middle|\, \alpha \geq x_1 \geq 0,\ \ \alpha \geq x_2 \geq 0,\ \ y \in [0,1],\ \ x_1 + x_2 \leq 2\alpha y\right\}.$$

If $(x_1, x_2, y) \in S'$, then $x_1 \leq \alpha y$ and $x_2 \leq \alpha y$. Consequently,

$$x_1 + x_2 \leq \alpha y + \alpha y = 2\alpha y.$$

This means that $(x_1, x_2, y) \in S$. Thus, $S' \subset S$.

The point $(x_1, x_2, y) = (2\alpha/3, \alpha/3, 1/2) \in S$ because

$$\tfrac{2}{3}\alpha + \tfrac{1}{3}\alpha = \alpha \leq 2\alpha\tfrac{1}{2}.$$

However, the point isn't in S' because $2\alpha/3 > \alpha(1/2)$. Thus S' is a strict subset S.

When y is binary, (4.23) and (4.24) define the same set of feasible points. Therefore (4.23) is a stronger formulation than (4.24) by definition.

Exercise 4.9 Solution. We initialize the algorithm in Step 0 by setting $S = \emptyset$, $\bar{S} = \{1, \ldots, 5\}$, $d(i) = \infty$, $i = 2, \ldots, 5$, $d(1) = 0$, $P(1) = 0$, and $\nu = 1$. We can't stop in Step 1 because $\nu < 5$. In Step 2, we select node $1 \in \operatorname{argmin}_{j \in \bar{S}} d(j)$. Step 3 updates $S = \{1\}$ and $\bar{S} = \{2, \ldots, 5\}$. Step 4 considers the three edges $E_1^+ = \{(1,2), (1,3), (1,4)\}$ and checks

$$j = 2: \quad d(2) = \infty; \quad d(1) + c_{12} = 1 \qquad \text{set: } P(2) = 1,\ \ d(2) = 1$$
$$j = 3: \quad d(3) = \infty; \quad d(1) + c_{13} = 2 \qquad \text{set: } P(3) = 1,\ \ d(3) = 2$$
$$j = 4: \quad d(4) = \infty; \quad d(1) + c_{14} = 3 \qquad \text{set: } P(4) = 1,\ \ d(4) = 3.$$

We're finished with the first iteration and set $\nu = 2$.

In Step 2, we select node $2 \in \operatorname{argmin}_{j \in \bar{S}} d(j)$. Step 3 updates $S = \{1,2\}$ and $\bar{S} = \{3,4,5\}$. Step 4 considers the edge $E_2^+ = \{(2,4)\}$ and checks

$$j = 4: \quad d(4) = 3; \quad d(2) + c_{24} = 2 \qquad \text{set: } P(4) = 2,\ \ d(4) = 2.$$

We're finished with the second iteration and set $\nu = 3$.

In Step 2, we select (arbitrarily between the two choices) node $4 \in \operatorname{argmin}_{j \in \bar{S}} d(j)$. Step 3 updates $S = \{1,2,4\}$ and $\bar{S} = \{3,5\}$. Step 4 considers $E_4^+ = \{(4,5)\}$ and checks

$$j = 5: \quad d(5) = \infty; \quad d(4) + c_{45} = 4 \qquad \text{set: } P(5) = 4,\ \ d(5) = 4.$$

We're finished with the third iteration and set $\nu = 4$.

In Step 2, we select node $3 \in \text{argmin}_{j\in\bar{S}} d(j)$. Step 3 updates $S = \{1, \ldots, 4\}$ and $\bar{S} = \{5\}$. Step 4 considers the two edges $E_3^+ = \{(3,4), (3,5)\}$ and checks

$$
\begin{aligned}
&j = 4: \quad d(4) = 2; \quad d(3) + c_{34} = 4 \qquad &\text{no update}\\
&j = 5: \quad d(5) = 4; \quad d(3) + c_{35} = 9 \qquad &\text{no update.}
\end{aligned}
$$

We're finished with the fourth iteration and set $\nu = 5$. The next iteration stops in Step 2. Since $d(5) = 4$, the minimum path has length 4. A path furnishing that length is recovered by working backward from vertex 5: $P(5) = 4, P(4) = 2, P(2) = 1$.

Exercise 4.10 Solution. The maximum flow from vertex 1 to vertex 8 is 30. In detail, the flow on edge (1,2) is 10, on edge (1,3) is 5, on edge (1,4) is 15, on edge (2,5) is 9, on edge (2,6) is 1, on edge (3,6) is 4, on edge (3,7) is 1, on edge (4,7) is 5, on edge (4,8) is 10, on edge (5,8) is 9, on edge (6,8) is 5, and on edge (7,8) is 6.

The solution makes sense because there's flow-balance at each node and the three edges emanating from vertex 1 are used at full capacity. Consequently, it isn't possible to improve the solution.

Exercise 4.11 Solution. For $T = 3$, we obtain $V_3(0) = 25$ and $V_3(s_3) = 0$ for $s_3 > 0$. At $t = 2$, we obtain

$$
\begin{aligned}
V_2(0) &= \min_{x_2\in\{0,1\}} 25 + 3x_2 + 25p_{00}(x_2) = 28\\
V_2(1) &= \min_{x_2\in\{0,1\}} 2 + 3x_2 + 25p_{10}(x_2) = 5\\
V_2(2) &= \min_{x_2\in\{0,1\}} 4 + 3x_2 + 25p_{20}(x_2) = 7\\
V_2(3) &= \min_{x_2\in\{0,1\}} 6 + 3x_2 + 25p_{30}(x_2) = 9\\
V_2(4) &= \min_{x_2\in\{0,1\}} 8 + 3x_2 + 25p_{40}(x_2) = 10.5\\
V_2(5) &= \min_{x_2\in\{0,1\}} 10 + 3x_2 = 10.
\end{aligned}
$$

At $t = 1$, we obtain

$$
\begin{aligned}
V_1(0) &= \min_{x_1\in\{0,1\}} 25 + 3x_1 + 28p_{00}(x_1) + 10p_{05}(x_1) = 38\\
V_1(1) &= \min_{x_1\in\{0,1\}} 2 + 3x_1 + 28p_{10}(x_1) + 5p_{11}(x_1) + 10p_{15}(x_1) = 15\\
V_1(2) &= \min_{x_1\in\{0,1\}} 4 + 3x_1 + 28p_{20}(x_1) + 5p_{21}(x_1) + 7p_{22}(x_1) + 10p_{25}(x_1) = 17\\
V_1(3) &= \min_{x_1\in\{0,1\}} 6 + 3x_1 + 28p_{30}(x_1) + 5p_{31}(x_1) + 7p_{32}(x_1) + 9p_{33}(x_1) + 10p_{35}(x_1)\\
&= 18.25\\
V_1(4) &= \min_{x_1\in\{0,1\}} 8 + 3x_1 + 28p_{40}(x_1) + 5p_{41}(x_1) + 7p_{42}(x_1) + 9p_{43}(x_1) + 10.5p_{44}(x_1)\\
&\quad + 10p_{45}(x_1) = 18.075\\
V_1(5) &= \min_{x_1\in\{0,1\}} 10 + 3x_1 + 5p_{51}(x_1) + 7p_{52}(x_1) + 9p_{53}(x_1) + 10.5p_{54}(x_1)\\
&\quad + 10p_{55}(x_1) = 18.825.
\end{aligned}
$$

At $t = 0$, we obtain

$$V_0(1) = \min_{x_0 \in \{0,1\}} 2 + 3x_0 + 38p_{10}(x_0) + 15p_{11}(x_0) + 18.825p_{15}(x_0) = 23.825.$$

Thus, an optimal policy produces a minimum value of 23.825. The specific details of the optimal policy follow by (4.15):

$$\pi_0(1) = 1,\ \ \pi_1(5) = 0,\ \ \pi_2(1) = 1,\ \ \pi_2(2) = 1,\ \ \pi_2(3) = 1,\ \ \pi_2(4) = 0,\ \ \pi_2(5) = 0.$$

Exercise 4.12 Solution. (a) We obtain

$$\begin{aligned}
V_2(0) &= \min_{x_2 \in \{0,1\}} 25 + 3x_2 + 30p_{00}(x_2) = 28\\
V_2(1) &= \min_{x_2 \in \{0,1\}} 2 + 3x_2 + 30p_{10}(x_2) = 5\\
V_2(2) &= \min_{x_2 \in \{0,1\}} 4 + 3x_2 + 30p_{20}(x_2) = 7\\
V_2(3) &= \min_{x_2 \in \{0,1\}} 6 + 3x_2 + 30p_{30}(x_2) = 9\\
V_2(4) &= \min_{x_2 \in \{0,1\}} 8 + 3x_2 + 30p_{40}(x_2) = 11\\
V_2(5) &= \min_{x_2 \in \{0,1\}} 10 + 3x_2 = 10.
\end{aligned}$$

(b) We obtain

$$\begin{aligned}
V_2(0) &= \min_{x_2 \in \{0,1\}} 25 + 3x_2 + 25p_{00}(x_2) = 28\\
V_2(1) &= \min_{x_2 \in \{0,1\}} 5 + 3x_2 + 25p_{10}(x_2) = 8\\
V_2(2) &= \min_{x_2 \in \{0,1\}} 10 + 3x_2 + 25p_{20}(x_2) = 13\\
V_2(3) &= \min_{x_2 \in \{0,1\}} 15 + 3x_2 + 25p_{30}(x_2) = 18\\
V_2(4) &= \min_{x_2 \in \{0,1\}} 20 + 3x_2 + 25p_{40}(x_2) = 22.5\\
V_2(5) &= \min_{x_2 \in \{0,1\}} 25 + 3x_2 = 25.
\end{aligned}$$

Exercise 5.1 Solution. (a) The Rockafellian encodes changes to the probability (or weights) of the different objective functions $f_1, \ldots, f_m$, with the perturbed probabilities being nonnegative but not necessarily summing to one. The unperturbed problem is

$$\underset{x}{\text{minimize}} \sum_{i=1}^{m} p_i f_i(x).$$

(b) The Rockafellian encodes changes to the right-hand side of inequality constraints. The unperturbed problem is

$$\underset{x}{\text{minimize}}\ f_0(x) \quad \text{subject to } g_i(x) \le 0,\ i = 1, \dots, q.$$

(c) The Rockafellian encodes changes to the right-hand side of equality constraints, with the perturbations being associated with a quadratic penalty scaled by θ. The unperturbed problem is

$$\underset{x}{\text{minimize}}\ f_0(x) \quad \text{subject to } f_i(x) = 0,\ i = 1, \dots, m.$$

Exercise 5.2 Solution. We utilize Theorem 5.3 and (5.1) to estimate how the min-value function given by

$$p(u) = \inf_x f(u, x)$$

changes near $u = 0$. Since $\operatorname{argmin}_x f(0, x) = \{1\}$, the subgradients in $\partial p(0)$ are the vectors y satisfying $(y, 0) \in \partial f(0, 1)$; see Theorem 5.3. We leverage Theorem 2.12 to compute

$$\partial f(0, 1) = \big(0,\ -\exp(-1)\big) + N_C(0, 1).$$

The normal cone $N_C(u, x)$ follows by Proposition 2.22 because the feasible set takes the form

$$C = \{w = (u, x) \mid Dw \le d\}, \ \text{ where } D = \begin{bmatrix} 0 & -1 \\ 1 & 1 \end{bmatrix} \ \text{ and } \ d = \begin{bmatrix} 1 \\ 1 \end{bmatrix}.$$

The constraint $-x \le 1$ isn't active at $x = 1$. Thus,

$$N_C(0, 1) = \{D^\top z \mid z_1 = 0, z_2 \ge 0\} = \big\{(z_2, z_2) \bigm| z_2 \ge 0\big\}.$$

We therefore obtain

$$\partial f(0, 1) = \big(0,\ -\exp(-1)\big) + \big\{(z_2, z_2) \bigm| z_2 \ge 0\big\} = \big\{(z_2, z_2 - \exp(-1)) \bigm| z_2 \ge 0\big\}.$$

Following Theorem 5.3, we seek vectors y such that $(y, 0) \in \partial f(0, 1)$. The first row in this expression says that $y = z_2$. The second row says that $0 = z_2 - \exp(-1)$. Thus, there's only one possibility: $y = \exp(-1)$. Using (5.1), we reach the conclusion that

$$p(u) \ge p(0) + yu = \exp(-1) + \exp(-1)u \approx 0.368(1 + u).$$

To check this result, we compute $p(u)$ directly and find that $p(u) = \exp(u - 1)$ for $u \in (-\infty, 2]$. Our estimate $0.368(1 + u)$ is 0.405 and 0.331 for $u = 0.1$ and $u = -0.1$, respectively. These are slightly off compared to $p(0.1) = 0.407$ and $p(-0.1) = 0.333$.

We also verifying that Theorem 5.3 applies. The Rockafellian is proper because its values aren't $-\infty$ and its epigraph isn't empty; the point

$$\big((0, 0),\ f(0, 0)\big) \in \operatorname{epi} f$$

since $f(0,0) = \exp(-0) + \iota_C(0,0) = 1 < \infty$. The Rockafellian is convex by Proposition 2.4 and the Hessian rule in Proposition 1.9; note that C is a polyhedron. The Rockafellian is also lsc because C contains all its boundary points and is therefore closed. For any scalars $\bar{u}$ and α, the level-set

$$\{x \mid f(u,x) \le \alpha\} = \{x \mid \exp(-x) \le \alpha\} \cap \{x \mid -x \le 1,\ u + x \le 1\} \subset [-1, 2 - \bar{u}]$$

whenever $|u - \bar{u}| \le 1$. Thus, f is level-bounded in x, locally uniformly in u.

Exercise 5.3 Solution. The Rockafellian relaxation is in this case

$$\underset{u,x}{\text{minimize}}\ f(u,x) - yu = (x-1)^2 + \iota_{\{0\}}(x+u) + \tfrac{1}{2}\theta u^2 - yu,$$

where y is the multiplier. The indictor function forces $u = -x$. Thus, the problem simplifies to

$$\underset{x}{\text{minimize}}\ (x-1)^2 + \tfrac{1}{2}\theta(-x)^2 + yx.$$

This problem is convex so it suffices, by the Fermat rule, to solve $2(x-1) + \theta x + y = 0$. We obtain

$$x^\star = \frac{2-y}{2+\theta}$$

and then $(u^\star, x^\star) = (-x^\star, x^\star)$ is a minimizer for the Rockafellian relaxation. The actual problem has (trivially) $x = 0$ as the minimizer, and we observe that $x^\star$ approaches that minimizer as θ becomes large. The convergence is "accelerated" if y is near 2.

The Rockafellian relaxation has minimum value

$$\psi(y) = \Big(\frac{y+\theta}{2+\theta}\Big)^2 + \tfrac{1}{2}\theta\Big(\frac{2-y}{2+\theta}\Big)^2 + y\frac{2-y}{2+\theta} = \frac{(-1-\theta/2)y^2 + (4+2\theta)y + \theta^2 + 2\theta}{(2+\theta)^2},$$

and this value approaches the minimum value of the actual problem, which is 1, as θ tends to infinity. Regardless of y and nonnegative θ, the minimum value of the Rockafellian relaxation isn't greater than 1 by (5.2). We verify this here by taking the derivative of ψ and solving

$$\psi'(y) = \frac{(-2-\theta)y + 4 + 2\theta}{(2+\theta)^2} = 0.$$

This produces $y = 2$ regardless of θ. Since $\psi(2) = 1$ and ψ is concave, we conclude that $y = 1$ is the maximizer of ψ and $\psi(y) \le 1$ for all y. We note that ψ is indeed the dual objective function produced by the Rockafellian f.

Exercise 5.4 Solution. The Lagrangian takes the form $l(x,y) = \exp(x) - xy$ for $y \ge 0$ and $l(x,y) = -\infty$ otherwise. Certainly, the actual minimum value

$$\inf_{x\ge 0} e^x = 1 = \exp(0) - 0 \cdot y \ge \inf_{x\in\mathbb{R}} \exp(x) - xy.$$

Thus, (5.2) holds. The Lagrangian for $y = 0.5, 1, 2$ is illustrated in Figure A.32, where $f_0(x) = \exp(x)$. We note that each $x < 0$ is infeasible so the comparison is between $l(x, y)$ and $f_0(x) + \iota_{[0,\infty)}(x) = \infty$, which holds trivially.

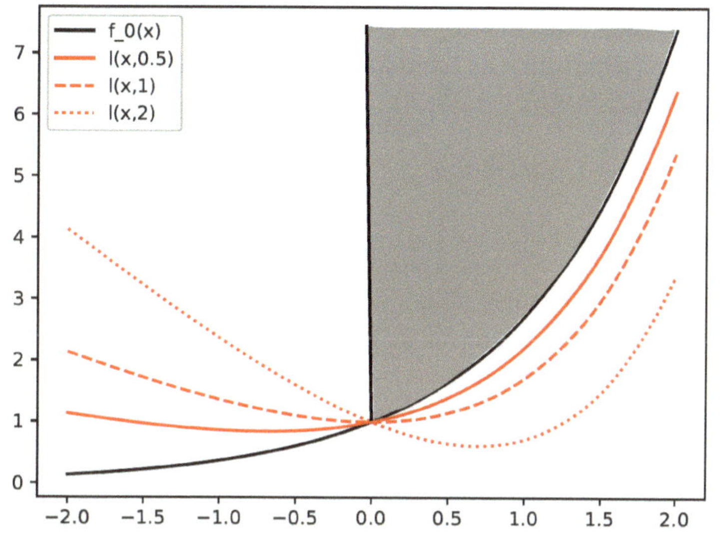

Fig. A.32: Graphs of $f_0(x)$, $l(x, 0.5)$, $l(x, 1)$, and $l(x, 2)$, with the epigraph of $f_0 + \iota_{[0,\infty)}$ being shaded in gray.

Exercise 5.5 Solution. For (a), we follow the framework of Example 5.9 and obtain that $c = (1, -2)$, $A = [2\ \ 1]$, and $b = 3$. Thus, the dual problem becomes

$$\underset{y \in \mathbb{R}}{\text{maximize}}\ 3y \ \text{ subject to }\ 2y \le 1,\ 1y \le -2.$$

For (b), we directly find that $y = -2$ is maximizer for the dual problem; its maximum value is -6. The minimum value of the actual problem is also -6 as provided by the minimizer $(0, 3)$. In this case, the relaxation underpinning the dual problem can be tuned to produce a value matching that of the actual problem.

Exercise 5.6 Solution. (a) We find that

$$A = \begin{bmatrix} 0 & 1 & 0 \end{bmatrix},\ \ b = 1,\ \ c = \begin{bmatrix} 1 \\ 1 \\ 1 \end{bmatrix}.$$

(b) Since the dual problem from Example 5.9 is to maximize $b^\top y$ subject $A^\top y \le c$, we obtain the dual problem

$$\underset{y \in \mathbb{R}}{\text{maximize}}\ y \ \text{ subject to }\ y \le 1.$$

(c) The maximum value of this dual problem is 1 because the maximizer is $y = 1$. (d) For $x \geq 0$, the specific Lagrangian takes the form

$$l(x, y) = x_1 + x_2 + x_3 + y(1 - x_2).$$

Thus, for $y = 1$, we obtain $l(x, 1) = x_1 + x_3 + 1$. The minimum value of $l(x, 1)$ over $x \geq 0$ is then 1. This is the same value as the maximum of the dual problem because $y = 1$ is actually the maximizer; see (b).

Exercise 5.7 Solution. (a) We define

$$f(x) = c^\top x + \iota_{[0,\infty)^n}(x) + \iota_{\{0\}^m}(b - Ax).$$

Minimizing f is equivalent to the actual problem. The first of the three terms is convex by linearity. The second term is convex as it's the indicator function of a convex set; see Proposition 2.4. The third term is convex by the composite rule of Proposition 1.8(d): the inside part $b - Ax$ is affine and the outer function $\iota_{\{0\}^m}$ is convex because it's the indicator function of a convex set. We can then invoke the sum rule in Proposition 1.8(a). (b) The minimizer is $x^\star = (1, 0)$ so the minimum value is 0. (A drawing of the feasible set might help you reach this conclusion.) (c) Maximize y subject to $y \leq 0$, $y \leq 1$. (d) Maximizer is $y^\star = 0$ so the maximum value is 0. (e) Using the specific numbers,

$$l(x, 0) = x_2 + y(1 - x_1 - x_2) = x_2.$$

The minimization of $l(x, 0)$ over $x \geq 0$ produces a minimum value of 0. This is the same as the maximum value of the dual problem because $y = 0$ is a maximizer for the dual problem.

Exercise 5.8 Solution. An example of when both the primal and the dual problems are infeasible is furnished by

$$\underset{x \in \mathbb{R}^4}{\text{minimize}}\ -x_1 - x_2 \ \text{ subject to } \ x_1 - x_2 - x_3 = 1,\ -x_1 + x_2 - x_4 = 1,\ x \geq 0$$

$$\underset{y \in \mathbb{R}^2}{\text{maximize}}\ y_1 + y_2 \ \text{ subject to } \ y_1 - y_2 \leq -1,\ -y_1 + y_2 \leq -1,\ -y_1 \leq 0,\ -y_2 \leq 0.$$

The primal problem is infeasible because the constraints state that

$$-x_1 + x_2 = -1 - x_3, \quad -x_1 + x_2 = 1 + x_4.$$

Consequently, the right-hand sides in these equations must be equal: $-1 - x_3 = 1 + x_4$. This equation simplifies to $x_3 + x_4 = -2$, which isn't possible for nonnegative x_3 and x_4.

The dual problem is infeasible because the two inequalities $y_1 - y_2 \leq -1$ and $-y_1 + y_2 \leq -1$ can't hold simultaneously.

Exercise 5.9 Solution. Following Example 5.5, the Lagrangian is given by

$$l(x,y) = \inf_u f(u,x) - y^\top u = f_0(x) + yg(x)$$
$$= 2x_1^2 + x_2^2 + 3x_3^2 + y(-2x_1 - 3x_2 - x_3 + 1)$$

for $y \geq 0$ and $l(x,y) = -\infty$ otherwise. The corresponding primal problem is

$$\underset{x\in\mathbb{R}^3}{\text{minimize}} \; \varphi(x) = \sup_y l(x,y),$$

which becomes specifically

$$\underset{x\in\mathbb{R}^3}{\text{minimize}} \; \sup_{y\geq 0} 2x_1^2 + x_2^2 + 3x_3^2 + y(-2x_1 - 3x_2 - x_3 + 1).$$

Since any x satisfying $-2x_1 - 3x_2 - x_3 + 1 > 0$ produces infinity during the maximization over y, we conclude that the primal problem is equivalent to

$$\underset{x\in\mathbb{R}^3}{\text{minimize}} \; 2x_1^2 + x_2^2 + 3x_3^2 \quad \text{subject to} \quad -2x_1 - 3x_2 - x_3 + 1 \leq 0,$$

which is the actual problem. We can find a minimizer by hand using the KKT condition in Theorem 2.23 or give the instance to a linear optimization solver. In either case, we obtain $x_1 = 0.088235$, $x_2 = 0.264706$, and $x_3 = 0.029412$, with minimum value 0.088235.

The corresponding dual problem is

$$\underset{y\in\mathbb{R}}{\text{maximize}} \; \psi(y) = \inf_x l(x,y),$$

which becomes specifically

$$\underset{y\geq 0}{\text{maximize}} \; \inf_x 2x_1^2 + x_2^2 + 3x_3^2 + y(-2x_1 - 3x_2 - x_3 + 1).$$

Since $l(x,y)$ is convex in x regardless of y, we can utilize the optimality condition in Theorem 1.6 to conclude that the minimizers of $l(x,y)$ with respect to x are characterized by the equations

$$4x_1 - 2y = 0, \quad 2x_2 - 3y = 0, \quad 6x_3 - y = 0.$$

Thus, $(y/2, 3y/2, y/6)$ is the unique minimizer of $l(x,y)$ with respect to x. This leads to

$$\psi(y) = 2\left(\tfrac{1}{2}y\right)^2 + \left(\tfrac{3}{2}y\right)^2 + 3\left(\tfrac{1}{6}y\right)^2 + \left(-2\tfrac{1}{2}y - 3\tfrac{3}{2}y - \tfrac{1}{6}y + 1\right)y = y - \frac{17}{6}y^2$$

when $y \geq 0$. Thus, the dual problem takes the explicit form

$$\underset{y\in\mathbb{R}}{\text{maximize}} \; y - \frac{17}{6}y^2 \quad \text{subject to} \quad y \geq 0.$$

Again Theorem 2.23 or a solver applied to the dual problem produces $y = 0.176471$, with maximum value 0.088235. We've verified strong duality because this value is the same as the minimum value found above.

For the assumptions in Theorem 5.13, the Rockafellian in this problem is proper because it's never $-\infty$ and its epigraph is nonempty. It's lsc because its epigraph is a closed set. It's convex because f_0 is convex by the Hessian rule in Theorem 1.9, $\iota_{(-\infty,0]}$ is convex by Proposition 2.4, and Theorem 1.8(a,d) applies. Theorem 5.13 also requires that the min-value function $p(u) = \inf_x f(u,x) < \infty$ for u a neighborhood of 0. This is the case because

$$g(1,1,1) + u = -2 - 3 - 1 + 1 + u = -5 + u \le 0 \quad \forall u \in [-5,5].$$

So $p(u) = \inf_x f(u,x) \le f_0(1,1,1) = 6 < \infty$ for such u.

Exercise 5.10 Solution. (a) $\sum_{(i,j)\in E} c_{ij}x_{ij}$. (b) The model becomes

$$\begin{aligned} &\underset{x\in\mathbb{R}^{10}}{\text{minimize}} \sum\nolimits_{(i,j)\in E} c_{ij}x_{ij} \\ &\text{subject to} \sum\nolimits_{j:(i,j)\in E} x_{ij} - \sum\nolimits_{j:(j,i)\in E} x_{ji} = b_i, \quad i = 1,\dots,7 \\ &\qquad\qquad x_{ij} \ge 0 \quad \forall (i,j) \in E, \end{aligned}$$

where $b_1 = 5, b_2 = 4, b_3 = b_4 = 0, b_5 = b_6 = b_7 = -3$. (c) The updated model becomes

$$\begin{aligned} &\underset{x\in\mathbb{R}^{10}}{\text{minimize}} \sum\nolimits_{(i,j)\in E} c_{ij}(1 + 9w_{ij})x_{ij} \\ &\text{subject to} \sum\nolimits_{j:(i,j)\in E} x_{ij} - \sum\nolimits_{j:(j,i)\in E} x_{ji} = b_i, \quad i = 1,\dots,7 \\ &\qquad\qquad x_{ij} \ge 0 \quad \forall (i,j) \in E. \end{aligned}$$

(d) The max-min problem becomes

$$\underset{w\in\mathbb{R}^{10}}{\text{maximize}}\ f(w) \ \text{subject to} \sum\nolimits_{(i,j)\in E} w_{ij} \le \alpha, \ \ w_{ij} \in \{0,1\} \ \ \forall (i,j) \in E,$$

where $f(w)$ is the minimum value from (c).

(e) Following Example 5.9, we see that the problem in (c) is of the form minimize $c^\top x$ subject to $Ax = b$, $x \ge 0$. It's apparent that the vector c has component $c_{ij}(1 + 9w_{ij})$ in our case and b is the vector of components b_i. The tricky part is to figure out the matrix A. It must have 7 rows and 10 columns because the problem in (c) has seven equality constraints and 10 variables. Let's organize the variables in the following order

$$x = (x_{13}, x_{14}, x_{23}, x_{24}, x_{35}, x_{36}, x_{37}, x_{45}, x_{46}, x_{47}).$$

Studying the equality constraints in more detail, we find that

$$A = \begin{bmatrix} 1 & 1 & 0 & 0 & 0 & 0 & 0 & 0 & 0 & 0 \\ 0 & 0 & 1 & 1 & 0 & 0 & 0 & 0 & 0 & 0 \\ -1 & 0 & -1 & 0 & 1 & 1 & 1 & 0 & 0 & 0 \\ 0 & -1 & 0 & -1 & 0 & 0 & 0 & 1 & 1 & 1 \\ 0 & 0 & 0 & 0 & -1 & 0 & 0 & -1 & 0 & 0 \\ 0 & 0 & 0 & 0 & 0 & -1 & 0 & 0 & -1 & 0 \\ 0 & 0 & 0 & 0 & 0 & 0 & -1 & 0 & 0 & -1 \end{bmatrix}$$

because then Ax produces the left-hand side in the seven equality constraints. According to Example 5.9, the dual problem corresponding to (c) is then

$$\underset{y \in \mathbb{R}^7}{\text{maximize}} \; b^\top y \;\text{ subject to }\; A^\top y \le c.$$

Writing this out explicitly, we obtain

$$\begin{aligned} \underset{y \in \mathbb{R}^7}{\text{maximize}} \; & 5y_1 + 4y_2 - 3y_5 - 3y_6 - 3y_7 \\ \text{subject to} \; & y_1 - y_3 \le c_{13}(1 + 9w_{13}) \\ & y_1 - y_4 \le c_{14}(1 + 9w_{14}) \\ & y_2 - y_3 \le c_{23}(1 + 9w_{23}) \\ & y_2 - y_4 \le c_{24}(1 + 9w_{24}) \\ & y_3 - y_5 \le c_{35}(1 + 9w_{35}) \\ & y_3 - y_6 \le c_{36}(1 + 9w_{36}) \\ & y_3 - y_7 \le c_{37}(1 + 9w_{37}) \\ & y_4 - y_5 \le c_{45}(1 + 9w_{45}) \\ & y_4 - y_6 \le c_{46}(1 + 9w_{46}) \\ & y_4 - y_7 \le c_{47}(1 + 9w_{47}). \end{aligned}$$

The maximum value here is identical to $f(w)$ by Example 5.11. Combining this maximization with the maximization in (d) and using a more economical sum-notation, we obtain

$$\begin{aligned} \underset{y,w}{\text{maximize}} \; & \sum_{i=1}^{7} b_i y_i \\ \text{subject to} \; & y_i - y_j \le c_{ij}(1 + 9w_{ij}) \quad \forall (i,j) \in E \\ & \sum_{(i,j) \in E} w_{ij} \le \alpha \\ & w_{ij} \in \{0, 1\} \quad \forall (i,j) \in E. \end{aligned}$$

Using a mixed-integer optimization solver, we obtain:

$\alpha = 0$: maximum value = 20.2
$\alpha = 1$: maximum value = 27.1 and attack $w_{13} = 1$
$\alpha = 2$: maximum value = 76.9 and attacks $w_{35} = 1$, $w_{45} = 1$
$\alpha = 3$: maximum value = 87.6 and attacks $w_{13} = 1$, $w_{35} = 1$, $w_{45} = 1$.

We observe that the best attack plans aren't necessarily nested as α increases: the single attack of (1,3) isn't included when two attacks are allowed but reemerge for three attacks. This is typically the case and makes "prioritized lists" of attacks potentially suboptimal.

To check how the best respond varies for the different α, we solve the model in (c) under each of the best attack and this produces the table:

	α			
	0	1	2	3
x_{13}	5.0	0.0	5.0	0.0
x_{14}	0.0	5.0	0.0	5.0
x_{23}	4.0	4.0	4.0	4.0
x_{24}	0.0	0.0	0.0	0.0
x_{35}	3.0	0.0	3.0	3.0
x_{36}	3.0	3.0	3.0	1.0
x_{37}	3.0	1.0	3.0	0.0
x_{45}	0.0	3.0	0.0	0.0
x_{46}	0.0	0.0	0.0	2.0
x_{47}	0.0	2.0	0.0	3.0

The second column gives the transportation plan under no attack. The cost is 20.2 but this worsens moderately to 27.1 under one attack at edge $(1, 3)$. The increase in cost is just moderate because the flow can be moved from edge (1,3) to (1,4) as we see in the third column of the table. The situation becomes more critical under two attacks; the cost increases dramatically to 76.9. The two most effective attacks are at (3,5) and (4,5), which completely isolate vertex 5. It's interesting that an attack occurs at (4,5), which carried no commodities when $\alpha = 0$. This means that it might be suboptimal to merely look at the current flow of commodities and attack edges with (large) flow. It might be better to "think ahead" about how the supply chain may operate after attacks. The last column gives the response to three attacks, which only causes a moderate further increase in cost.

References

1. R. K. Ahuja, T. L. Magnanti, and J. B. Orlin. *Network Flows: Theory, Algorithms, and Applications*. Pearson, 1993.
2. A. Beck. *First-Order Methods in Optimization*. SIAM, 2017.
3. D. P. Bertsekas. *Dynamic Programming and Optimal Control. Approximate Dynamic Programming. Volume 2*. Athena Scientific, 4. edition, 2012.
4. D. P. Bertsekas. *Dynamic Programming and Optimal Control. Volume 1*. Athena Scientific, 4. edition, 2017.
5. D. P. Bertsekas and J. N. Tsitsiklis. *Neuro-Dynamic Programming*. Athena Scientific, 1996.
6. D.P. Bertsekas. *Convex Optimization Theory*. Athena Scientific, 2009.
7. D. Bertsimas and J. N. Tsitsiklis. *Introduction to Linear Programming*. Athena Scientific, 1997.
8. G. G. Brown and G. W. Graves. Real-time dispatch of petroleum tank trucks. *Management Science*, 27(1):19–32, 1981.
9. G. C. Calafiore and L. El Ghaoui. *Optimization Models*. Cambridge University Press, 2014.
10. W. M. Carlyle, J. O. Royset, and R. K. Wood. Lagrangian relaxation and enumeration for solving constrained shortest-path problems. *Networks*, 52(4):256–270, 2008.
11. A. R. Ferguson and G. B. Dantzig. The allocation of aircraft to routes. an example of linear programming under uncertain demand. *Management Science*, 3(1):45–73, 1956.
12. G. Garrigos and R. M. Gower. Handbook of convergence theorems for (stochastic) gradient methods. *Preprint arXiv:2301.11235*, 2023.
13. W. E. Hart, C. D. Laird, J.-P. Watson, D. L Woodruff, G. A. Hackebeil, B. L. Nicholson, and J. D. Siirola. *Pyomo - Optimization Modeling in Python*. Springer, 3. edition, 2021.
14. J. L. Higle and S. Wallace. Sensitivity analysis and uncertainty in linear programming. *Interfaces*, 33(4):53–60, 2003.
15. V. Katz. *A History of Mathematics*. Addison Wesley, 2008.
16. A. J. King and S. W. Wallace. *Modeling with Stochastic Programming*. Springer, 2. edition, 2024.
17. D. P. Kingma and L. J. Ba. ADAM: A method for stochastic optimization. In *International Conference on Learning Representations*, pages 1–13, 2015.
18. G. Lan. *First-Order and Stochastic Optimization Methods for Machine Learning*. Springer, 2020.
19. M. Mahsuli. *Probabilistic Models, Methods, and Software for Evaluating Risk to Civil Infrastructure*. Ph.d. thesis, University of British Columbia, 2012.
20. A. Martin. General mixed integer programming: Computational issues for branch-and-cut algorithms. In *Computational Combinatorial Optimization*, volume 2241 of *Lecture Notes in Computer Science*, pages 1–25. Springer, Berlin, Heidelberg, 2001.
21. Y. Nesterov. *Introductory Lectures on Convex Optimization*. Springer, 2004.
22. J. Nocedal and S. Wright. *Numerical Optimization*. Springer, 2006.
23. P. M. Pardalos and O. A. Prokopyev, editors. *Encyclopedia of Optimization*. Springer Cham, 2020.
24. W. B. Powell. *Reinforcement learning and stochastic optimization: A unified framework for sequential decisions*. John Wiley & Sons, 2022.
25. R. T. Rockafellar and R. J-B Wets. *Variational Analysis*. Springer, 1998 (2009, 3. printing).
26. J. O. Royset. Risk-adaptive approaches to stochastic optimization: A survey. *SIAM Review*, 67(1):3–70, 2025.

J. O. Royset, Optimization Concepts, *Springer Series in Operations Research and Financial Engineering*,
https://doi.org/10.1007/978-3-032-36819-5

27. J. O. Royset, W. M. Carlyle, and R. K. Wood. Routing military aircraft with a constrained shortest-path algorithm. *Military Operations Research*, 14(3):31–52, 2009.
28. J. O. Royset and R. J-B Wets. Multivariate epi-splines and evolving function identification problems. *Set-Valued and Variational Analysis*, 24(4):517–545, 2016. Erratum, pp. 547-549.
29. J. O. Royset and R. J-B Wets. On univariate function identification problems. *Mathematical Programming B*, 168(1-2):449–474, 2018.
30. J. O. Royset and R. J-B Wets. *An Optimization Primer*. Springer, 2021.
31. J. O. Royset and R. K. Wood. Solving the bi-objective maximum-flow network-interdiction problem. *INFORMS J. Computing*, 19(2):175–184, 2007.
32. A. Ruszczynski. *Nonlinear Optimization*. Princeton University Press, 2006.
33. L. D. Stone, J. O. Royset, and A. R. Washburn. *Search for Moving Targets*. Springer, 2016.
34. E. D. Taillard. *Design of Heuristic Algorithms for Hard Optimization*. Springer, 2022.
35. M. D. Teter, A. M. Newman, and M. Weiss. Consistent notation for presenting complex optimization models in technical writing. *Surveys in Operations Research and Management Science*, 21(1):1–17, 2016.
36. M. D. Teter, J. O. Royset, and A. M. Newman. Modeling uncertainty of expert elicitation for use in risk-based optimization. *Annals of Operations Research*, 280(1-2):189–210, 2019.
37. L. A. Wolsey and G. L. Nemhauser. *Integer and Combinatorial Optimization*. Wiley, 1999.

Index

J. O. Royset, Optimization Concepts, *Springer Series in Operations Research and Financial Engineering*,
https://doi.org/10.1007/978-3-032-36819-5

FSC
www.fsc.org
MIX
Papier | Fördert gute Waldnutzung
FSC® C083411

Kolibri 360 GmbH
Ferdinand-Jühlke-Straße 7
99095 Erfurt, Deutschland
produktsicherheit@kolibri360.de